COMPUTATIONAL APPROACHES TO BIOCHEMICAL REACTIVITY

Understanding Chemical Reactivity

Volume 19

The titles published in this series are listed at the end of this volume.

Computational Approaches to Biochemical Reactivity

edited by

Gábor Náray-Szabó

Department of Theoretical Chemistry,
Eötvös Loránd University,
Budapest, Hungary

and

Arieh Warshel

Department of Chemistry,
University of Southern California,
Los Angeles, California, U.S.A.

Springer-Science+Business Media, B.V.

A C.I.P. Catalogue record for this book is available from the Library of Congress

ISBN 978-1-4020-0415-5 ISBN 978-0-306-46934-3 (eBook)
DOI 10.1007/978-0-306-46934-3

Printed on acid-free paper

TABLE OF CONTENTS

PREFACE

A quantitative description of the action of enzymes and other biological systems is both a challenge and a fundamental requirement for further progress in our understanding of biochemical processes. This can help in practical design of new drugs and in the development of artificial enzymes as well as in fundamental understanding of the factors that control the activity of biological systems. Structural and biochemical studies have yielded major insights about the action of biological molecules and the mechanism of enzymatic reactions. However it is not entirely clear how to use this important information in a consistent and quantitative analysis of the factors that are responsible for rate acceleration in enzyme active sites. The problem is associated with the fact that reaction rates are determined by energetics (i.e. activation energies) and the available experimental methods by themselves cannot provide a correlation between structure and energy. Even mutations of specific active site residues, which are extremely useful, cannot tell us about the totality of the interaction between the active site and the substrate. In fact, short of inventing experiments that allow one to measure the forces in enzyme active sites it is hard to see how can one use a direct experimental approach to unambiguously correlate the structure and function of enzymes. In fact, in view of the complexity of biological systems it seems that only computers can handle the task of providing a quantitative structure-function correlation.

The use of computer modelling in examining the microscopic nature of enzymatic reactions is relatively young and this book provides a glimpse at the current state of this fast growing field. Although the first hybrid quantum mechanical/molecular mechanical (QM/MM) study of enzymatic reactions was reported two decades ago this is clearly not a mature field and many of the strategies used are not properly developed and no general consensus has been established with regards to the optimal strategy. Moreover, it is clear that many studies are still missing crucial points in their attempt to model biological processes. Many of the problems are due to the complexity of enzyme-substrate systems and the fact that the strategies developed for QM calculations of isolated molecules in the gas phase are not adequate for studies of enzymatic reactions. The same is true for other chemical concepts that should be re-evaluated when applied to complex, non-homogeneous systems.

This book presents different approaches that can be useful in theoretical treatments of biological activities. In doing so we try to bring together parts of the overall picture of what is needed in order to model and analyse the energetics and kinetics of enzymatic reactions. As editors, we do not necessarily fully agree with the philosophy of each chapter. However, we believe that presenting different approaches is an optimal way of exposing the reader to the current state of the field and for reaching scientific consensus. Chapter 1 considers the general issue of modelling of chemical processes in solution emphasising continuum approaches. Solvation energies provide the essential connection between gas phase QM studies and the energetics of processes in condensed phase. In fact, we chose this as the opening chapter since one of the main problems in elucidating the origin of enzyme catalysis has been associated with the difficulties of estimating solvation free energies. Chapter 2 presents attempts to advance the accuracy of the QM parts of reactivity studies by representing the solute using ab initio methods. Such methods will eventually become the methods of choice and early exploration of their performance is crucial for the development of the field. Chapter 3 reviews combined QM/MM calculations of chemical reactions in solutions and enzymes considering some of the currently used methods. Here the emphasis is on the crucial aspect of combining the quantum mechanical and classical regions. Chapter 4 considers MM and molecular dynamic (MD) approaches. Such approaches are essential for representing the conformational energies of biological molecules and can be used for example in assessing the importance of strain effects or other *ground-state* properties. This chapter also presents attempts to use ground-state MD in studies of mechanistic issues. Here, it might be useful to caution that definitive information about different mechanisms can only be obtained by going beyond such approaches and considering the quantum mechanical changes that necessarily take place in chemical reactions. Chapter 5 considers calculations of electrostatic energies in proteins. This aspect is an essential part of analysing of the energetics of enzymatic processes because without reliable ways of estimating electrostatic energies it is impossible to ask any quantitative question about enzyme catalysis. The approaches considered in Chapter 1 are not always applicable to studies of electrostatic energies in proteins and one has to be familiar with calculations both in proteins and solutions if gaining a clear understanding of this challenging field is the objective. Chapter 6 considers the general issue of the catalytic power of enzymes and demonstrate that electrostatic energies are the most important factor in enzyme catalysis. This point is illustrated by both quantitative calculations and simple molecular electrostatic potential calculations where it is shown that enzymes provide a complementary environment to the charge distribution of their transition states. Chapter 7 considers in detail the important class of proteases and reviews the current theoretical effort in this specific case. Chapter 8 presents quantitative calculations of enzymatic reactions that and focuses on studies of proton transfer reactions using the empirical valence bond (EVB) method. This chapter illustrates and explains the catalytic role of the enzyme in providing electrostatic stabilisation to high energy intermediates and in reducing reaction reorganisation energies. At last, Chapter 9 deals with protein-ligand interactions that can be treated by using methods described in the previous chapters. Quantitative understanding of such interactions is of primary importance in rational drug design.

While the chapters presented here reflect different aspects and opinions it is useful to emphasise some points that might not be obvious to the readers. These points are important since the current status of the field is somewhat confusing and some readers might be overwhelmed by the technological aspects rather than by logical considerations of the energetics. Thus we will outline below some of the main problems that should be considered in a critical examination whether a given approach for studies of enzymatic reactions is really useful. We start stating what should have been obvious by now, that calculations of enzymatic reactions must reflect the energetics of the complete enzyme-substrate-solvent system. Thus calculations of subsystems in the gas phase or even calculations that involve a few amino acids cannot be used to draw any quantitative conclusion about enzyme mechanism. That is, the gradual "build-up" process must involve an increasing sophistication of describing the *complete* system, rather than adding different physical parts to a rigorous but incomplete description. This might not be so clear to readers who are familiar with the use of accurate gas phase calculations and prefer rigorous treatments of isolated subsystems over an approximate but reasonable treatment of the whole system. However enzyme modelling does not lend itself to incremental studies where one can learn by considering parts of the system in a step by step process. For example, if a reaction involves a formation of an ion pair the error of not including the surrounding solvent can amount to 40 kcal/mol regardless of how accurate is the treatment of the reacting fragments. The problem of incomplete description cannot be over-emphasised since it can lead to major conceptual problems, such as concluding that a helix macro-dipole accounts for the catalytic effect of an enzyme while using unsolved protein as a model for the given enzyme. On the other hand, correct calculations might indicate that the solvent around the protein screens the helix effect and even leads to less stabilisation than that provided by the solvent for the reference reaction in water. Similarly, modelling an enzymatic reaction with an unscreened metal ion can lead one to believe that this ion alone provides enormous catalytic effect, but in reality the field of the ion might be largely screened.

Another problem with regards to modelling of enzymatic reactions is the recent tendency to believe that ground state MD simulations can provide concrete information about enzyme mechanism and catalysis. This assumption is unjustified since ground state dynamics cannot tell us much about the probability of reaching the transition state in different feasible mechanisms. Thus, for example, finding a proton near a proton acceptor does not mean that the barrier for proton transfer is reduced by the given active site. Finally we would like to warn about the use of combined QM/MM approaches. Here one might assume that since the complete enzyme substrate system can be considered the results can be trusted blindly. This is unfortunately incorrect. First, many such studies do not consider the solvent molecules in and around the protein, and thus may lead to enormous errors. Secondly, even approaches that include the solvent molecules are likely to provide irrelevant results unless the free energy of the system is evaluated by reliable free energy perturbation (FEP) or related approach. Using energy minimisation in an enzyme active site might be quite ineffective. Even the use of convergent free energy calculations does not guarantee the accuracy of the calculated activation free energies since the given QM method might not be reliable enough. Thus, unless the method can reproduce the correct energetics in reference solution reactions it is unlikely to reproduce correctly the energetics of enzymatic reactions. Thus we believe that any approach that is used in studies of enzymatic reactions must be able to accurately reproduce electrostatic energies in enzymes (e.g.

accurate pKa's) and accurate energetics in solutions, otherwise, such methods cannot be considered as quantitative.

The perspective given above might look somewhat critical and almost pessimistic. However, most of the warnings given here are related only to the current and short term status of the field. There is no doubt that once grown out of its infancy, computer modelling will provide the most powerful way of using structural and biochemical information in quantitative description of biological reactions. We believe that this maturation will occur in the next several years and will involve a major progress in the use of theoretical methods in studies of enzymatic reactions and related processes. We hope that this book will contribute to this progress.

The Editors

QUANTUM MECHANICAL MODELS FOR REACTIONS IN SOLUTION

J. TOMASI, B. MENNUCCI
Dipartimento di Chimica e Chimica Industriale, Università di Pisa
Via Risorgimento 35, Pisa, Italy
R. CAMMI
Dipartimento di Chimica, Università di Parma
Viale delle Scienze 1, Parma, Italy
AND
M. COSSI
Dipartimento di Chimica, Università "Federico II" di Napoli
Via Mezzocannone 4, Napoli, Italy

1. Naive picture of liquids and chemical reactions in liquids.

Models in theoretical chemistry are often quite complex, but at the same time they are always based on simple and naive pictures of the real systems and the processes which are the object of modelling.

To gain a better understanding of a given model, with its subtleties and characterizing features, it is often convenient to go back to basic naive pictures. Also the opposite way, i.e. contrasting different naive pictures for the same problem, may be of some help in the appreciation of a model. Simple pictures emphasize different aspects of the problem, and their comparison is of great help in grasping both merits and limits of the theoretical and computational methods proposed in scientific literature.

We shall start with a couple of such naive models for the liquid state, and for reactions occurring in solution. A molecular liquid in macroscopic equilibrium may be viewed as a large assembly of molecules incessantly colliding, and exchanging energy among collision partners and among in-

G. Náray-Szabó and A. Warshel (eds.), Computational Approaches to Biochemical Reactivity, 1–102.

ternal degrees of freedom. A limited number of collisions leads to more drastic effects, perturbing the internal electronic distribution of collision partners, and causing the formation of molecules with a different chemical composition.

This model of the liquid will be characterized by some macroscopic quantities, to be selected among those considered by classical equilibrium thermodynamics to define a system, such as the temperature T and the density ρ. This macroscopic characterization should be accompanied by a microscopic description of the collisions. As we are interested in chemical reactions, one is sorely tempted to discard the enormous number of non–reactive collisions. This temptation is strenghtened by the fact that reactive collisions often regard molecules constituting a minor component of the solution, at low-molar ratio, i.e. the solute. The perspective of such a drastic reduction of the complexity of the model is tempered by another naive consideration, namely that reactive collisions may interest several molecular partners, so that for a nominal two body reaction: $A + B \rightarrow$ products, it may be possible that other molecules, in particular solvent molecules, could play an active role in the reaction.

This is the naive picture on which many tentative models of chemical reactions used in the past were based. The material model is reduced to the minimal reacting system (A+B in the example presented above) and supplemented by a limited number of solvent molecules (S). Such material model may be studied in detail with quantum mechanical methods if A and B are of modest size, and the number of S molecules is kept within narrow limits. Some computational problems arise when the size of reactants increases, and these problems have been, and still are, the object of active research. This model is clearly unsatisfactory. It may be supplemented by a thermal bath which enables the description of energy fluxes from the microscopic to the outer medium, and *vice versa*, but this coupling is not sufficient to bring the model in line with chemical intuition and experimental evidence.

Now we proceed to consider another naive picture of liquid systems. A liquid system is disordered on a large scale, but more ordered locally. The properties of the liquid may be understood by looking at this local order, and examining how it fades away at larger distances. The local order is due to the microscopic characteristics of the intermolecular interaction potential. By introducing interaction potentials of different type in the computational machinery of the corresponding theoretical model, and starting, for example, from short–range repulsive potentials and then adding appropriate medium and long–range terms, one may learn a lot about the properties of the liquid. Using more and more realistic interaction potentials, one has the perspective of gaining a sufficiently accurate description of the liquid.

However, it is hard to introduce chemical reactions in this naive picture.

The relatively rare reactive events have an extremely low statistical weight. To study them one has to force the model, bringing into contact two local structures based on molecules A and B, in our example, and then studying the evolution of such local structures in the whole liquid. We are so led once again to consider a microscopic event, i.e. the chemical reaction, which is now set in a wider and more detailed model for the liquid.

The main theoretical approach to describe the reactive event is still quantum mechanics (QM). Alternative semiclassical models can be used only after an accurate calibration on very similar processes studied at the appropriate QM level. However, in this case, things are more complex than in the previous model. The description of local liquid structures, and their decay at long distances, cannot be made at the QM level. Severe limitations are necessary: if we confine ourselves to the most used methods, computer simulations using Monte Carlo or molecular dynamics techniques (standard references are provided by Valleau and Whittington (1977) and Valleau and Torrie (1977) for MC, and by Kushick and Berne (1977) for MD, while more recent developments can be found in Beveridge and Jorgensen (1986), and Allen and Tildesley (1987)) we recognize these limitations in the use of semiclassical two–body potentials (many–body potentials increase computational times beyond acceptable values). There is now experience and availability of computational resource which are sufficient to make the derivation of a two body potential a feasible task if the two partners are at a fixed geometry. In chemical reactions the change of internal geometry, and of electronic structure, is a basic aspect that cannot be grossly approximated. Therefore the solute–solvent interaction potential must be re-evaluated for a sufficient number of nuclear conformations of the reactive subsystem A–B, with the additional problem, hard to be solved, that the charge distribution of A–B, and then its interaction potential with a solvent molecule S, critically depends on the interactions with the other S molecules nearby. In addition the introduction of explicit solvent molecules in the reactive system (say an A–B–S_n subsystem) is not easy.

There are of course methods which attempt to tackle these problems in their complexity, and avoid some of the approximations we have described; we quote, as an outstanding example, the Car–Parrinello approach (Car and Parrinello, 1985), that in recent extensions has aimed to give a coherent QM description of such models (Laasonen *et al.*, 1993; Fois *et al.*, 1994). However, these approaches are still in their infancy, and it is advisable to look at other models for liquids and chemical reactions in solutions.

The models we are considering now are less related to naive descriptions. They take some elements from the two previous models, and attempt a synthesis. In this synthesis emphasis is laid on efficiency (combined with accuracy) in computational applications, but another aspect has to be considered, namely the flexibility of the models, i.e. their capability to include, when necessary, additional details as well as to describe reacting molecular systems more deeply.

As we have seen, both models considered in the previous pages lead to the definition of a microscopic portion of the whole liquid system, the larger portion of the liquid being treated differently. We may rationalize this point by introducing, in the quantum mechanical language, an effective Hamiltonian of the subsystem (A–B–S_n), where the Hamiltonian of the isolated system $M = A$–B ($\hat{\mathcal{H}}_M^o$) is supplemented by an effective solute–solvent interaction potential ($\hat{\mathcal{V}}_{int}$):

$$\hat{\mathcal{H}}_M^{eff} = \hat{\mathcal{H}}_M^o + \hat{\mathcal{V}}_{int} \tag{1}$$

$\hat{\mathcal{V}}_{int}$ aims to describe the interaction of M with the local solvent structure, envisaged in the second naive picture of liquids and hence bearing in action the concept of average interaction, as well as the non–reactive collisions, envisaged in the first solution picture and hence introducing the concept of solvent fluctuations.

$\hat{\mathcal{V}}_{int}$ may be modeled in many different ways. One of the extreme examples is the solvation model proposed years ago by Klopman (1967), which is quoted here to show the flexibility of this approach, and not to suggest its use (the limits of this model have been known since a long time). In this model each nucleus of M is provided with an extra phantom charge (the solvaton), which introduces, *via* Coulombic interactions, a modification of the solute electronic wavefunction and of the expectation value of the energy, E_M, mimicking solvent effects.

The expressions of $\hat{\mathcal{V}}_{int}$ which are now in use belong to two categories: expressions based on a discrete distribution of the solvent, and expressions based on continuous distributions. The first approach leads to quite different methods. We quote here as examples the combined quantum mechanics/molecular mechanics approach (QM/MM) which introduces in the quantum formulation computer simulation procedures for the solvent (see Gao, 1995, for a recent review), and the Langevin dipole model developed by Warshel (Warshel, 1991), which fits the gap between discrete and continuum approaches. We shall come back to the abundant literature on this subject later.

In the second approach one has to define the status of the continuum solvent distribution. If the distribution corresponds to an average of the possible solvent conformations, given for example by Monte Carlo or by RISM calculation, (Chandler and Andersen, 1972; Hirata and Rossky, 1981; Rossky, 1985) $\hat{\mathcal{V}}_{int}$ may be assimilated to a free energy. With other solvent distributions the thermodynamic status of $\hat{\mathcal{V}}_{int}$ may be different according to the imposing conditions. In our exposition we shall mainly rely on the second model, namely effective Hamiltonians based on continuous solvent distributions: EH–CSD.

There are several versions of EH–CSD models. To make the exposition less cumbersome, in the next pages we shall only summarize one version, that was elaborated in Pisa and known with the acronym PCM (Polarizable Continuum Model) (Miertuš *et al.*, 1981; Miertuš and Tomasi, 1982). We shall consider other versions later, and the differences with respect to PCM will be highlighted. Other approaches, based on effective Hamiltonians expressed in terms of discrete solvent distributions, EH–DSD, or not relying on effective Hamiltonians, will also be considered.

Limitations of space make impossible a thorough consideration of all the topics here mentioned; some suggestions of further reading will be given at appropriate places. For the same reason other topics will never be mentioned: we suggest a recent book (Simkin and Sheiker, 1995) to gain a broader view on quantum chemical problems in solution and suggestions for further readings. Also some recent collective books, namely 'Structure and Reactivity in Aqueous Solution' (Cramer and Truhlar, 1994) and 'Quantitative Treatments of Solute/Solvent Interactions' (Politzer and Murray, 1994), with their collection of reviews and original papers written by eminent specialists, are recommended.

2. A phenomenological partition of the solute–solvent interactions.

To build up an appropriate version of the EH–CSD model we have to be more precise in defining which energy terms must be included in the model. To this end we may start defining a phenomenological partition of a quantity having a precise thermodynamic meaning. We shall select the Gibbs solvation free energy of a solute M, and follow Ben-Naim's definition of the solvation process (Ben–Naim 1974; 1987). In this framework, the solvation process is defined in terms of the work spent in transferring the solute M from a fixed position in the ideal gas phase to a fixed position in the liquid S. This work, $W(M/S)$, called "coupling work", is the basic ingredient of the solvation free energy:

$$\Delta G_{sol} = W(M/S) + RT \ln \left(\frac{q_{rot,g} q_{vib,g}}{q_{rot,s} q_{vib,s}} \right) - RT \ln \left(\frac{n_{M,g} \Lambda^3_{M,g}}{n_{M,s} \Lambda^3_{M,s}} \right) \quad (2)$$

where $q_{rot,g}$, $q_{vib,g}$, $q_{rot,s}$, $q_{vib,s}$ are the microscopic partition functions for the rotation and the vibration of M in the gas phase and in solution. $\Lambda_{M,g}$ and $\Lambda_{M,s}$ are the momentum partition functions, and $n_{M,g}$ and $n_{M,s}$ the number densities of M in the two phases. There is an additional $P\Delta\mathcal{V}$ term, here neglected since it is quite small in normal cases. The sum of the first two terms of eq.(2) is indicated by Ben Naim with the symbol ΔG^*. The last term in eq.(2) is called "liberation free energy".

Ben-Naim's definition has many merits: it is not limited to dilute solutions, it avoids some assumptions about the structure of the liquid, it allows to use microscopical molecular partition functions; moreover, keeping M fixed in both phases is quite useful in order to implement this approach in a computationally transparent QM procedure. The liberation free energy may be discarded when examining infinite isotropic solutions, but it must be reconsidered when M is placed near a solution boundary.

We may now introduce a phenomenological partition of $W(M/S)$. The analogy of $W(M/S)$ with the few body intermolecular ΔE, extensively studied in QM models, could suggest the use of one of the numerous ΔE decompositions available in literature. In the past we used, with good results, the following partition (Kitaura and Morokuma, 1976):

$$\Delta E(A/B) = \Delta E_{coul} + \Delta E_{pol} + \Delta E_{exch} + \Delta E_{dis} + \Delta E_{chtr} \qquad (3)$$

where the interaction is resolved into Coulombic, polarization, exchange, dispersion, and charge–transfer terms; however, its direct adaptation to $W(M/S)$, assimilating M to A and S to B presents some inconveniences. Some analogous considerations apply to other partitions of $\Delta E(A/B)$.

In the EH–CSD approach it is not convenient to decouple electrostatic terms into rigid Coulombic and polarization contributions: the effective Hamiltonian leads to compute these two terms together. Exchange repulsive terms are hardly computed when the second partner of the interaction is a liquid; they may be obtained with delicate simulation procedures, and it is convenient to decouple them into two contributions, namely the work spent to form a cavity of a suitable shape and an additional repulsion contribution. Dispersion contributions may be kept: we shall examine this term in more detail later. Charge–transfer contributions are damped in liquids; their inclusion could introduce additional problems in the definition of $\hat{\mathcal{V}}_{int}$ *via* continuous solvent distributions. It is advisable to neglect them, as it is done in the interaction potentials used in simulations; with the present approach it is possible to describe the charge transfer effect by "enlarging" the solute: $M \rightarrow M \cdot S_n$.

The phenomenological partition of $W(M/S)$ we consider computationally convenient is:

$$W(M/S) = \Delta G_{el} + \Delta G_{cav} + \Delta G_{dis} + \Delta G_{rep} \qquad (4)$$

and then, for the free solvation energy:

$$\Delta G_{sol} = \Delta G_{el} + \Delta G_{cav} + \Delta G_{dis} + \Delta G_{rep} + \Delta G_{Mm} \qquad (5)$$

where the subscripts stand for electrostatic, cavitation, dispersion, and repulsion terms, respectively. In the last term we have collected all the contributions explicitely reported in eq.(2) and concerning the nuclear motions of M.

3. The free energy hypersurface.

Ab initio results represent a benchmark for all studies on chemical reactions. It is thus convenient to reformulate the phenomenological description of the solvation energy, given in eq.(5), introducing an "absolute" reference energy, similar to that used in *ab initio* calculations *in vacuo*.

In the continuum solvent distribution models this reference energy corresponds to non–interacting nuclei and electrons at rest (in the number which is necessary to build up the "solute" M) supplemented by the unperturbed liquid phase. This is not a simple energy shift of ΔG_{sol} as given by eq.(5), in fact, we introduce here a supplementary term corresponding to the energy of assembling electrons and nuclei, these last at a given geometry, in the solution. We define:

$$G = G_{el} + G_{dis} + G_{rep} + G_{cav} + G_{Mm} \tag{6}$$

where in G_{el} we collect all terms of electrostatic origin, i.e. the work spent to assemble nuclei and electrons of M at the chosen nuclear geometry, and the electrostatic solute–solvent contributions to the free energy.

In our definition of reference energy we have implicitly assumed the validity of the Born–Oppenheimer approximation: the phenomenological partition of ΔG_{sol}, and the statistical mechanics considerations leading to eq.(2), also assume the separation of nuclear and electronic motions of the solute. The first four terms of the right–side of eq.(6) define a hypersurface, $G(\mathbf{R})$, in the space spanned by the nuclear coordinates of M, which is the analog of the potential energy hypersurface, $E^o(\mathbf{R})$, of the same molecular system *in vacuo*. The last term of eq.(6) also depends, in an indirect way, on the nuclear conformation $\mathbf{R}$: its expression may be easily derived from eq.(2). The $E^o(\mathbf{R})$ surface is provided with an analogous term. Zero–point vibrational contributions are included in the G_{Mm} term (and, in analogy, in the E^o_{Mm} term for the *in vacuo* case).

We may thus compare two energy hypersurfaces, $G(\mathbf{R})$ and $E^o(\mathbf{R})$, or, if it is allowed by the adopted computational procedure, avoid a separate calculation of $E^o(\mathbf{R})$ and rely for our studies on the $G(\mathbf{R})$ surface solely. $G(\mathbf{R})$ may be viewed as the sum of separate contributions:

$$G(\mathbf{R}) = G_{el}(\mathbf{R}) + G_{dis}(\mathbf{R}) + G_{rep}(\mathbf{R}) + G_{cav}(\mathbf{R}) \tag{7}$$

In the computational practice, the various components of eq.(7) are often calculated separately.

The most important one is $G_{el}(\mathbf{R})$. It is easy to recover from it the electrostatic contribution to the solvation energy, ΔG_{el}:

$$G_{el}(\mathbf{R}) = E^o(\mathbf{R}) + \Delta G_{el}(\mathbf{R}) \tag{8}$$

In such a way one may pass from the *ab initio* formulation to semiempirical or semiclassical formulations. It is sufficient to replace $E^o(\mathbf{R})$ with another semiempirical or semiclassical surface $E^{app}(\mathbf{R})$ (there is no more need of an 'absolute' zero of energy); ΔG_{el} depends on the solute charge distribution (and on its electric response function, i.e. polarizability), and *ab initio* calculations are not strictly necessary. However, it has to be remebered that in chemical reactions the description of the solute charge distribution and of its response function must be checked with great care.

The same concepts may be recast in a different form. We may define a solvation free energy hypersurface:

$$\Delta G_{sol}(\mathbf{R}) = \Delta G_{el}(\mathbf{R}) + G_{cav}(\mathbf{R}) + G_{dis}(\mathbf{R}) + G_{rep}(\mathbf{R}) \tag{9}$$

so that

$$G(\mathbf{R}) = E^*(\mathbf{R}) + \Delta G_{sol}(\mathbf{R}) \tag{10}$$

Definition (10) of $G(\mathbf{R})$ is of little use in *ab initio* computations, as the procedure directly gives $G(\mathbf{R}) = E^o(\mathbf{R}) + \Delta G_{sol}(\mathbf{R})$. This definition is useful in other approaches, where $E^*(\mathbf{R}) = E^{app}(\mathbf{R})$ is computed with some approximate methods and the attention is focussed on $\Delta G_{sol}(\mathbf{R})$, which may be independently computed by means of *ad hoc* procedures.

The absolute values of ΔG_{el}, G_{cav}, and G_{dis} are often comparable, while G_{rep} is noticeably smaller, and often discarded. In most cases, the shape of $\Delta G_{sol}(\mathbf{R})$ is determined by $\Delta G_{el}(\mathbf{R})$; one has to consider bulky hydrocarbons to find cases in which the shape of $\Delta G_{sol}(\mathbf{R})$ is dictated by $G_{cav}(\mathbf{R}) + G_{dis}(\mathbf{R})$ (which actually are of opposite sign). For this reason several computational methods discard all non–electrostatic components, devoting all efforts to the computation of ΔG_{el} and E^o. However, the numerical experience derived from the more recent calculations shows that contributions due to $G_{cav} + G_{dis} + G_{rep}$ are not negligible. Anyway it is safer to compute all $G(\mathbf{R})$ components.

4. A sketch of the use of $G(\mathbf{R})$ to study reactions in solution.

Before entering in a more detailed description of the computational aspects of the model, we report a concise outline of the most important ways according to which one may use $G(\mathbf{R})$ values in the study of chemical reactions.

The real nature of the model, a microscopic system treated at a fine level of description and supplemented with an effective solute–solvent potential, makes it evident that much matter is in common with the analogous problem of using $E^o(\mathbf{R})$ values for reactions *in vacuo*. The only things that are missing in problems regarding reactions *in vacuo* are the contributions related to the interaction potential $\hat{\mathcal{V}}_{int}$. However, their presence requires the consideration of further aspects of the computational problem, that may result to be critical in assessing the soundness of the study. This is the reason why we put this Section before those devoted to methods, in order to highlight the specific computational points deserving more attention when applications to chemical reactions are envisaged.

The information on chemical reactions one may draw from calculations can be divided into two broad classes, i.e. reaction equilibria and reaction mechanisms. Mechanisms, in turn, may be considered either at a static level or including dynamical aspects. It is convenient to treat these items separately.

4.1. REACTION EQUILIBRIA.

To study reaction equilibria we simply need to know the values of $G(\mathbf{R})$ corresponding to two local minima (reagents and products). If we confine our consideration to cases in which there is no change in the number of solute molecules, i.e. to molecular processes involving changes of conformation, or bond connectivity ($A \rightleftharpoons B$), the desired quantity $\Delta G^o(A \rightarrow B)$ may be computed in two ways, according to the following scheme:

$$
\begin{array}{ccccc}
 & A(g) & \xrightarrow{\Delta G^o_g(A \rightarrow B)} & B(g) & \\
\Delta G^o_{sol}(A) & \downarrow & & \downarrow & \Delta G^o_{sol}(B) \\
 & A(s) & \xrightarrow{\Delta G^o_s(A \rightarrow B)} & B(s) &
\end{array}
$$

$$\Delta G^o_s(A \rightarrow B) = G_s(B) - G_s(A) = \Delta G^o_g(A \rightarrow B) + \Delta G^o_{sol}(B) - \Delta G^o_{sol}(A) \tag{11}$$

This scheme has been reported to emphasize some points which deserve attention in performing calculation. Firstly, any effort to improve the quality of the solvation energies, $\Delta G^o_{sol}(M)$, is meaningless unless it is accompanied by a parallel effort in giving a good description of the energy difference $\Delta G^o_g(A \rightarrow B)$ (and, conversely, good ΔG^o_g calculations are meaningless if the parallel ΔG^o_{sol} calculations are not able to describe the difference between reagent and product value at a comparable level of accuracy). Secondly, minima for A and B in the gas phase may refer to geometries which are somewhat different from the corresponding minima in solution: the use of rigid geometries (computed *in vacuo*) may be another source of errors.

The scheme may be extended to the case of multiple minima by using a standard Boltzmann formalism. Attention must be paid here to the correct evaluation of the ΔG_{Mm} contributions (see equations (2) and (5) in Section 2). The shape of the $G(\mathbf{R})$ surface (in solution) around these minima may be different from that of the corresponding portion of the $E^o(\mathbf{R})$ surface (*in vacuo*) at such an extent as to give significant changes in the final $\Delta G(A \rightarrow B)$ values. Generally speaking, for compounds with polar groups dissolved in water, the $G(\mathbf{R})$ surface is flatter than $E^o(\mathbf{R})$, giving thus larger entropic contributions which are not well described by the usual first–order approximation of internal motions (the opposite may also occurs, some local minima of $G(\mathbf{R})$ may be deeper than the corresponding minima of $E^o(R)$). In solvents with a low dielectric constant, the $\Delta G_{dis-rep}$ contribution may play an essential role (see, for example, the limitations of an only–electrostatic solvation model in the study of conformational equilibria in acetonitrile which have been recently stressed by Varnek *et al.* (1995)).

A typical example in which all these points can be found is the study of equilibria of amino acids. In these cases the conformational equilibria are combined with intramolecular proton transfer. The comparison between neutral (**1**) and zwitterionic (**2**) forms

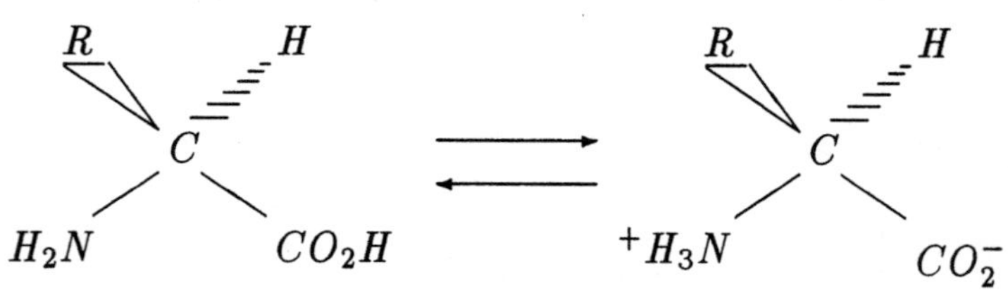

1 **2**

must take into account the effects due to the large charge separation in (**2**), the basis set, the changes in internal geometry going from (**1**) to (**2**) and from gas phase to solution, etc. (Bonaccorsi *et al.*, 1984b).

If we proceed now to consider reaction equilibria involving changes in the number of solute molecules ($A + B \rightleftharpoons C$, for example), other problems have to be taken into account. A scheme analogous to that reported for the $A \rightleftharpoons B$ process is reported below

$$\begin{array}{ccc}
A(g) + B(g) & \xrightarrow{\Delta G^o_g(A+B\rightarrow C)} & C(g) \\
\Delta G^o_{sol}(A) \downarrow \quad \downarrow \Delta G^o_{sol}(B) & & \downarrow \Delta G^o_{sol}(C) \\
A(s) + B(s) & \xrightarrow{\Delta G^o_s(A+B\rightarrow C)} & C(s)
\end{array}$$

In the association process some degrees of freedom of the reacting system change their nature (from translation and rotation to internal motions). Statistical thermodynamics suggests us the procedures to be used in gas phase calculations; application to processes in solution requires a careful analysis. The additional internal motions are in general quite floppy, and their separation from rotational motions of the whole C is a delicate task.

A typical case is the process of dimer formation in amides. Calculations *in vacuo* lead to planar dimers, thanks to the association force provided by intramolecular hydrogen bonds: the stabilization energy ΔE^o *in vacuo* may be of the order of -20 kcal/mol (the values obviously depend on the quality of the calculation and of the chemical composition of the dimer). In water, the corresponding ΔG_{el} value is of the order of -1 kcal/mol, and the inclusion of the other terms leads to a positive association free energy $\Delta G_s(2A \rightarrow A_2)$, with large estimated error bars (Biagi *et al.*, 1989). In the formamide dimer case, the favoured interaction in water exhibits the two monomers face–to–face. These results, which have been cursorily reported here, are in agreement with experimental values and with the Monte Carlo simulations performed by Jorgensen (1989a). We would like to shortly remark that the mean force integration technique used by Jorgensen to get ΔG values can be used only when there are good reasons to assume that the integration may be limited to just one or two coordinates.

This example emphasizes an important difference between reactions in low–pressure gas phase and in solution, which has been considered in the naive pictures introduced in Section 1. A close contact between two solute molecule, eventually leading to a chemical reaction, is always a substitution process, in which a portion of the first solvation shell molecules is replaced by the second solute. This point is of particular relevance in the study of reaction mechanisms, but it must be taken into account even when this study is limited to the assessment of equilibrium constants.

Another typical case which deserves to be mentioned is the complexion of a ligand L with a metal cation M^{n+}. Cations have all tightly–bound water molecules belonging to the first solvation shell. Microscopic models composed by "bare" L and M^{n+} are not sufficient: at least the first solvation shell of M^{n+} must be included, $M^{n+} \cdot mS$ (Floris *et al.*, 1995; Pappalardo *et al.*, 1993, 1995). When the tight complex is formed, a portion of these m molecules of solvent loses its specific role and becomes a component of the larger assembly of solvent molecules, with an evidently higher mobility which fades away in the bulk solvent. This means that for the evaluation of ΔG_{Mm} one has to consider a different chemical composition before and after the reaction, i.e. different numbers of degrees of freedom and different partition functions. The computational method must be able to include (or to eliminate) in the microscopic part of the material model a limited number of solvent molecules, not engaged in specific interactions, without altering the shape and the values of the pertinent portion of the $G(\mathbf{R})$ surface too much.

The analysis could be extended to other processes, as acid–base equilibria and reactions related to intermolecular proton transfer, or involving a change of electronic state, but what has been said is sufficient to convey the essential message, i.e. the determination of the energetic balance of a reaction, and of the equilibrium between reagents and products, albeit conceptually simple, requires a serious consideration of the computational tools one has to select.

4.2. REACTION MECHANISMS.

Nowadays the study of a reaction mechanism may be done by performing a well determined sequence of computational steps: we define this sequence as the canonical approach to the study of chemical reactions. At first, one has to define the geometry of reagents and products, then that of other locally stable intermediates, especially those acting as precursors of the true reaction process, and finally that of the transition state or states and of the reaction intermediates, if any. The determination of these geometries will of course be accompanied by the computation of the relative energies. All the points on the potential energy hypersurface we have mentioned are stationary points, defined by the condition:

$$g(\mathbf{R}_{st}) = grad\, E(\mathbf{R}_{st}) = 0 \tag{12}$$

The gradient function is thus defined:

$$\begin{aligned} grad\, E(\mathbf{R}_p) &= \{q_1, q_2, \ldots.., q_{3N}\}^{\dagger} \left\{ \frac{\partial E(\mathbf{R})}{\partial q_1}, \frac{\partial E(\mathbf{R})}{\partial q_2}, \ldots.., \frac{\partial E(\mathbf{R})}{\partial q_{3N}} \right\}_{\mathbf{R}=\mathbf{R}_p} \\ &= \sum_i q_i \left. \frac{\partial E(\mathbf{R})}{\partial q_i} \right|_{\mathbf{R}=\mathbf{R}_p} \end{aligned} \tag{13}$$

where $\{q_1, \ldots..\}$ and $\{\frac{\partial E(\mathbf{R})}{\partial q_1}, \ldots..\}$ are the column matrices of the unit vectors on the $3N$ nuclear coordinates, and of the partial derivatives of $E(\mathbf{R})$, respectively. The values of $g(\mathbf{R})$ define a vectorial field which accompanies the scalar field $E(\mathbf{R})$.

The difference among stationary points is given by the local curvature of $E(\mathbf{R})$, which is related to the eigenvalues of the Hessian matrix, $\mathbf{H}(\mathbf{R})$, i.e. the matrix of the second derivatives of the energy:

$$\mathrm{H}(\mathbf{R}_p)_{ij} = \left.\frac{\partial^2 E(\mathbf{R})}{\partial q_i \partial q_j}\right|_{\mathbf{R}=\mathbf{R}_p} \tag{14}$$

Since the potential energy of a molecule in a homogeneous medium (either a *vacuum* or an isotropic solution) is invariant with respect to transitions and/or rotations of the whole system, the $\mathbf{H}$ spectrum always exhibits six (five for linear molecules) zero eigenvalues. The characteristics of the stationary points are determined by the number ν of negative eigenvalues of $\mathbf{H}$. We are here interested in the cases in which $\nu = 0$, or $\nu = 1$. In the first case we have a local minimum, and the energy increases for infinitesimal displacements in all directions; in the second case the stationary point is a saddle point of the first type (SP1) and the displacements parallel to the negative eigenvalue of $\mathbf{H}(\mathbf{R}_{st})$ correspond to an energy decrease.

We have here summarized some points of the mathematical analysis of potential energy hypersurfaces: many other properties are of interest for the study of chemical reactions, and the interested reader can look up in a number of accurate monographs (e.g. Mezey, 1987; Heidrich *et al.*, 1991) which resume the abundant literature.

From our short summary it comes out that the calculation of energy derivatives with respect to the nuclear coordinates is an essential point in the characterization of stationary points. Actually, the calculation of derivatives is also a decisive tool in the search for the location of these stationary points. There is a large, and still fast growing, number of reviews surveying the formal and computational aspects of this problem (Schlegel, 1987; Bernardi and Robb, 1987; Dunning 1990; Schlick 1992; McKee and Page, 1993).

In recent years, the elaboration of efficient methods to compute analytical energy derivatives has made it possible to apply the canonical approach for the study of reaction mechanisms to chemical systems of sizeable dimension (for reviews, see Jorgensen and Simons, 1986; Pulay, 1987; Helgaker and Jorgensen, 1988; Yamagouchi *et al.*, 1994). As a matter of fact, there is a gain of almost three orders of magnitude in computing first–order derivatives with analytical formulas with respect to finite difference methods. This technical achievement allows the use of sophisticated and efficient methods in the search for minima or SP1 saddle points (in general, it is advisable to use different methods for minima and saddle points). We shall not enter into more details, what said being sufficient to discuss the application of this first step of the canonical approach to reaction mechanisms in solution.

In principle there are no differences in applying this strategy to $G(\mathbf{R})$ (eq.7) instead of $E(\mathbf{R})$. On the contrary, from a practical point of view, the differences are important. All the EH–CSD methods are characterized by the presence of boundary conditions defining the portion of space where there is no solvent (in many methods it is called the cavity hosting the solute). A good model must have a cavity well tailored to the solute shape, and the evaluation of the derivatives $\partial G(\mathbf{R})/\partial q_i$ and $\partial^2 G(\mathbf{R})/\partial q_i \partial q_j$ must include the calculation of partial derivatives of the boundary conditions.

The formulation and the computer implementation of the complete analytic expressions of $G(\mathbf{R})$ derivatives is a challenging task, which has only recently been considered with the due attention. Computer codes are at present less efficient than those for the derivatives of $E(\mathbf{R})$; in those codes we know better there is at least an order of magnitude of difference: in other words if the ratio of computational times of analytical and finite difference methods for $E(\mathbf{R})$, $\frac{\partial E}{\partial q_i}/\frac{\Delta E}{\delta q_i}$, is 0.005, the analogous ratio $\frac{\partial G}{\partial q_i}/\frac{\Delta G}{\delta q_i}$ is 0.05, or worse. The analytical calculation of the $G(\mathbf{R})$ derivatives is more efficient when the cavity has a regular shape (sphere, ellipsoid). There are EH–CSD methods (e.g. Rinaldi *et al.*, 1992) which update the shape of the ellipsoids along the search path, exploiting this feature of regular cavities. The calculation of the derivatives is even faster when the cavity is kept fixed. However, geometry optimization may be severely distorted when programs which are only provided with this option are used.

The precise location of minima, and especially of SP1 points on the $E(\mathbf{R})$ surface is often a delicate task: according to the local features of the potential energy shape, different computational strategies may exhibit strong differences in their effectiveness. The experience so far gathered in the location of stationary points on $G(\mathbf{R})$ surfaces indicates that there is often a loss of efficiency in the last steps of the search; this could be due to the elements of discreteness that all the EH–CSD methods have (finite boundary elements, discrete integration grids, truncated multipole expansions). Maybe *ad hoc* search strategies with a more holistic character could be convenient.

An even more important point is the correct definition of the chemical composition of the "solute" determining the dimension of the nuclear conformation space. The problem is similar to the one we have already discussed for chemical equilibria involving metal cations. In many cases the solvent, always present, may act as a catalyst. In the study of chemical reactions in solution, we consider it important to reach reliable conclusions about the role of the solvent, whose molecules may give a non–specific assistance to the reaction and, in some cases, a limited number of them may also play a specific role. We stress here our belief that the occurrence of hydrogen bonds between solute and solvent molecules does not imply that these solvent molecules must be included in the molecularly treated "solute": a good continuum model is able to treat hydrogen bonding effects without resorting to a molecular description of both partners.

Evidence of a specific (catalytic) role of one, or few, solvent molecules is to be found in the proximity of the transition state (TS): the examination of solvation clusters of reagents or products, in general, gives little information. The strategy we suggest is based on a progressive enlargement of the number of solvent molecules included in the "solute" when the geometry of the relevant SP1 point is searched. In our experience it is easy to discern when the last added solvent molecule does not play an active role any longer. When the correct number of solvent molecules for the description of the TS is determined, it is convenient to keep the same chemical composition of the "solute" for the determination of the other stationary points: in some cases the structure of the preliminary complexes and intermediates depends on some specific roles played by these solvent molecules. To adopt this strategy one has to use a solvation procedure able to describe the system at a comparable degree of precision when one solvent molecule, which does not play a specific role, is added, or removed from the molecularly treated "solute".

The following step in the canonical procedure is the determination of the reaction path (RP). In its definition all the dynamical effects are discarded: the pictorial image of the RP as the hypothetical trajectory on the potential energy surface followed by a molasses drop, starting from SP1 and reaching the minimum well describes the differential equations one has to solve to obtain the RP. The references given at the beginning of this subsection, to which we can add Kraka and Dunning (1990), may be used as a guide for the specialized literature on this subject. Actually, there is not a unique definition of the RP (or of the corresponding curvilinear coordinate, often called reaction coordinate, RC). The most used definition is due to Fukui (1970), and is generally named intrinsic reaction coordinate (IRC). The IRC satisfies the following system of differential equations:

$$\frac{d}{ds}\mathbf{X}(s) = -\frac{\mathbf{g}(s)}{||\mathbf{g}||} = \mathbf{V}(s) \qquad (15)$$

where $\mathbf{X}(s)$ is a vector of mass-weighted Cartesian coordinates expressed as a function of the distance (arc length) parameter s, $\mathbf{g}(s)$ is the corresponding gradient and $||\mathbf{g}|| = \sqrt{\mathbf{g}^\dagger \mathbf{g}}$ is the norm.

The determination of the RP is computationally intensive, and hardly feasible (at least at the *ab initio* level) if analytical expressions of the gradient are not available.

Now if we go on to consider the calculation of the IRC on $G(\mathbf{R})$ surfaces, the delicate point is the evaluation of the portion of this coordinate (and of the corresponding energy profile) when the solute boundary conditions (i.e. the cavity surface) show an abrupt change of connectivity. These changes are always present in bimolecular interactions, such as $A{+}B \rightarrow TS \rightarrow C{+}D$, when the RC coordinate describes the formation of the preliminary $A \cdot B$ complex and the separation of the final $C \cdot D$ complex. When the size of the reacting system is bigger, there may be other points on the RP showing local changes of connectivity. The solvation procedure to be used must be able to describe these effects. We remember that the physically most suitable definitions of the cavity surface are those given by the Solvent Accessible Surface, $\mathcal{S}_{SA}$, introduced by Lee and Richards (1971) (that may present changes of local connectivity even for changes in molecular conformations leaving the bond connectivity unchanged) and by the Molecular Surface, MS, introduced by Richards (1977). In the following we shall adopt for the last a different name, Solvent Excluding Surface, $\mathcal{S}_{SE}$.

Once the energy profile on the RP and the corresponding changes in the internal geometry of the system have been determined, the first goal in the canonical description of the mechanism is reached. The further steps begin to consider dynamical aspects of the reaction. We shall consider these points later, after some comments on the use of the information so far collected.

The publication of numerical data regarding RP and its energy profile is merely a report of what has been found. It may be sufficient for many purposes, but the ultimate goal of theoretical chemistry is to give an interpretation of what has been computationally found. The interpretation of the studied phenomenon must be based on clear and simple physical and chemical concepts. In order to perform this analysis we need to define and use appropriate theoretical and computational tools.

The literature offers quite a large number of tools satisfying these basic requirements, but to consider all of them adequately would require a separate review. We shall limit ourselves to a few remarks, mainly regarding application to reactions in solution.

Many tools regard subunits of the reacting system: for example, they refer to the analysis of the energy of specific molecular orbitals, to the shape of localized charge distributions (under the form of localized orbital distribution, values of atomic charges), to local properties such as the stiffness of a specific bond, or the classical forces between two molecular groups, etc. In solution, there is an additional component to be considered, namely the solvent. The elaboration of descriptive/interpretative tools exploiting this new dimension is still in its infancy. A lot of new ideas may be tested, with the perspective of exploiting the personal ingenuity and ability to get algorithms of wide application.

One example may suffice. In the EH–CSD version that our group has introduced (it will be documented in Section 6), we are deliberately making a systematic use of cavity surface distribution of the components of the solvation effects (electrostatic apparent charges, dispersion–repulsion elements, etc.). These local components may be further partitioned according to the source from which they derive (for example, apparent charges due to different chemical groups), and used to elaborate tools. In fact, if we condense the information they carry to a limited number of scalar quantities, the resulting tools may assess the influence of specific solvent/molecular group interactions in facilitating or impeding the reaction course (for an example, see Coitiño *et al.*, 1994).

The following steps in the study of molecular mechanisms cannot claim the status of canonical procedure, being the number of applications by far smaller, and limited to prototypical examples. We shall consider here two methods, both emphasizing the role of the RC previously defined.

First, we would like to mention the reaction path Hamiltonian approach (RPH) proposed by Miller *et al.* (1980). In this approach the information enclosed in the RP is supplemented by some information about the shape of the PES on the $3N-7$ coordinates which are perpendicular to each point of s, and are described in the harmonic approximation. The PES thus assumes the form:

$$E(s, Q_1, \ldots\ldots, Q_{3N-7}) = E(s) + \sum_{k=1}^{3N-7} \frac{1}{2}\nu_k^2 Q_k^2(s) \tag{16}$$

where Q_k are normal coordinates and ν_k the corresponding harmonic frequencies. To use the Hamiltonian associated to this definition of the potential energy one has to consider coupling terms among the motions along s and the $Q_k(s)$ coordinates. These couplings are called curvature coupling coefficients because they measure to what extent the trajectory may curve in a particular transverse coordinate.

The method is quite computer–demanding: one has to compute and to diagonalize the Hessian matrix on a sizeable number of points on the RP coordinate, and, in addition, to compute the curvature factors, which require the knowledge of the third derivatives of $E(\mathbf{R})$. Application to polyatomic systems may present some problems, as in the case of hydrogen–transfer reactions characterized by the presence of a heavy/light–heavy atom system. To overcome these problems an alternative RPH formulation has been proposed, which is based on a straight–line, the least motion path that interpolates between reactant and product geometries (Ruf and Miller, 1988; Miller, 1988). Being the methodology dependent on the definition of the RP, more exploratory studies are needed before assessing a "canonical" RPH procedure to be used in polyatomic systems.

Another approach which exploits RC data, supplemented by a description of the perpendicular modes, is the variational transition state theory (VTST), whose aim is to improve the results given by the the conventional transition state theory (TST). The basic idea is simple: conventional TST gives but an upper limit to the classical rate constant, because it assumes that at the saddle point there are no recrossing trajectories. Truhlar and coworkers (Truhlar and Garret, 1984; Truhlar *et al.*, 1985; 1987) developed new methods for the detection of the separating surface with a minimal number or recrossing trajectories (the "true" TS), and also including tunneling quantum effects, which make the VTST a practical method for the computation of rate constants.

The application of the VTST to reactions in solution has to face several computational problems, of the type we have discussed for the canonical static description of the reaction mechanism. Moreover, there is another important problem to face, not present when this approach is applied to reaction *in vacuo*, which literally adds "new dimensions" to the model. We shall consider now this point, even if shortly.

The potential energy surface used in solution, $G(\mathbf{R})$, is related to an effective Hamiltonian containing a solute–solvent interaction term, $\hat{\mathcal{V}}_{int}$. In the implementation of the EH–CSD model, that will be examined in Section 6, use is made of the equilibrium solute–solvent potential. There are good reasons to do so; however, when the attention is shifted to a dynamical problem, we have to be careful in the definition of $\hat{\mathcal{V}}_{int}$. This operator may be formally related to a response function $\vec{\mathcal{R}}$ which depends on time. For simplicity's sake, we may replace here $\vec{\mathcal{R}}$ with the polarization vector $\vec{P}$, which actually is the most important component of $\vec{\mathcal{R}}$ (another important contribution is related to G_{dis}). For the calculation of G_{el} (see eq.7), we resort to a static value, while for dynamic calculations we have to use a $\vec{P}(t)$ function: quantum electrodynamics offers the theoretical framework for the calculation of $\vec{P}$ as well as of $\vec{\mathcal{R}}$. The strict quantum electrodynamical approach is not practical, hence one usually resorts to simple naive models.

The use of a continuum distribution of the solvent, and of its response function, does not eliminate the fact that the solvent is made up of distinct molecules, each one composed by a given set of nuclei held together by the electrons. These molecules clearly translate, rotate, vibrate, and are subject to continuous changes in their electronic charge distribution. When the solvent suffers an external, time–dependent perturbation – and the internal dynamics of a subsystem undergoing a reaction is a perturbation of this kind – the equilibrium is troubled and the various degrees of freedom of the solvent react to this change with specific relaxation times τ_k. This naive picture gives a physical basis to the phenomenological expansion of

$\vec{P}$ over normal modes, with frequency ω_k

$$\vec{P} = \sum_{\omega_k} \vec{P}_{\omega_k} \tag{17}$$

The relevant frequencies ω_k (the summation may also be continuous) are related to the inverse of the relaxation times τ_k (Levich, 1966). In many cases, it may be acceptable to use a simpler expression of this expansion (the so–called Pekar separation: Pekar, 1951)

$$\vec{P} = \vec{P}_{\text{slow}} + \vec{P}_{\text{fast}} \tag{18}$$

where $\vec{P}_{\text{fast}}$ regards electronic relaxations that in this model are assumed to be immediate, while $\vec{P}_{\text{slow}}$ collects all the other relaxation phenomena.

Thus, if a charged particle moves fast enough in the medium, it will experience a retarding force (friction) due to the fact that, during its trajectory, the orientation and the position of the solvent molecules are not in equilibrium with respect to their actual position: this effect, which is expressed by $\vec{P}_{\text{slow}}$, is called dielectric friction. In addition to this, we may invoke another naive picture. During its motion a particle (which, for simplicity's sake, may be assumed as uncharged, but with a non–zero collisional diameter) collides with solvent molecules, and thus experiences a different retarding force, i.e. the mechanical friction.

For these reasons we cannot use $G(\mathbf{R})$ as a rigid support for dynamical studies of trajectories of representative points. $G(\mathbf{R})$ has to be modified, at every point of each trajectory, and these modifications depend on the nature of the system, on the microscopic properties of the solution, and on the dynamical parameters of the trajectories themselves. This rather formidable task may be simplified in several ways: we consider it convenient to treat this problem in a separate Section. It is sufficient to add here that one possible way is the introduction into $G(\mathbf{R})$ of some extra coordinates, which reflect the effects of these retarding forces. These coordinates, collectively called solvent coordinates (nothing to do with the coordinates of the extra solvent molecules added to the "solute") are here indicated by $\mathbf{S}$, and define a hypersurface of greater dimensionality, $G(\mathbf{R}\oplus\mathbf{S})$. To show how this approach of expanding the coordinate space may be successfully exploited, we refer here to the proposals made by Truhlar *et al.* (1993). Their formulation, that just lets these solvent coordinates partecipate in the reaction path, allows to extend the algorithms and concepts of the above mentioned variational transition state theory to molecules in solution.

As a last feature, it is worth mentioning another solvent effect of general occurrence. The solvent molecules possess an intrinsic thermal motion which may induce local fluctuations in their equilibrium distribution around the solute, and the effect of these fluctuations on the single molecular reaction act may be reflected in macroscopic quantities such as the reaction rate or the apparent reaction barrier. Solvent fluctuations can be introduced in the study of chemical ractions in many ways: this subject, as well as the others we have mentioned here as general flashes on methods for describing dynamical aspects of reactions, will be resumed with more details and references in the next Section.

5. Dynamical aspects of reactions in solution.

The possible role of solvent dynamics in influencing reaction in solutions has recently received considerable scrutiny. We cannot exhaustively review the impressive number of recent methodological and applicative contributions in this field, which have been supported and stimulated by new experimental evidence based on innovative techniques, and by the increasing reliability of molecular dynamics and MC simulations. Following the approach used in the previous Section to treat the static description of the solvent, we shall focus our review about dynamical aspects almost entirely on methods in which the continuum model plays a key role.

The goal of this Section is to offer a perspective on solvation dynamics starting from the classic investigation developed before 1970s until recent developments in the basic underlying theory. It is beyond our scope to provide a full overview of these developments, but the interested reader can find more details and references in many papers. Here we shall confine ourselves to quote Hynes (1985) for an excellent introduction to the subject, and Weaver (1992) for a review on recent progress, but the list should be much longer, especially if one wants to examine this field in a systematic way.

5.1. HISTORICAL OVERVIEW.

In the standard Transition State Theory (TST) the general reaction is viewed as a passage over a mean potential barrier located between reactant and product potential wells. It is often stated that TST is exact under a certain dynamical condition, i.e. a trajectory initially crossing the barrier top or transition state surface S from the side of reactants to the side of products must proceed to products without recrossing S. This no–recrossing condition makes the TST rate constant an upper limit.

Of course, TST is sometimes incorrect even in gases (see, for example, the well known breakdown of TST in its standard form exhibited by activated unimolecular reactions); in a solvent, this approach can fail due to different reasons, such as the retarding effects or collisionally induced recrossing. All these sources of breakdown of the Transition State Theory have been discussed by Kramers (1940). In his treatment the reaction system is considered as an effective particle of mass μ crossing a one–dimensional potential barrier. This motion is described by the stochastic Langevin equation (LE):

$$\mu\ddot{x} = F - \mu\zeta\dot{x} + R \tag{19}$$

The force arising from the potential is F, while R is a gaussian random force. The net effect of the "collisions", i.e. dynamical interactions between the particle and solvent molecules, is thus approximately accounted for by the frictional, or damping force, $F_d = -\mu\zeta\dot{x}$, where ζ is a friction constant related to the time correlation of the random force:

$$\zeta = (\beta/\mu)\int_0^\infty dt < RR(t) > \tag{20}$$

where the brackets denote a solvent average.

The potential, $U(x)$, in the barrier region is approximated to an inverted parabola with a frequency w_b related to the barrier curvature ($w_b = [\mu^{-1}|\partial^2 U(0)/\partial^2 x|]^{1/2}$). Actually, Kramers solves the steady–state Fokker–Planck equation associated with eq.(19) to find the following rate constant:

$$k = k^{TST}\kappa^{KR} \tag{21}$$

$$\kappa^{KR} = [1 + (\zeta/2w_b)^2]^{1/2} - (\zeta/2w_b) \tag{22}$$

where k^{TST} is the TST rate constant of the model.

Kramers' result predicts a reduction of the rate constant from its TST value due to collisions with solvent molecules, as it is shown by the Kramers transmission coefficient, $\kappa^{KR} \leq 1$. Only if $\zeta/w_b \ll 1$, $\kappa^{KR} \to 1$ and TST value is obtained. In this case, the solvent friction is ineffective in inducing recollisions, and the reaction is a "free" passage across S from an equilibrium distribution.

The characteristic time scale for the motion of the particle in the parabolic top barrier is the inverse barrier frequency, w_b^{-1}: the sharper is the barrier, the faster is the motion. Typically, atom transfer barrier are quite sharp; therefore the key time scale is very short, and the short–time solvent response becomes relevant instead of the long–time overall response given by the ζ used in Kramers' theory (see eq.(20)). To account for this critical feature of reaction problems, Grote and Hynes (1980) introduce the generalized Langevin equation (GLE):

$$\mu\ddot{x}(t) = F(t) - \mu\int_0^t d\tau\zeta(\tau)\dot{x}(t-\tau) + R(t) \tag{23}$$

where the time dependent friction coefficient is governed by the fluctuating forces on the coordinate x through their time correlation function:

$$\zeta(t) = (\beta/\mu) < RR(t) > \tag{24}$$

With the assumption of a parabolic reaction barrier in the top region, the reaction transmission coefficient, i.e. the ratio of the actual rate constant to its TST value, $\kappa = k/k^{TST}$, is found to be the ratio of the reactive frequency λ_r to the mean barrier frequency:

$$\kappa = \lambda_r / w_b \quad (25)$$

$$\lambda_r = w_b^2[\lambda_r + \hat{\zeta}(\lambda_r)]^{-1} \quad (26)$$

$$\hat{\zeta}(\lambda_r) = \int_0^\infty dt\, e^{-\lambda_r t}\zeta(t) \quad (27)$$

Equations (25) and (26) are Grote-Hynes key results. They show that κ is determined by λ_r and that, in its turn, this reactive frequency λ_r is determined both by the barrier frequency w_b and by the Laplace transform frequency component of the friction (see eq.(27)).

Many qualitative aspects of the rate constant follow from the self–consistent eq.(26). If the friction is very weak ($\hat{\zeta}(\lambda_r) \ll \lambda_r$), trajectories across the barrier are negligibly perturbed by collisions with solvent molecules and eq.(26) gives $\lambda_r \cong w_b$. Thus, the reactive frequency is just the barrier frequency and the TST result is obtained ($\kappa = 1$). In the opposite limit of large friction ($\hat{\zeta}(\lambda_r) \gg \lambda_r$), barrier region trajectories are strongly and continually perturbed by solvent collisions. Since the solvent is rapid on the time scale of λ_r^{-1}, the frequency dependence of $\hat{\zeta}(\lambda_r)$ can be ignored, and Grote–Hynes equations are reduced to the Kramers theory result (eq.(22)), in which zero frequency friction, or friction constant, ζ, is given by $\hat{\zeta}(0)$.

The Grote–Hynes theory has been found to be in good agreement with the numerically determined rate constant in several computer simulation studies (see for example Gertner *et al.*, 1989; Ciccotti *et al.*, 1989, and references therein). In recent years, it has been demonstrated that the GLE-based Grote–Hynes theory is equivalent to *multidimensional* TST which includes fluctuations of the environmental force modes through an effective harmonic bath bilinearly coupled to the system coordinate (Pollack, 1986). While it seems evident that the GLE in eq.(23) is a powerful way to approximately describe the frictional influence of the evironment on the reaction coordinate during an activated rate process, this form of the GLE rests on the assumption that the friction coefficient $\zeta(t)$ is independent from the motion of the reaction coordinate. As discussed by several authors (Lindberg and Cortes, 1984; Krishnan *et al.*, 1992; Straub *et al.*, 1990), a spatial indipendence of the friction may not be a very accurate representation of the dynamics of realistic Hamiltonians. In light of this fact, it is more appropriate to recast the expression for the friction in the spatially dependent form:

$$\zeta(t; x) = (\beta/\mu) < RR(t) >_x \quad (28)$$

where $< RR(t) >_x$ is the bath averaged correlation function for the fluctuating force exerted by the bath modes on the reaction coordinate, x, with x being "clamped" at a particular position. In the Grote–Hynes theory, the friction coefficient to be used in eq.(23), is usually calculated with the reaction coordinate clamped at the top of the barrier along x. If eq.(28) exhibits a significant spatial dependence, however, the use of eq.(23) to describe realistic condensed phase activated dynamics is not rigorously justified and might not provide a good approximation to the true dynamics in some instance. Noteworthy efforts to include spatially dependent diffusion in limiting equations for the rate constant are those of Northrup and Hynes (1979,1980), and later Gavish (1980) and Grote and Hynes (1980), who derived an expression for the high friction diffusive limit for reactive systems characterized by a spatially dependent diffusion coefficient. Another effort which treats both space and time dependent friction in the overdamped regime of the barrier crossing dynamics is the "effective Grote–Hynes" theory by Voth (1992). Yet another effort is the subject of a paper of Pollack and coworkers (Haynes *et al.*, 1994), wherein a complete theory for the barrier crossing rate with space and time dependent friction of arbitrary strength is presented. In this paper the theory is specialized for the case of the uniform coupling model, in which the friction is represented as a product of two functions which are independently spatially dependent and time dependent. This work and related studies indicate that the effect of the spatially dependent friction leads to not trivial deviations from the Grote–Hynes theory, and that, consequently, care must be exercised in the application of the latter to any experimental systems.

5.2. SOLVENT COORDINATE.

We have seen that dynamical solvent effects in the friction can lead to a breakdown of TST. As stressed above, this is also a breakdown in the equilibrium solvation assumption for the transition state and configurations in its neighborhood. In fact, the standard TST view is a special one–dimensional equilibrium perspective, i.e. a mean potential curve for the reacting species is visualized and no friction of any sort is considered. The solvent influence can be felt solely *via* this potential, hence it is assumed that for each configuration of the reacting species, the solvent is equilibrated. On the contrary, the discussion above about Kramers and Grote–Hynes theories has documented the importance of nonequilibrium solvation effects in a frictional language.

Actually, there is another equivalent perspective of the problem, in which an additional solvent–dependent coordinate is introduced. The advantage here is that the solvent participation in the reaction can be examined explicitely; indeed, the reaction coordinate, and the free–energy hypersurface, depend on both the reactive species coordinate and the solvent coordinate. A limiting case may be considered here as an example. In the outer–sphere electron–transfer (ET) reaction mechanism between two spherical ions, there is no change in the first solvation shell radii during the ET process, according to Marcus' model (Marcus, 1956, 1960, 1963, 1965). The barrier for the electron transfer appears in a "solvent coordinate", s, which measures deviations from the equilibrium distribution of the solvent. The specification of the s coordinate has been left very vague until Zusman (1980), who identified it in the context of a continuum solvent model as:

$$s = \int_{\text{solv}} d\vec{r}(\vec{E}^o_R - \vec{E}^o_P) \cdot [\vec{P}(\vec{r}) - \vec{P}^{eq}_R(\vec{r})] \tag{29}$$

Here $\vec{E}^o_i(\vec{r})$ are the bare or *vacuum* electric fields of the reactant (R) and product (P) states. $\vec{P}^{eq}(\vec{r})$ denotes the equilibrium solvent polarization for the R state, while $\vec{P}(\vec{r})$ is the actual time–dependent polarization.

Here, what said in Section 4.2 about time–dependent perturbations is worth recalling, trying to give a more detailed analysis. The best approach to treat problems characterized by the presence of a $\vec{P}(t)$ function is provided by the quantum electrodynamics theories, where $\vec{P}$ is described in terms of an expansion over normal modes of the dielectric polarization. This model can be simplified by considering only two terms, often called the "fast" and the "slow" contribution to $\vec{P}$ (the Pekar separation introduced in eq.(18) of Section 4.2):

$$\vec{P} = \vec{P}_{\text{fast}} + \vec{P}_{\text{slow}} \tag{30}$$

The fast component is clearly related to electronic polarization, $\vec{P}_{\text{fast}} = \vec{P}_{el}$, while the slow component, connected to nuclear motions of the solvent molecules, is often called the "orientational" polarization ($\vec{P}_{\text{slow}} = \vec{P}_{or}$), or "inertial" component ($\vec{P}_{\text{slow}} = \vec{P}_{in}$). This simplified model has been developed and applied by many authors: we shall recall here Marcus (see the papers already quoted), who first had the idea of using $\vec{P}_{\text{slow}}$ as a dynamical coordinate. For description of solvent dynamical coordinates in discrete solvent models see Warshel (1982) and other papers quoted in Section 9.

An important step in the definition of combined solute and solvent coordinates is presented in some papers by van der Zwan and Hynes (1984) and particularly by Lee and Hynes (1988), who have shown how Marcus' treatment of nonequilibrium polarization effects may be accomplished for model charge–transfer reactions (e.g. proton transfer, hydride transfer, and S_N2 reactions) in a generalized continuum description of nondissipative polar solvents. In their formalism, a single effective solvent coordinate, s, is defined in terms of changes of the "slow" (here called "heavy particle atomic") polarization of the dielectric medium, $\vec{P}_h$, which is modeled in terms of variations of charge e_α on the atom (or moiety) α as a function of a single solute reaction coordinate x:

$$s = s^{eq} + A(x) \int d\vec{r}[\vec{P}_h(\vec{r}|x) - \vec{P}_h^{eq}(\vec{r}|x)] \cdot \Delta\vec{E}^o(\vec{r}|x) \tag{31}$$

$$\Delta\vec{E}^o(\vec{r}|x) = \sum_\alpha \frac{\vec{r} - \vec{r}_\alpha(x)}{|\vec{r} - \vec{r}_\alpha(x)|^2}(e_\alpha^P - e_\alpha^R) \tag{32}$$

where $\vec{P}_h^{eq}$ is the "slow" polarization in equilibrium with the reaction system charge distribution $\{e_\alpha(x)\}$ (for which $s = s^{eq}$), and $A(x)$ is a function of dielectric constants, static and high frequency, and of the vacuum field $|\Delta\vec{E}^o(\vec{r}|x)|$ (indices R and P of eq.(32) stand for reactant and product states, respectively).

For polyatomic complex systems the use of a single dynamical coordinate s, as in Marcus' theory of ET reactions and in the following generalizations described above, may not be sufficient. The extension of these formalisms to many coordinates have been exploited by several groups. In the already quoted work on the extension of VTST to reactions in solution, Truhlar *et al.* (1993), generalize Lee and Hynes' formalism defining a solvent coordinate y_i for each internal coordinate of the solute q_j:

$$y_i = y_i^{eq} + A'(x) \sum_j \int d\vec{r}[\vec{P}_h(\vec{r}) - \vec{P}_h^{eq}(\vec{r}|\mathbf{q})] \cdot \frac{\partial P_h^{eq}(\vec{r}|\mathbf{q})}{\partial q_j} \tag{33}$$

Like the single solvent coordinate of Lee and Hynes, each y_i is determined from the component of the deviation of the nonequilibrium "slow" polarization from its equilibrium value that is parallel to a reference vector. However, in this case the reference vectors are locally determined from small displacements of the solute geometries rather than globally determined from the changes of charge distributions from reactants to products (see eqs. 31–32). The advantage of Truhlar's treatment with respect to that of Lee and Hynes is that it can also model reactions involving motion of charges through the dielectric media with little or no changes in the atomic charges themselves, or reactions affected by internal motion of the solute other than only the reaction coordinate x; hence this model is suitable for a treatment of larger classes of reactions.

A different set of dynamical variables can be given by the use of continuum models based on the apparent surface charge approach (ASC). In the PCM (Aguilar *et al.*, 1993b; Cammi and Tomasi, 1995a) the set of coordinates is reduced to a discrete number, related to the cavity shape and to the set of surface charges $\{q_k\}$ whose dimension is given by the number of the representative points selected on the solute surface (see Section 6 for a detailed description of the PCM model). Actually, this set is previously partitioned (as already seen in eq.(30) for $\vec{P}$) into "slow" and "fast" components related to relaxation processes occurring in solution with different times. The same considerations made for the polarization vector are still valid for charges q_k. Their "slow" components are related to the orientational relaxations modes of the solvent, and easily identified as pertinent dynamical coordinates. As a second possible application, the time evolution of the cavity shape has been considered in the form of a delay with respect to the solute transformations (Aguilar *et al.*, 1993a).

Another proposal based on the ASC approach is given by Basilevsky's group. In the work on the S$_N$2 reaction $Cl^- + CH_3Cl \rightarrow ClCH_3 + Cl^-$ (Basilevsky *et al.*, 1993), the set of surface charges $\{q_i(\alpha)\}$ and the solute charge distributions $\rho_\alpha(\vec{r})$ are computed in a self–consistent way for each value x_α of the reaction coordinate, so that they are always mutually equilibrated. Then a solute charge distribution $\rho_{\alpha\beta}(\vec{r})$ is calculated for the solute configuration x_α in the field created by surface charges $\{q_i(\beta)\}$, with $\beta \neq \alpha$. This non–equilibrium situation generates a two–dimensional free energy surface $G(\alpha, \beta)$, which can be rewritten in terms of the "discrete medium coordinate" which is defined, also in this case, by the inertial (previously indicated as "slow") component of the surface charges. To proceed further, the authors introduce two limit cases, as it is customarily done in order to examine dynamical phenomena. The first one , the so–called "Born–Oppenheimer" (BO) approximation, corresponds to the infinitely fast motion of the medium electrons on the solute electronic time scale, while the other one, called Self–Consistent (SC) limit, represents the opposite situation. In a more recent paper (Basilevsky *et al.*, 1995), this description has been improved using a configuration interaction (CI) formulation of the solute wavefunction. The corresponding decomposition of the solute charge density leads to the definition of a larger set of dynamical variables, Y_{ab}:

$$Y_{ab} = -\int d^3r \int_S d^2s \frac{\sigma_{\text{slow}}(\vec{s})\rho_{ab}(\vec{r})}{|\vec{s} - \vec{r}|} \tag{34}$$

where ρ_{ab} is a single state, or transition density involving the configurations a and b (these indices run on all the configurations included in the calculation).

The use of a CI description of the solute wavefunction, instead of MO treatment, has allowed a further development of the continuum medium model. In fact, in its traditional versions, solvent electrons are considered in terms of the "fast" polarization field defined by the solute charge density. The limitations of this "classical" description become clear if we consider the time scale tipically involved in the motions of these electrons. As a matter of fact, electronic relaxation times are generally of the order of 10^{-16}–10^{-15} s. Several advanced treatments with an explicit quantum consideration of the medium electrons have been reported (Gehlen *et al.*, 1992; Kim and Hynes, 1992; Basilevsky and Chudinov, 1992; Kuznestov, 1992). The most elaborate theory introduced for this purpose is that of Kim and Hynes (1992). Here, the full quantization of the "fast" (or electronic) polarization, $\vec{P}_{el}$, is effected *via* a multiconfigurational self-consistent (MCSCF) representation of the solute–solvent wavefunction:

$$\Psi = \sum_i c_i \left|i\right\rangle \left|P_{el}^{(i)}\right\rangle \tag{35}$$

where $|i\rangle$ denotes the electron localized states of the solute (in a diabatic Valence Bond state framework) and $\left|P_{el}^{(i)}\right\rangle$ is an arbitrary coherent state for the solvent electronic polarization, which will be determined by some specific optimization conditions. At first the model was limited to the two–state formulation ($i = 1, 2$ in eq.(35)), but recently it has been generalized to account for more than two solute VB states (Bianco *et al.*, 1994).

This analysis also allows a new reformulation of the two extreme models, SC and BO, that were previously defined. In fact, when $\vec{P}_{el}$ is much faster than solute electrons (BO limit), each $\left|P_{el}^{(i)}\right\rangle$ is determined solely by the single charge–localized solute state $|i\rangle$; whereas in the opposite limit, where $\vec{P}_{el}$ is much slower, the electronic polarization becomes equilibrated to the smeared–out solute charge distribution, e.g. it "sees" an average of all the states $|i\rangle$. Kim–Hynes' theory states that the electronically adiabatic ground–state free energy is bound by these two limits for a large variety of reactions. In the last few years the approach has been applied to a rapidly growing number of chemical problems. We shall quote here some applications to S_N1 (Mathis *et al.*, 1993; Kim *et al.*, 1993), electron transfer (Mathis and Hynes, 1994), and S_N2 (Mathis *et al.*, 1994) reactions. A recent review on the results so far obtained following the Kim–Hynes' approach (Hynes *et al.*, 1994) gives a clear indication of its potentiality. The VB diabatic model can be used in discrete solvent models too, for example Warshel (1982), this point will be reconsidered in Section 9.

To end this Section, we would like to go back to what we had just mentioned in Section 4.2 about dynamical effects of stochastic solvent fluctuations on chemical reactions. As already said, solvent fluctuations can be introduced in many effective ways; as an example we quote an attempt to model them, *via* the continuum PCM approach (Bianco *et al.*, 1992) using a S_N2 reaction as test case. In this formulation a fluctuation of a reasonable magnitude is modelled as a time–dependent change in the polarization charges $\{q_k\}$ on the surface of the cavity surrounding the solute. It is worth noting that the introduction of a single fluctuation does not allow to consider the computed solvation energy as a free energy (as it is the case for the static model), but now an average over fluctuations must be performed. This example has to be considered as a first attempt for a non–equilibrium description of a reaction including fluctuations: of course much work has still to be done, and other, maybe more effective, ways can be introduced in order to get a faithful description of solvent fluctuations.

More recently, Rivail's group (Ruiz–López *et al.*, 1995) has developed a model which analyses non–equilibrium effects on chemical reactions, using the SCRF multipole expansion method (MPE) (see Section 8.2). The role of solvent fluctuations is emphasized in this study and explicit use of a solvent dynamical coordinate is made. This coordinate is related to the dipole contribution to the reaction field, thanks to the partition of this field into separate multipole components which is more immediate in MPE approaches. This simple model is appropriate to the S_N2 reaction considered as a test case, but extension to higher multipole moments or to multicenter expansions is feasible. There is thus the prospect of a hierarchy of models of increasing complexity, which, if systematically applied, may give important information on dynamical effects on more complex reactions.

In this Section we have discussed some important aspects of the dynamical problems, and reported some attempts to model the large variety of dynamical phenomena taking place in any kind of reaction in solution. However, the most convenient way of merging the various solvent effects (frictions, fluctuations, etc.) into a unified computational scheme is still under development. Only the progress of the research will show if simple strategies are possible and what features, not considered until now, are necessary to improve our understanding of the dynamics of chemical reactions.

6. Methods for the evaluation of $G(\mathbf{R})$ in the PCM formalism.

After the parenthesis on general aspects of reactions in solution, in this Section we shall go back to methods and evaluate the free energy hypersurface $G(\mathbf{R})$ defined in eq.(7). Our analysis will consider each component of $G(\mathbf{R})$, eq.(7), separately, using the formalism of the PCM method developed in Pisa as reference. The combination of two or more terms in the same calculation will be examined at the appropriate places.

6.1. ELECTROSTATIC TERM G_{EL}.

We have to solve the following Schrödinger equation

$$(\hat{\mathcal{H}}_M^o + \hat{\mathcal{V}}_{el})\Psi = E\Psi \tag{36}$$

where the Hamiltonian is similar to that of eq.(1) but where $\hat{\mathcal{V}}_{int}$ is replaced by $\hat{\mathcal{V}}_{el}$.

In the continuum solvent distribution models, $\hat{\mathcal{V}}_{el}$ is evaluated by resorting to the description of the solvent as a dielectric medium. This medium may be modeled in many different ways, being the continuous methods quite flexible. We shall consider the simplest model only, i.e. an infinite linear isotropic dielectric, characterized by a scalar dielectric constant ε. The interested reader can refer to a recent review (Tomasi and Persico, 1994) for the literature regarding more detailed and more specialistic models. However, the basic model we are considering here is sufficient to treat almost all chemical reactions occurring in bulk homogeneous solutions.

In the PCM procedure $\hat{\mathcal{V}}_{el}$ is expressed in terms of an apparent charge distribution $\sigma(\vec{s})$ which is spread on the cavity surface (Apparent Surface Charge method, ASC):

$$\sigma(\vec{s}_k) = \frac{\varepsilon - 1}{4\pi\varepsilon} \vec{\nabla}\Phi_{tot}(\vec{s}_k) \cdot \hat{n}_k \tag{37}$$

where $\hat{n}_k$ is the unit vector, normal to the cavity $\mathcal{S}$ at the point $\vec{s}_k$ and pointing outwards.

The potential Φ_{tot} derives from all the sources present in the model, i.e. the solute and the apparent surface charges. Factor $1/\varepsilon$ of eq.(37) reflects the boundary conditions of the electrostatic problem. The inclusion of this factor means that the gradient at $\vec{s}_k$ is computed in the inner part of the surface element (the dielectric constant is ε for the medium and 1 for the inner cavity space).

In principle, the ASC method gives an exact solution of the electrostatic problem for this model. In practice, the fact that the source of $\hat{\mathcal{V}}_{el}$ is confined to a close surface makes the numerical solution of the problem easier. The continuous charge distribution $\sigma(\vec{s})$ is replaced by a set of point charges $\{q_k\}$, each placed at the center $\vec{s}_k$ of an element of the cavity surface (called tessera) having an area $\Delta\mathcal{S}_k$:

$$q_k = \Delta\mathcal{S}_k \, \sigma(\vec{s}_k) \tag{38}$$

with $\sum_k \Delta\mathcal{S}_k$ equal to the cavity surface $\mathcal{S}$.

Using this discretization of the charge distribution, we can write:

$$\hat{\mathcal{V}}_{el} = \sum_k \frac{q_k}{|\vec{r} - \vec{s}_k|} \tag{39}$$

It is important to remark that this formulation for $\hat{\mathcal{V}}_{el}$ makes it possible to treat solvent–separated subsystems (for example, the reactants A and B) at the same level as the composite molecular system (i.e. A–B in our example). The possibility of describing portions of the overall hypersurface corresponding to solvent–separated reagents or products is a feature which is not present in other methods. The generalization of formulas (37)–(38) to the case of many boundary surfaces may be found in Bonaccorsi *et al.* (1986). Analogous expressions have been considered by Sakurai and coworkers in a cognate formulation of the PCM method (Hoshi *et al.*, 1987; 1988).

This approach, which is based on a discretization of σ on the cavity surface $\mathcal{S}$, is often referred to as the Boundary Element Method (BEM). There is an abundant literature about BEM in the field of applied engineering and physics (Banerjee and Butterfield 1981; Beskos 1987). In this application of the BEM approach we need to specify how the surface $\mathcal{S}$ is defined and partitioned in tesserae, and how $\hat{\mathcal{V}}_{el}$ is computed.

6.1.1. *Definition of the cavity and of its tessellation.*

Solvation energy is rather sensitive to the shape and the volume of the cavity. Also the shape of the $G(\mathbf{R})$ surface may be deformed by the choice of a badly balanced cavity.

In the PCM procedure we use cavities built according to the GEPOL method (Pascual-Ahuir *et al.*, 1987; 1994; Pascual-Ahuir and Silla, 1990; Silla *et al.*, 1991), which has been originally devised just as a supplement of the PCM. In this method each atom (or small group of atoms, like CH_3) is provided with a sphere of suitable radius. The union of these spheres gives the "first–generation" cavity, with surface $\mathcal{S}_I$. For solutes of small size $\mathcal{S}_I$ is sufficient, but when the complexity of the solute increases, there will be regions of the outer space where the solvent cannot enter. In these cases one may resort to the definition of Solvent Excluding Surface ($\mathcal{S}_{SE}$) first defined by Richards (1977). In GEPOL $\mathcal{S}_{SE}$, and the corresponding volume $\mathcal{V}_{SE}$, is defined by introducing additional spheres (second–generation) which are no longer centered on the nuclei. Each sphere, either atomic or additional, is partitioned into 60 equivalent curvilinear triangles, with their vertices corresponding to the ones of a pentakisdodecahedron inscribed in it. A considerable number of these triangles are deleted being completely contained in the volume of the cavity. A number of triangles is partially exposed to the continuum medium: they are replaced by surfaces having an appropriate shape. Thus, each sphere contributes to the definition of the whole cavity with a number of tesserae equal or less than 60, these tesserae being spherical triangles or spherical polyhedra.

Other ASC solvation methods use Connolly's algorithm (Connolly, 1983; 1985) to define $\mathcal{S}_{SE}$ and to provide a tessellation of the surface. GEPOL gives surface and volume values which are more accurate and at a lower cost than Connolly's algorithm (Pascual-Ahuir and Silla 1990; Pascual-Ahuir *et al.*, 1994; Silla *et al.*, 1991). Recently we have introduced a modification in GEPOL to make it more suited for the calculation of analytical derivatives of $G(\mathbf{R})$ (Cammi and Tomasi, 1994; Cossi *et al.*, 1996a).

6.1.2. *Calculation of $\hat{\mathcal{V}}_{el}$ and obtainment of G_{el}.*

The calculation of $\hat{\mathcal{V}}_{el}$ is linked to the solution of the Schrödinger equation (36), since $\hat{\mathcal{V}}_{el}$ depends on the solute charge distribution ρ_M (electrons and nuclei). It is convenient to divide ρ_M into electronic and nuclear contributions:

$$\rho_M = \rho_{M,n} - \rho_{M,e} \tag{40}$$

with

$$\rho_{M,e}(\vec{r}_1;\mathbf{R}) = \int |\Psi(\mathbf{r};\mathbf{R})|^2 \, d\vec{r}_2.....d\vec{r}_n \tag{41}$$

$$\rho_{M,n}(\vec{r}_1;\mathbf{R}) = \sum_{\alpha}^{N} Z_\alpha \delta(\vec{r}_1 - \vec{R}_\alpha) \tag{42}$$

If we work at a fixed nuclear conformation (clamped nuclei), $\hat{\mathcal{V}}_{el}$ will be a function of $\rho_{M,e}$, i.e. $\hat{\mathcal{V}}_{el}(\Psi\Psi^*)$. Schrödinger equation (36) is not linear as the Hamiltonian depends on the eigenfunction Ψ. To solve this equation three methods have been devised, i.e.: a) iterative solution, b) closure solution, and c) matrix inversion. All methods are of current us. According to the type of problem, it is convenient to adopt a different procedure.

a) Iterative solution.

The potential Φ_{tot} may be decomposed into components due to the solute M, and to the apparent polarization charges:

$$\Phi_{tot}(\vec{r}) = \Phi_M(\vec{r}) + \Phi_\sigma(\vec{r}) \tag{43}$$

Φ_M in turn may be decomposed into a nuclear and an electronic contribution:

$$\Phi_M(\vec{r}) = \Phi_{M,n}(\vec{r};\mathbf{R}) + \Phi_{M,e}(\vec{r};\Psi) \tag{44}$$

with

$$\Phi_{M,n}(\vec{r}) = \int \frac{\rho_{M,n}(\vec{r}\,')}{|\vec{r} - \vec{r}\,'|} d\vec{r}\,' \tag{45}$$

$$\Phi_{M,e}(\vec{r}) = -\int \frac{\rho_{M,e}(\vec{r}\,')}{|\vec{r} - \vec{r}\,'|} d\vec{r}\,' \tag{46}$$

In eq.(44) we have indicated the parametric dependences which are directly relevant to the discussion. In the clamped nuclei approximation, $\Phi_{M,n}(\vec{r};\mathbf{R})$

is constant during the whole process, while $\Phi_{M,e}(\vec{r};\Psi)$ changes when Ψ is modified by the solvent polarization effects.

The self–consistency between Ψ and $\Phi_{M,e}$ can be achieved through an iterative cycle: at the first step $\Phi_{M,e}$ and $\hat{V}_{el}$ are calculated using the solute wavefunction obtained *in vacuo*. A new Ψ is then determined through eq.(36), and used to re-calculate $\Phi_{M,e}$; the cycle is repeated until reaching convergency (usually 3-5 iterations are sufficient). Another iterative cycle, nested to the one discussed above, is necessary to get apparent polarization charges in equilibrium with the total electrostatic potential, eq.(43). Keeping Φ_M frozen, the apparent charges are calculated starting with only the solute term in Φ_{tot} of eq.(37), in order to obtain a first guess for σ. The latter is then iteratively renewed by also adding the contribution Φ_σ due to the apparent charges themselves. This inner cycle, which is repeated at each outer cycle on Ψ, usually gets convergency in 4-5 iterations.

The original procedure (Miertuš *et al.*, 1981) was expressed in the restricted Hartree–Fock (HF) formalism. Extensions to other levels of the quantum theory are easy, and there are versions of the PCM program accepting UHF, ROHF, MP*n*, CASSCF, SDCI, MRCI, and CASSCF–CI levels of quantum molecular theory. The extension to semiempirical quantum methods has been elaborated by several groups. We quote here Miertuš *et al.* (1988) for CNDO methods, and Luque and Orozco (Luque *et al.*, 1993; 1995; Negre *et al.*, 1992) for PM3 and AM1 version. Both of them follow this approach. The computational codes are of public domain. The PM3 version has been included in MOPAC package (Stewart, 1990).

We shall consider later other quantum versions introducing some changes in this scheme, as well as semiempirical versions which do not use the outer cycle related to the solution of the Schrödinger equation.

b) Closure solution.

The apparent charges at the end of each inner cycle of the iterative version presented above may be expressed in a compact form through a combination of series: two approximations have been proposed which replace the whole polarization cycle with simple expressions for the charges. Following them we can write:

$$q_k = (1 - A_k)^{-1}\Big[q_k^0 - \sum_{l \neq k} q_l^0 B_{k,l}(1 - A_l)^{-1}\Big] \tag{47}$$

and

$$q_k = (1 - A_k)^{-1}\Big(q_k^0 - \sum_{l \neq k} B_{k,l}(1 - A_l)^{-1}\Big[q_l^0 - \sum_{n \neq l} B_{l,n}(1 - A_n)^{-1}\Big]\Big) \tag{48}$$

In these two equations, q_i^0 is the apparent charge calculated on tessera i, by considering only the electrostatic potential Φ_M due to the solute in eq.(37), and

$$A_k = \frac{\varepsilon - 1}{2\varepsilon}(1 - \xi_k) \tag{49}$$

$$B_{k,l} = \frac{\varepsilon - 1}{4\pi\varepsilon}\Delta\mathcal{S}_k \frac{(\vec{s}_k - \vec{s}_l)}{|\vec{s}_k - \vec{s}_l|^3} \cdot \hat{n}_k \tag{50}$$

where ξ_k is a factor taking into account the curvature of the surface on tessera k (Miertuš *et al.*, 1981; Tomasi and Persico, 1994).

Eqs. (47) and (48), called CLS–1 and CLS–2 respectively (Coitiño *et al.*, 1995a), bracket the correct value and represent good approximations to the set of charges $\{q_k\}$. Moreover, the related program is computationally less demanding than the iterative version, and it is quite effective, especially when applied to a semiempirical formulation (Pomelli and Tomasi, 1995).

c) Matrix inversion procedure.

The expressions giving the set of final $\{q_k\}$ charges may be formally put in the form of linear equations. Using a matrix notation we have:

$$\mathbf{D}\mathbf{\Sigma}^{-1}\mathbf{q} = -\mathbf{E}_n \tag{51}$$

D is a square matrix with dimensions equal to the number of tesserae and elements defined as:

$$\mathbf{D}_{kk} = \frac{4\pi\varepsilon}{\varepsilon - 1}(1 - A_k) = \frac{4\pi\varepsilon}{\varepsilon - 1} - 2\pi(1 - \xi_k) \tag{52}$$

$$\mathbf{D}_{kl} = \frac{4\pi\varepsilon}{\varepsilon - 1}B_{k,l} = \Delta\mathcal{S}_l \frac{(\vec{s}_k - \vec{s}_l)}{|\vec{s}_k - \vec{s}_l|^3} \cdot \hat{n}_k \tag{53}$$

$\mathbf{\Sigma}$ is a diagonal matrix, containing the areas of the tesserae: $\mathbf{\Sigma}_{kk} = \Delta\mathcal{S}_k$. Both **D** and $\mathbf{\Sigma}$ depend only on geometrical parameters defining the cavity, and its partition. The dielectric constant ε is present in the diagonal elements of **D**. The column matrices **q** and $\mathbf{E}_n$ contain the unknown charges, and the normal components of the solute electric field at the center of each tessera, respectively. In particular we have:

$$\mathbf{E}_k = -\vec{\nabla}\Phi_M(\vec{s}_k) \cdot \hat{n}_k \tag{54}$$

This matrix formulation may be used in the iterative procedure by replacing the inner cycle with the solution of linear equation system of eq.(51) (Coitiño *et al.*, 1995a). However, this approach could be too cumbersome; a more interesting application is the direct minimization of the free energy functional. We need to make a digression here.

When the Hamiltonian is non–linear, but contains a term depending on some power t of the wavefunction, i.e. a term that may be expressed as

$$\hat{\mathcal{V}}_{eff} = \hat{\mathcal{A}}(\Psi\Psi^*)^t \tag{55}$$

where $\hat{\mathcal{A}}$ is a suitable integral operator, the correct form of the functional which is subject to variational optimization is

$$\mathcal{J} = \frac{\left\langle \Psi \left| \hat{\mathcal{H}}_M^o + \frac{1}{t+1}\hat{\mathcal{V}}_{eff} \right| \Psi \right\rangle}{\langle \Psi | \Psi \rangle} \tag{56}$$

A demonstration has been given by Sanhueza *et al.* (1979). In our case we are using models in which the response of the medium is linear. Therefore the operator $\hat{\mathcal{A}}$ is reduced to a Green operator.

On the other hand, the expectation value of the Schrödinger equation (36)

$$E = \frac{\left\langle \Psi \left| \hat{\mathcal{H}}_M^o + \hat{\mathcal{V}}_{el}(\Psi\Psi^*) \right| \Psi \right\rangle}{\langle \Psi | \Psi \rangle} \tag{57}$$

has not the status of free energy. The electrostatic free energy G_{el} is given by

$$G_{el} = \frac{\left\langle \Psi \left| \hat{\mathcal{H}}_M^o + \frac{1}{2}\hat{\mathcal{V}}_{el}(\Psi\Psi^*) \right| \Psi \right\rangle}{\langle \Psi | \Psi \rangle} + \widetilde{V}_{NN} \tag{58}$$

The $\widetilde{V}_{NN}$ term collects contributions that are not related to the electronic wavefunction:

$$\widetilde{V}_{NN} = V_{NN} + \frac{1}{2}U_{NN} \tag{59}$$

where V_{NN} is the nuclear repulsion energy of the solute M, and U_{NN} is the interaction between the component of the reaction field generated by the nuclear part of Φ_M (see eq.(44)) and the nuclear charges themselves.

Eq. (58) has been derived by Yomosa (1974) and can be demonstrated in different ways, see Tomasi and Persico (1994) for some derivations. This fact has been commented by several authors. The reader is referred to the clear exposition given by Tapia (1982). By comparing eq.(56) with eq.(58), one immediately recognizes that the functional $\mathcal{J}$ is exactly G_{el}, being t=1. This fact may be exploited to elaborate a direct procedure to compute G_{el}. If we confine ourselves to the case of an SCF description of the problem (but extension to other levels of the quantum molecular theory is immediate) the variational optimization process may be reduced to the solution of the following HF equation (which is expressed, for convenience, in a matricial form)

$$\widetilde{\mathbf{F}}\mathbf{C} = \varepsilon\ \mathbf{S}\mathbf{C} \tag{60}$$

where

$$\widetilde{\mathbf{F}} = \widetilde{\mathbf{h}} + \widetilde{\mathbf{G}}(\mathbf{P}) \tag{61}$$

Here, one may recognize the usual matrix expression for the HF solution *in vacuo*: $\widetilde{\mathbf{h}}$ collects one electron matrix element expressed in terms of an expansion basis set $\{\chi\}$ and $\widetilde{\mathbf{G}}(\mathbf{P})$ the two–electron matrix elements. $\widetilde{\mathbf{G}}(\mathbf{P})$ depends on the one–electron density matrix

$$\mathbf{P} = \mathbf{CnC}^{\dagger} \tag{62}$$

where $\mathbf{n}$ is the orbital occupation number diagonal matrix.

The tilde put on $\mathbf{h}$ and $\mathbf{G}$ means that these matrices contain terms which account for the presence of the solvent, and hence are not present in the usual formulation given for an isolated molecule. In fact, we have:

$$\widetilde{\mathbf{h}} = \mathbf{h} + \frac{1}{2}(\mathbf{j} + \mathbf{y}) \tag{63}$$

and

$$\widetilde{\mathbf{G}}(\mathbf{P}) = \mathbf{G}(\mathbf{P}) + \mathbf{X}(\mathbf{P}) \tag{64}$$

$\mathbf{h}$ is the usual one–electron operator matrix representation (on the basis set $\{\chi\}$) for M *in vacuo*. $\mathbf{j}$ is the matrix collecting the interaction between solute electronic elementary charge distribution $\chi_\mu \chi_\nu^*$ and the apparent charge distribution generated by the solute nuclear charge density, $\rho_{M,n}$, and conversely $\mathbf{y}$ is the matrix collecting the interactions between all the nuclear charges of M and the components of the apparent charges having their source in the solute electronic charge distribution $\rho_{M,e}$.

$\mathbf{G}(\mathbf{P})$ collects the two–electron integrals used in normal *in vacuo* calculations, and $\mathbf{X}(\mathbf{P})$ the integrals describing the interactions between the elementary charge distribution, $\chi_\mu \chi_\nu^*$, and the apparent charges generated by solute $\rho_{M,e}$. The elements of the latter matrix are one–electron integrals.

The explicit expressions of the $\mathbf{j}$, $\mathbf{y}$, and $\mathbf{X}$ matrices can be found elsewhere (Cammi and Tomasi, 1995b). They have been obtained by making use of eq.(51).

When self–consistency conditions are achieved, we obtain from eq.(60) and eq.(58) the following expression for G_{el}

$$G_{el} = tr\mathbf{P}\widetilde{\mathbf{h}} + \frac{1}{2}tr\mathbf{P}\widetilde{\mathbf{G}}(\mathbf{P}) + \widetilde{V}_{NN} \tag{65}$$

This expression may be compared with the one obtained by making use of the iterative procedure. From eq.(36) we have:

$$E = \left\langle \Psi \left| \hat{\mathcal{H}}_M^o + \hat{\mathcal{V}}_{el} \right| \Psi \right\rangle \tag{66}$$

and

$$G_{el} = E - \frac{1}{2}\left\langle \Psi \left| \hat{\mathcal{V}}_{el} \right| \Psi \right\rangle - \frac{1}{2}\int \rho_{M,n} \hat{\mathcal{V}}_{el} d\vec{r} + V_{NN} + U_{NN} \tag{67}$$

Using again the HF matrix formulation, we have:

$$E = tr\mathbf{P}(\mathbf{h} + \mathbf{V}_\sigma) + \frac{1}{2}tr\mathbf{P}\mathbf{G}(\mathbf{P}) \tag{68}$$

and

$$G_{el} = E - \frac{1}{2}tr\mathbf{P}\mathbf{V}_\sigma + \widetilde{V}_{NN} \tag{69}$$

Here $\mathbf{h}$ and $\mathbf{G}$ are the one and two–electron matrices used *in vacuo*, and $\mathbf{V}_\sigma$ collects the one–electron integrals related to the potential generated by the apparent surface charges $\{q_k\}$ defined in Section 6.1. Expressions (65) and (69) are formally equivalent, and give exactly the same results when properly applied (Cammi and Tomasi, 1995b).

We have opened a page to show the relationship between iterative and direct procedures, and repeated again simple things which had been said in previous papers, because there has been some confusion about this subject in the literature of the past years. The final message is simple, that is iterative and direct procedures are not alternative approaches in setting up the ASC continuum model and they give the same results. On the contrary, the closure formulation is an approximation. The difference lies in the computational efficiency. Here, several parameters have to be considered, related to the characteristics of the computer and to the size of the problem (number of basis set functions and of tesserae). The paper we have quoted here several times (Tomasi and Persico, 1994) has been written two years ago. Meanwhile we have improved the efficiency of the direct PCM version, which is now more efficient than the iterative one, that has not been modified.

6.2. DISPERSION TERM G_{DIS}.

In the previous Subsection we have introduced three PCM versions differing in the technique adopted for the calculation of ΔG_{el}, and there are several others which consider different models for the solvent, such as liquid crystals, confined liquids, etc.. Most of such PCM procedures use an approach based on pair–potential effective energies for the calculation of G_{dis}. However, there are other PCM versions based on a reaction field approach.

6.2.1. *Pair potential approach.*

In the pair potential approach presently used in the PCM , the dispersion potential between two molecules (the solute M and a solvent molecule S) is expressed with the atom–atom approximation, with a truncated expression in terms of powers of $1/r$. If m and s are two atoms belonging to M and

S, respectively, we have:

$$U_{MS} = \sum_m \sum_s \sum_{k=6,8,10,\cdots} d_{ms}^{(k)} \, r_{ms}^{-k} \tag{70}$$

Introducing now continuous distribution functions for each m-s couple:

$$\rho_{ms}(\vec{r}_{ms}) = N_s \, \rho_S \, g_{ms}(\vec{r}_{ms}) \tag{71}$$

where N_s is the number of atoms of type s present in each solvent molecule, ρ_S is the macroscopic density of the solvent and $g_{ms}(\vec{r}_{ms})$ is a correlation function depending on the position of s with respect to m, we may write the average solute–solvent dispersion energy:

$$\langle E_{dis} \rangle = \int \cdots \int U(\Omega)\rho(\Omega)d\Omega \tag{72}$$

in a form quite effective from the computational view point. In eq.(72) we call Ω the set of all the coordinates, indicating the relative positions of the components of the system (M is considered fixed in an external reference frame) and $\rho(\Omega)$ is the proper combination of $\rho_{ms}(\vec{r}_{ms})$ functions. The evaluation of the integral in eq.(72) is a quite complex task.

The details of the formal elaboration may be found in the source papers (Floris *et al.*, 1989; 1991; 1993). Here we report and comment the final expression

$$\langle E_{dis} \rangle = \rho_S \sum_{s \in S} N_s \sum_{m \in M} \sum_{t=1}^{NT} \sum_k \vec{A}_{ms}^{(k)} \cdot \hat{n}_t \, \Delta S_t \tag{73}$$

The multidimensional integral (72) is reduced to a simple summation over the centers t of the cavity tesserae (their number is NT) of the normal component of the simple auxiliary function:

$$\vec{A}_{ms}^{(k)} = \frac{d_{ms}^{(k)}}{(k-3)r_{ms}^k} \, \vec{r}_{ms} \tag{74}$$

The introduction of $\vec{A}_{ms}^{(k)}$ replaces the integration over the whole solvent volume, and reduces the evaluation of (72) to a surface integral, further reduced to a discrete summation, as in all BEM methods. The simple analytical expression (74) for $\vec{A}_{ms}^{(k)}$ has been obtained by using the so-called "uniform approximation" for the solute–solvent distribution functions, i.e.:

$$\begin{aligned} g_{ms}(\vec{r}_{ms}) &= 1 \quad \text{for } r_{ms} \text{ outside the cavity} \\ g_{ms}(\vec{r}_{ms}) &= 0 \quad \text{for } r_{ms} \text{ inside the cavity} \end{aligned} \tag{75}$$

It is worth noting that the radii of the spheres defining the cavity for the calculation of G_{dis} are not necessarily the same used for G_{el} (Bonacccorsi *et al.*, 1990).

To adopt more realistic $g_{ms}(\vec{r})$ functions, derived either from experiment, simulations or by RISM perturbation theory treatment, one has to use more complex mathematics, as detailed in Floris *et al.* (1993). Note that the use of the uniform distribution makes it possible to assimilate the internal energy to the free energy, then eq.(73) gives the G_{dis} value we were looking for.

6.2.2. *Reaction field approach.*
The second approach for the evaluation of G_{dis} is based on the concept of reaction field we have already expressed in the case of G_{el}. This approach has been pioneered by Linder (1967) and introduced in general continuum solvation model by Rinaldi *et al.* (1986) in Nancy. The insertion of this procedure for the evaluation of G_{dis} in PCM has been done by Aguilar and Olivares del Valle (1989) in Badajoz. Before giving a short sketch of this approach, we would like to point out that it allows both the calculation of G_{dis} and the definition of an effective operator $\hat{\mathcal{V}}_{dis}$. The latter will be inserted in the Schrödinger equation to complement the electrostatic effective operator $\hat{\mathcal{V}}_{el}$.

This method is based on the London approximation of dispersion forces, namely in terms of the solute polarizability tensor $\boldsymbol{\alpha}$, and of the mean excitation energies of solute and solvent molecules, and leads to the compact formula:

$$\hat{\mathcal{V}}_{dis} = -\frac{1}{8}\frac{\bar{E}_M \bar{E}_S}{\bar{E}_M + \bar{E}_S}\sum_p \sum_q g_{pq}(\epsilon_\infty)\alpha_{pq} \tag{76}$$

where $\bar{E}_M$ and $\bar{E}_S$ are the mean excitation energies of solute and solvent molecules, indexes p and q refer to the components (l_1, m_1) and (l_2, m_2) of a spherical tensor expansion, and $g_{pq}(\epsilon_\infty)$ are the components (extrapolated for the dielectric constant at infinite frequency) of the dipole reaction tensor, defined by the relation

$$\mathbf{g} \cdot \vec{\mu} = \vec{R} \tag{77}$$

where $\vec{\mu}$ is the solute dipole and $\vec{R}$ the reaction field, resulting from the polarization of the solvent.

The differences between the Nancy and Badajoz approaches are related to the different general strategies of the solvation procedures followed by the two groups. We shall comment later the whole SCRF computational scheme developed by Rivail and coworkers. Here we limit ourselves to say that they use multipolar expansions. The multipole approach implicit in eq.(76) may be developed without any change. The calculation done by Rinaldi *et al.* (1986) refer to the use of the dipole term ($l = 1$) alone, but the extension to higher terms is straightforward.

Olivares del Valle and Aguilar (1993) further elaborated the expression (76), obtaining an expression of $\hat{\mathcal{V}}_{dis}$ in terms of apparent charges $\{q_{dis}\}$ placed on the cavity surface tesserae, like the charges used to describe $\hat{\mathcal{V}}_{el}$. The iterative PCM procedure is extended to include a third iterative cycle regarding the definition of $\hat{\mathcal{V}}_{dis}$. To summarize, the Hamiltonian used for the final calculation of Ψ may be written

$$\hat{\mathcal{H}}_M = \hat{\mathcal{H}}_M^o + \hat{\mathcal{V}}_{el}^{(kl)} + \hat{\mathcal{V}}_{dis}^{(klm)} \tag{78}$$

where the superscripts indicate the number of cycles needed to get the desired degree of self–consistency. The flowchart of this procedure is given in the already quoted paper by Olivares del Valle and Aguilar (1993). These authors also include many body corrections for the correlation energy at MP3 level, with some options for the calculations with or without any dispersion term in the Hamiltonian.

The reaction field approach is more congruent to the general picture used in continuum models than the pair potential method. However, the results are quite similar for simple solutes at the equilibrium geometry. The dipole approximation used in Nancy and Badajoz approaches has some limitations, that could be partly removed. We do not want to start a discussion about these technical aspects, and we would like only to point out that we are planning the implementations of another formulation, according to the lines shown by Amovilli (1994). We stress that the study of reaction mechanisms in solution requires the use of methods able to treat "supersolutes" ($M{\cdot}S_n$), on a wide range of nuclear geometries, including geometries where phenomena of bond formation/disruption are important. Neither the pair potential nor the reaction field approach in the dipole–London formulation seem adequate to accurately describe G_{dis} at such geometries.

6.3. REPULSION TERM G_{REP}.

This term is often neglected in many computational procedures, even when they are derived from the original PCM. Its absolute value at equilibrium geometries is actually quite small. We judge important to include it, in order to complete the computational scheme, and also because its presence makes a variational optimization of the cavity volume, when needed, possible.

The computational procedure now in use in the PCM scheme is similar to the one adopted for the dispersion term according to the pair potential method; the relevant formulas may be found in Floris *et al.* (1991) and in Tomasi and Persico (1994).

Though computed in similar ways, dispersion and repulsion terms have a different physical origin. The analysis and the descriptions of the repulsion energy referred to bimolecular systems cannot be directly translated into efficient computational procedures for continuum models. Several alternatives are possible to develop more detailed and efficient procedures, but the repulsion term seems less sensitive to bond formation/disruption than dispersion one. For this reason there is not an urgent need to replace the description now in use with more sophisticated versions.

6.4. CAVITATION TERM G_{CAV}.

We have already stressed that the definition of the shape and size of the cavity is a critical step in the evaluation of G_{el} as well as of G_{dis}. The cavity used to compute G_{cav} must be congruent with those used for the other contributions, although, as already remarked, there is no reason to constrain the model to the use of the same cavity for all energy terms.

The main method used in PCM for G_{cav} is based on Pierotti's elaboration (Pierotti 1963; 1965; 1976) of the Scaled Particle Theory which was introduced by Reiss *et al.* (1959, 1960). This model which belongs to the family of physical descriptions of liquid systems defines G_{cav} as an expansion in powers of R_{MS}, i.e. the radius of a sphere excluding the centers of solvent molecules:

$$G_{cav} = K_0 + K_1 R_{MS} + K_2 R_{MS}^2 + K_3 R_{MS}^3 \tag{79}$$

The coefficients K_n are expressed in terms of the bulk properties of the solution (temperature, pressure, number density) and of structural properties of the solvent (molecular radius R_S). The cavity radius is defined by the relation $R_{MS} = R_M + R_S$.

Pierotti's cavities always have a spherical shape. There have been many debates about whether eq.(79) may be applied to spherical cavities in water, where molecules exhibit a remarkable self–association, i.e. a behaviour which is quite different from that of hard spheres. The answer is positive. Recent unpublished Monte Carlo calculations performed by our group (Floris *et al.*, 1996) confirm and extend previous molecular dynamics simulations by Postma *et al.* (1982) showing that Pierotti's formula is quite good also for spherical cavities in water.

In general, solutes have a non spherical shape. To extend Pierotti's formula to the case of non–spherical cavities, defined in terms of overlapping spheres, one uses the following expression, proposed by Claverie's group (Langlet *et al.*, 1988)

$$G_{cav} = \sum_i \frac{A_i}{4\pi R_i^2} G_{cav}(R_i) \tag{80}$$

where each sphere of radius R_i contributes with a weight proportional to the surface exposed to the solvent. The validity of this approximation is now under scrutiny, on the basis of computer simulations both in water and in other solvents. Expressions (79) and (80) exploit cavities defined in terms of contact between solute and solvent spheres (i.e. $\mathcal{S}_{SA}$ surfaces). If one assumes van der Waals' values to describe these contacts, one arrives to cavities slightly different from those used for G_{el}. We recall that they are defined in terms of van der Waals' radii multiplied by a scaling factor $\simeq 1.2$, and, in most cases, supplemented by other spheres that are not centered on atoms ($\mathcal{S}_{SE}$ surfaces).

Another simple procedure included in PCM programs for the calculation of G_{cav} is based on the use of experimental values of the surface tension γ. The formula we have implemented is due to Sinanoglu (1967, 1974, 1981)

$$G_{cav} = k_S^g\,(\mathcal{V}_M/\mathcal{V}_S)\,\gamma\mathcal{S} \tag{81}$$

here $\mathcal{S}$ is the cavity area (the formula has been developed for spherical cavities and then extended to an arbitrary shape). The correction factor $k_S^g\,(\mathcal{V}_M/\mathcal{V}_S)$ depends on the ratio of the molecular volumes

$$k_S^g\,(\mathcal{V}_M/\mathcal{V}_S) = 1 + \left(\frac{\mathcal{V}_M}{\mathcal{V}_S}\right)^{2/3}(k^g(1) - 1) \tag{82}$$

and the factor $k^g(1)$ is estimated from solubility data referred to the solvent. Sinanoglu has given several different versions of these formulas, but no detailed information about the numerical definition of $k^g(1)$. Thus users have to confine themselves to the solvents listed in Sinanoglu's papers. Those who are interested to expand this listing could start from the treatment of microscopic curvature effects on surface tension that was given by Kirkwood and Buff (1950) and Buff (1955) and refined by others (e.g. Choi *et al.*, 1970).

For spherical cavities of small size, Pierotti's and Sinanoglu's formulas give quite similar results, if the solvent is water, and not too different results for other solvents (a couple of examples may be found in Tomasi and Persico, 1994). In our opinion Pierotti's results are more reliable, at least for water. The two approaches give diverging results for cavitation enthalpy and entropy. Neither formulation probably gives the correct trend for these thermodynamical functions.

Some PCM calculations (e.g. Bachs *et al.*, 1994) exploit a numerical correlation between G_{cav} (and G_{dis}) and the volume and surface of the solute, drawn from Floris and Tomasi's calculations (1989). The Scaled Particle Theory formulation for a spherical cavity is based on the volume. Pierotti's expansion leads to a formula where both volume and surface terms are present. Sinanoglu's formula is based on the surface, but with a factor depending on the volume. Monte Carlo and molecular dynamics simulations for spherical cavities in water give trends of G_{cav} which are well correlated with both surface and volume (a little better with volume, actually). Also concerning G_{dis} (and G_{rep}) there are almost equal correlations with the volume and the surface of the cavity for several sets of small molecules. These correlations have been calculated by means of the pair potential method, but, as we already said, G_{dis} values computed with the first term of the pair potential expansion ($d^{(6)}_{ms} r^{-6}_{ms}$) are similar to those found with the dipole reaction field approach, and the two successive terms ($d^{(8)}_{ms} r^{-8}_{ms}$ and $d^{(10)}_{ms} r^{-10}_{ms}$) cancel almost completely the opposite–sign G_{rep} contribution. The definition of the pair potential method gives us a hint about the identification of the correct trend of this quantity with respect to geometrical solute parameters. The presence of rapidly decaying interaction potentials (from r^{-6} to higher inverse powers of the distance) shows that only a portion of the solute lying within a limited layer around the molecular surface is effective in the determination of energy contributions. Such layer is of a depth of the order of two bond distances. For small or long linear molecules there will be a good correlation with the molecular volume as well as with the surface, while for larger solutes the correlation with the surface will be even better.

6.5. BEYOND THE HARTREE–FOCK APPROXIMATION.

In some points of the previous analysis we have used formulations based on the Hartree–Fock (HF) expression of the quantum problem, mainly for simplicity of exposition. As a matter of fact, there are no formal reasons to limit continuum solvent approach to the HF level. Actually, PCM solvation procedures have been extended to MCSCF (Aguilar *et al.*, 1993b), CI (Persico and Tomasi, 1984), MBPT (Olivares del Valle *et al.*, 1991, 1993), CASSCF, MR–SDCI (Aguilar *et al.*, 1993b), DFT (Fortunelli and Tomasi, 1994) levels of the quantum description. The other continuum solvation methods have, at least in principle, the same flexibility in the definition of the quantum theory level to be used in computations.

Continuum effective Hamiltonian needs a definition of the electronic charge distribution $\rho_{M,e}$. All quantum methods giving this quantity can be used, whereas other methods must be suitably modified. Quantum methods are not limited to those based on a canonical molecular orbital formulation. Valence Bond (VB) and related methods may be employed. The interpretation of reaction mechanisms in the gas phase greatly benefits by the shift from one description to another (e.g. from MO to VB). The same techniques can be applied to continuum effective Hamiltonians. We only mention this point here, which would deserve a more detailed discussion.

In this Section we shall shortly consider two points: the use of the density functional (DFT), and the use of Moller Plesset perturbation picture (MPn). DFT has gained a wide popolarity for the calculation of the isolate ground state molecules. The considerable wealth of information gathered in the last two years allows us to appreciate both merits and defects of the various computational options available in the DFT framework, when applied to specific systems and problems. Several continuum *ab initio* solvation methods are now provided with options allowing calculations with the DFT formalism. We shall review other methods in a next Section. For the PCM procedure we mention a preliminary report (Fortunelli and Tomasi, 1994) showing that gradient–corrected DFT values of the solvation energy ΔG_{sol} of small solutes are better than the corresponding HF values (and similar to the MP2 ones), with reduced computational costs. We have found worse performances using other versions of the DFT method.

When one adopts a method starting from an SCF description with further corrections to describe electron correlation, one has to deal with a combination of effects. Electron correlation modifies the solute charge distribution ρ_M^{SCF} and consequently the interaction potential $\Phi_\sigma(\rho_M)$. On the other hand, the polarization induced by the solvent modifies electron correlation effects. Olivares del Valle and coworkers (1991, 1993) proposed an iterative computational scheme to introduce both effects in the PCM procedure. The first step, called PTE, is based on the solution of the PCM equation at the SCF level (it contains $\Phi_\sigma(\rho_M^{SCF})$) and then an MP$n$ correction for the energy E is computed. Going further, the density function ρ_M^{MPn} so obtained is used to update $\Phi_\sigma(\rho_M)$. The final value is called PTDE. To study better the relative importance of the effects, another approximation, called PTD, uses the MPn corrected wavefunction *in vacuo* to define $\Phi_\sigma(\rho_M^{0,MPn})$ and to compute the corresponding value of ΔG_{el}. An analysis given by Fortunelli (1995a) considers the extension of the problem to other PT correlation models, where there is no wavefunction corresponding to the energy (i.e. coupled–cluster (CC) and quadratic configuration interaction (QCI) models), and shows that both PTE and PTDE methods give comparable ΔG_{el} results. To explain this similarity between PTE and PTDE methods, Fortunelli makes use of arguments drawn from the response theory. This has led to a simplification of the PCM formalism (Fortunelli, 1995b), which is computationally quite effective, without any detriment to the quality of the results. Computational improvements in routine PCM calculations at the SCF correlated to DFT levels are thus expected.

Coming back to the question of MP calculation, Ángyán's recent analysis on the MP2 approximation is worth mentioning (Ángyán, 1995). He shows that MP2 energy can be obtained by using SCF orbitals optimized with the self–consistent continuum procedure, and hence the SCF reaction field, i.e. the PTE approximation. PTDE at MP2 level introduces higher-order corrections. Things are not so simple when we pass to higher MP orders but, in current calculations on reaction mechanisms, electron correlation effects are evaluated at the lowest order only, and so the empirical practice of using PTE calculation finds a theoretical support in this analysis.

6.6. APPROXIMATE AND NON–QUANTUM PROCEDURES.

The PCM has been conceived for quantum calculations, according to the scheme shown in the previous Sections. At the same time, it has been conceived as a means to extend to solvation problems a strategy based on a "semiclassical approach". In past years, we found it convenient to treat on the same footing intermolecular interactions *in vacuo*, chemical substitution effects, internal geometry relaxation, and electronic excitation.

We quote a few review papers from our group (Tomasi *et al.*, 1991; Bonaccorsi *et al.*, 1984a; Alagona *et al.*, 1986) giving a fuller account of this procedure. The problem of getting a rationale of the chemical effects that can be reduced to interactions among molecular subunits is of basic importance in theoretical chemistry, and several other authors have proposed some models and theories addressed for this aim. Since they have some points in common with our approach, they are referred to in the above quoted review papers.

The application of this strategy to solvation problems means to introduce in the full quantum description of the problem a hierarchy of approximations which, step by step, leads to simpler algorithms. At each step the loss of information, and the errors introduced are controlled. The sequence of approximations we shall summarize here has been accompanied by a set of protocols to check the quality of the results. Our exposition will be rather concise. It is a summary of those given in our recent review on continuum solvation (Tomasi and Persico, 1994). All information about the quality of the results may be found there, as well as in the source papers.

Approximation 1 – The full PCM description of the system. The procedure gives G_{el}, from which the contribution to the free energy of solvation may be obtained;

$$\Delta G_{el,1} = G_{el} - E^o \tag{83}$$

Approximation 2 – Starting from this approximation there is no longer a direct calculation of G_{el}. The attention is focussed on $\Delta G_{el,n}$. At level 2 the expression is:

$$\Delta G_{el,2} = \frac{1}{2} \int \rho_M(\vec{r}) \Phi_\sigma(\vec{r}) d\vec{r} \tag{84}$$

We are using here the fully relaxed molecular density function and the corresponding $\Phi_\sigma(\rho_M)$ potential. This approximation, which corresponds to neglect solvent effects on the average kinetic energy of the electrons, may be justified by invoking the virial theorem.

Approximation 3 – The solute charge distribution computed *in vacuo*, ρ^o_M, is here used:

$$\Delta G_{el,3} = \frac{1}{2} \int \rho^o_M(\vec{r}) \Phi^o_\sigma(\vec{r}) d\vec{r} \tag{85}$$

This approximation is the most detailed version of the class of rigid models, disregarding solvent polarization effects. We shall encounter other rigid models at approximation 6.

Approximation 4 – *Ab initio* calculations are no longer used, starting from here. The solute charge distribution is expressed in terms of group contributions. Each group g is described in terms of a prototype, drawn from a set of calculations on simpler molecules containing the same group (the parent molecules). Each prototytpe is a neutral subunit, defined in terms of a suitable set of localized orbitals accompanied by a proper share of nuclear charges. The prototype gives a charge distribution which, by definition and by construction, is independent on the chemical environment and on external factors. This charge distribution is called $\rho(g, 0)$, where index 0 emphasizes its independence on the medium. Actually, it may be modified, *via* classical polarization effects, by the electric fields generated by molecular environment (M), and by the solvent (σ). In this approximation we use prototypes modified by both fields. The resulting charge distribution is:

$$\rho_{M,4} = \sum_g \rho(g/M, \sigma) \tag{86}$$

and the solution free energy becomes:

$$\Delta G_{el,4} = \frac{1}{2} \int \rho_{M,4}(\vec{r}) \Phi_{\sigma,4}(\vec{r}) d\vec{r} \tag{87}$$

E^o must be computed by means of a different method. For large solutes, molecular mechanics methods could be appropriate, when dealing with molecular conformation problems. The description of chemical reactions requires a separate discussion.

Approximation 5 – The prototype functions are polarized by the electric field due to the molecular remainder M, and not by the solvent:

$$\rho_{M,5} = \sum_g \rho(g/M, 0) \tag{88}$$

This approximation may be classified as a rigid solute model:

$$\Delta G_{el,5} = \frac{1}{2} \int \rho_{M,5}(\vec{r}) \Phi_{\sigma,5}(\vec{r}) d\vec{r} \tag{89}$$

Approximation 6 – The prototype functions are used without any modification:

$$\rho_{M,6} = \sum_g \rho(g/0) \tag{90}$$

and

$$\Delta G_{el,6} = \frac{1}{2} \int \rho_{M,6}(\vec{r}) \Phi_{\sigma,6}(\vec{r}) d\vec{r} \tag{91}$$

Approximation 7 – The prototype charge distributions are replaced by simpler models. In approximations 4–6, the $\rho(g/0)$ functions are expressed on the same basis set as the parent molecules used for their definition, hence for molecule M the use of a portion (diagonal block) of the one–electron penetration integrals over the whole basis set is required. Simpler descriptions, which have been here introduced, avoid the evaluation of these matrix elements. There are many methods to define local multipole developments for molecular fragments: a review is given in Tomasi *et al.* (1991). Starting from the prototype distributions $\rho(g/0)$, we have checked the quality of results obtained by replacing the diffuse electron distribution of each $\rho(g/0)$ by a set of point charges (monopoles), with integral values -1 or -2, fixed in positions to be determined by the conservation of the group dipole distribution.

Approximation 8 – The final step in this sequence of operations is taken when charges placed on atoms only are used. The natural extension of the hierarchy summarized in the previous steps consists in the use of potential derived atomic charges (PD charges), obtained by using $\rho_{M,6}$ or $\rho_{M,7}$. PD charges have been introduced by Momany (1978) following a previous proposal (Alagona *et al.*, 1972) in which there were more charges than nuclei. The procedure is based on a numerical fitting of the molecular electrostatic potential (MEP) given by these charges, with respect to the MEP due to the reference charge distribution. Since Momany's definition of PD charges, there has been a great interest in this approach, especially for molecular recognition, summarized in some reviews (see, e.g. Tomasi *et al.*, 1991). Fitting of PD charges addressed to solvation procedures are simpler than fitting addressed to molecular recognition, because it is sufficient to reproduce the good shape of the molecular electric field (MEF) across the cavity surface.

In this sequence of approximations we have not considered other solvation terms, such as G_{cav} and G_{dis}. They may be computed apart, with a low computational effort.

We have taken some space to describe some approaches that could be considered of little use in the study of chemical reactions. In fact, the first approximations in the sequence we have outlined lead to minor reductions in the computational cost. They have been defined, and examined to control the whole sequence. When we examine approximations that are more effective in reducing computational times, we have to draw the E^o hypersurface from other sources. To obtain the portion of the PES describing a chemical reaction *via* simple methods, like Molecular Mechanics, is still a delicate task, for which more efforts are needed.

However there is another quite promising approach. For reactions involving large molecules, it is often possible to enucleate a small portion, in which quantum processes leading to bond formation and disruption occur, while the largest portion of the molecular system acts in a more indirect way, and to describe it with the aid of a classical model. The use of hybrid QM/MM models are addressed to exploit these considerations. They will be examined in Section 9.2.

Enzymatic reactions will be considered in another chapter of this book. We shall only remark that continuum ASC models, and in particular the PCM, may be of some help in studying reactions involving large molecules. In the set of approximations we have considered no additional changes in the PCM formalism were introduced. In particular, the same definition of cavity in terms of interlocking spheres, its partition in tesserae resulting from the inscription of a pentakisdodecahedron in each sphere, and the self–polarization of apparent charges, have been used.

In order to get practical algorithms to be used in hybrid procedures, these points should be re-examined. When solutes containing hundreds or thousands of atoms are considered the usual definition of cavity and tesserae (and the related calculation of G_{cav} and G_{dis}) is too costly. A new definition of the cavity surface and of its tesselation has been recently proposed. According to the first checks it gives an excellent correlation with PCM values for cavity volume, and surface, G_{dis}, G_{cav}, ΔG_{el} (Approximation 8), ΔG_{sol} over a set of molecules ranging from 3 to more than 3000 atoms (from water to HIV protease+inhibitor). A preliminary report on this procedure may be found in a paper by Cammi *et al.* (1995a), a more detailed report is under elaboration (Pomelli and Tomasi, 1995).

This procedure is based on the use of unique sphere, with its tesserae defined in terms of a number of spherical triangles (from 60 to 940) derived by partition of the pentakisdodecahedron faces. The sphere is deformed to mimick the molecular shape, taking into account solvent excluded volume. The number of tesserae is automatically increased only in the portions of the surfaces where it is needed, keeping the shape of a curvilinear triangle. The gain in computational times reaches a $\sim$ 100 factor for the largest molecules.

To further reduce computational times, it is convenient to introduce appropriate cutoff in the evaluation of the separate elements of the procedure. This has been done for G_{dis} (Floris *et al.*, 1991), and work is in progress for

ΔG_{el}. G_{dis} is given as a sum of elements, one for each tessera. The effectors (or sources) of each element are the the atom–atom potentials which decay with the distance as r^{-6}. The introduction of a cutoff radius equal to 4.5 Å allows one to compute G_{dis} with a 99% precision. Things are more complex for ΔG_{el} for which there are two kind of sources. The expression of ΔG_{el} is given by the sum of products $q_k \Phi_M(\vec{r}_k)$, where q_k are the apparent charges (one for each tessera) and $\Phi_M(\vec{r}_k)$ is the solute electrostatic potential at $\vec{r}_k$. There will be sources for the apparent charges (the elements of the solute charge distribution) as well as for Φ_M (the elements of the solute charge distribution again, but with a different decay rate). Cutoffs may be introduced for both elements. We are still examining the best strategy to implement in computational codes.

7. Analytical derivatives of $G(\mathbf{R})$ in the PCM formalism.

The study of reaction mechanisms by means of theoretical methods heavily depends on the availability of analytical expressions of the free energy derivatives with respect to nuclear coordinates. We shall summarize here the status of the art for the PCM method developed in Pisa. Other approaches will be considered later in Section 8.

To simplify the notation we shall follow the convention used in the specialized literature, indicating partial derivatives with upper indices: X^α and $X^{\alpha\beta}$ will stand for $\frac{\partial X}{\partial \alpha}$ and $\frac{\partial^2 X}{\partial \alpha \partial \beta}$, respectively. In this Section α and β will be cartesian coordinates of a nucleus of the solute. We shall neglect here the cases in which α, β are parameters of different nature.

We recall that we have to consider all the components of G, thus, for example, the first derivative with respect to α will be

$$G^\alpha = G^\alpha_{el} + G^\alpha_{cav} + G^\alpha_{dis} + G^\alpha_{rep} \tag{92}$$

and a similar expression applies for the second derivative $G^{\alpha\beta}$.

The direct PCM procedure has been implemented to get analytical expressions of G^α and $G^{\alpha\beta}$.

7.1. G^α_{EL} AND $G^{\alpha\beta}_{EL}$.

The formal derivation of these quantities from the starting equation (67) follows the same lines which were worked out for molecules *in vacuo*. The subject is thoroughly treated in several excellent monographs (Helgaker and Jorgensen, 1988; Jorgensen *et al.*, 1986; Pulay, 1987; Yamagouchi *et al.*, 1994) and we may limit ourselves to report here the general formulas used at the SCF level.

For the first derivative we have:

$$G_{el}^{\alpha} = tr\mathbf{P}[\mathbf{h}^{\alpha} + \frac{1}{2}(\mathbf{j}^{\alpha} + \mathbf{y}^{\alpha})] + \frac{1}{2}tr\mathbf{P}[\mathbf{G}^{\alpha}(\mathbf{P}) + \mathbf{X}^{\alpha}(\mathbf{P})] - tr\mathbf{S}^{\alpha}\widetilde{\mathbf{W}} + \widetilde{V}_{NN}^{\alpha} \quad (93)$$

We have here introduced the auxiliary $\widetilde{\mathbf{W}}$ matrix, analogous to the $\mathbf{W}$ matrix defined *in vacuo* and defined as:

$$\widetilde{\mathbf{W}} = \mathbf{P}\widetilde{\mathbf{F}}\mathbf{P} \quad (94)$$

It should be remarked that in the expression of G_{el}^{α} the derivative of the density matrix $\mathbf{P}^{\alpha}$ does not appear.

For the second derivative we have:

$$\begin{aligned} G_{el}^{\alpha\beta} = \; & tr\mathbf{P}[\mathbf{h}^{\alpha\beta} + \tfrac{(\mathbf{j}^{\alpha\beta}+\mathbf{y}^{\alpha\beta})}{2}] + \tfrac{1}{2}tr\mathbf{P}[\mathbf{G}^{\alpha\beta}(\mathbf{P}) + \mathbf{X}^{\alpha\beta}(\mathbf{P})] - tr\mathbf{S}^{\alpha\beta}\widetilde{\mathbf{W}} + \\ & \widetilde{V}_{NN}^{\alpha\beta} + tr\mathbf{P}^{\beta}[\mathbf{h}^{\alpha} + \tfrac{(\mathbf{j}^{\alpha}+\mathbf{y}^{\alpha})}{2} + \mathbf{G}^{\alpha}(\mathbf{P}) + \mathbf{X}^{\alpha}(\mathbf{P})] - tr\mathbf{S}^{\alpha}\widetilde{\mathbf{W}}^{\beta} \quad (95) \end{aligned}$$

with

$$\widetilde{\mathbf{W}}^{\beta} = \mathbf{P}^{\beta}\widetilde{\mathbf{F}}\mathbf{P} + \mathbf{P}\widetilde{\mathbf{F}}^{(\beta)}\mathbf{P} + \mathbf{P}\widetilde{\mathbf{G}}(\mathbf{P}^{\beta})\mathbf{P} + \mathbf{P}\widetilde{\mathbf{F}}\mathbf{P}^{\beta} \quad (96)$$

$\mathbf{P}^{\beta}$ is obtained by solving the coupled Hartree–Fock equation:

$$\underline{\widetilde{\mathbf{F}}}\,\underline{\mathbf{P}}^{\beta}_{ov} - \underline{\mathbf{P}}^{\beta}_{ov}\underline{\widetilde{\mathbf{F}}} + \underline{\widetilde{\mathbf{G}}}(\underline{\mathbf{P}}^{\beta}_{ov}) = \underline{\widetilde{\mathbf{F}}}^{(\beta)}_{ov} - \underline{\widetilde{\mathbf{G}}}(\underline{\mathbf{S}}^{\beta}_{oo}) - \underline{\widetilde{\mathbf{F}}}\,\underline{\mathbf{S}}^{\beta}_{ov} \quad (97)$$

expressed in terms of occupied (o) and virtual (v) MO orbitals (this change of the basis is indicated by an underline). $\underline{\widetilde{\mathbf{F}}}^{(\beta)}$ is the Fock matrix formed with the derivative integrals.

It would be tedious to report all the detailed expressions obtained when these general formulas are applied to the cases in which α and β are nuclear cartesian coordinates. They may be found in the reference paper (Cammi and Tomasi, 1994), it should be noticed that there are several misprints in that paper.

Technical details are not necessary here, with the exception of one point which we shall discuss later. Suffice it to say that the formidable formal apparatus corresponds to the calculation of a few sets of additional one–electron integrals and to their manipulation with the usual matrix techniques.

Analytical gradient calculations are quite effective when compared to the finite difference calculations. A factor of $\sim$ 20 applies in the present PCM formulation (Cossi *et al.*, 1995). Although effective, this increment of efficiency is smaller, by a factor 10 or more, than the analogous speed up found for the calculations of gradients *in vacuo*.

To obtain G^{α} values one has to compute partial derivatives with respect to the area of tesserae, and to other parameters related to tesselation of the cavity surface. In fact, we consider it necessary for a complete definition of G^{α} to include these derivatives. However, the other possible alternative which computes G^{α} at a fixed cavity could also be explored.

It is worth noting that the derivatives of the tesserae's geometrical elements are heavily affected by the additional spheres added to take into account the solvent excluding surface (see Section 6.1.1). The analytical derivatives of the position and size of these spheres with respect to solute nuclear coordinates have been also implemented (Cossi *et al.*, 1996a). We have reported these short comments to point out that substantial improvements of the effectiveness of analytical gradient depends on a change of strategy for the calculation of surface partial derivatives. Several options are possible: to reduce the number of tesserae, especially on the small additional spheres, to change the definition of the surface.

A final comment regards the latter point. The cavity can be definied according to different options, each one accompanied by its tesselation:

1) the GEPOL program (Pascual-Ahuir *et al.* 1987; 1994; Silla *et al.*, 1991), in the version exploiting Gauss–Bonnet theorem (called GEPOL–GB, Cossi *et al.*, 1996a), which is at present the most effective one for analytical derivatives: all the surface elements are convex, and this feature simplifies the calculation of derivatives;

2) the programs elaborated by Connolly (1983, 1985): they are of wide use in molecular graphics, but present problems for their use in BEM procedures, due to the appearance of "holes" and "intersecting triangles". The more complex procedure suggested by Juffer *et al.* (1991) has not been tested for the calculation of derivatives. In addition, Connolly's definition leads to a combination of convex, concave and saddle–shaped portions of the cavity which are potential troubles for the calculation of derivatives;

3) a definition based on the value of some relevant molecular properties. More than ten years ago we checked the technique proposed by Francl *et al.* (1984) and a few years later the one proposed by Bader *et al.* (1987) to define isodensity surfaces, and we found both of them computationally too expensive for the computer facilities available at that moment. The isodensity surface is now introduced in the new GAUSSIAN94 program (Frisch *et al.*, 1994). An independent version has been recently proposed by Kölle and Jug (1995);

4) the new definition of surface given by Pomelli and Tomasi (1995), shortly described in Section 6.6. This definition leads to a smaller number of tesserae, all of triangular shape, either convex or concave. The program for the evaluation of partial derivatives is still under testing and it is premature to make comments about merits and defects of this approach, which apparently has a remarkable computational efficiency.

This is the list of possible alternatives we have considered. To dispel erroneous impressions we stress that we have considered a number of possible ways to improve a program (Cossi *et al.*, 1995, 1996a) that gives G^{α}_{el} at a good computational efficiency, without introducing any approximation, just to produce something better.

7.2. G^{α}_{DIS}.

The analytical calculation of G^{α}_{dis} with the pair–potential method requires the use of the partial derivatives with respect to tesserae's elements we have discussed in the previous Section. The complete analytical derivation of G^{α}_{dis} formulas is given in Cossi *et al.* (1996a). The computational cost is minimal.

7.3. G^{α}_{CAV}.

Formulas are given in the already quoted Cossi *et al.*'s paper (1996a). To compute this contribution with Pierotti–Claverie's formula (eq. 80), the tessellation of the cavity portions is not compulsory. However, to exploit the subroutines used for G^{α}_{dis} and G^{α}_{el}, we can use the partial surface calculations *via* summation over the tesserae. Also the total volume $\mathcal{V}^{\alpha}_{M}$ is computed by using a BEM procedure, i.e. exploiting the definition of tesserae. According to the definition of the SPT cavity, there is no need of introducing GEPOL additional spheres (the $\mathcal{S}_{AS}$ surface is here used). The computational cost of G^{α}_{cav} is minimal.

8. Other continuum solvation methods.

In the two previous Sections we have described several points of methodological and computational interest for the calculation of solvation effects using the PCM procedures as reference. We have considered it convenient to focus our attention on a single procedure to make the exposition more clear. We have selected the PCM approach for two reasons: first, it is the procedure we know better, second, it has been elaborated with a number of details large enough to show all the relevant points.

The exposition of the PCM procedures has been limited to the aspects we consider more important for the study of chemical reactions. We would like to point out here that there are other extensions that use more specialistic medium models (for example non–homogeneous, Cossi *et al.*, (1994), and anisotropic solvents, Mennucci *et al.* (1995, 1996)) or that are addressed to the study of specific classes of phenomena (for example electric and magnetic susceptivity, non linear optics: Cammi *et al.*, 1995b, 1996). After saying this about PCM, we would like to dispel any false impression that solvation problems, and in particular reactions, can be treated with the PCM only. There are numerous valid alternatives, each one having specific points of strength and of weakness, when compared with PCM. Readers who are also potential users must be guided in the forest of specialized literature, to make a reasonable choice of the method to be adopted in the study of a specific problem.

To this end, we complement the two previous Sections by indicating other procedures, and by adding comments which are more addressed to highlight points of interest for potential users rather than to specify further technical details.

We shall start from methods similar to that previously described, characterized by the use of the apparent surface charge (ASC) description of the electrostatic interaction term $\hat{\mathcal{V}}_{el}$, passing then to consider other continuum methods, which use a different description of $\hat{\mathcal{V}}_{el}$. To complete the exposition we shall introduce, where appropriate, methods not based on the solution of a Schrödinger equation, and hence not belonging to the category of continuum effective Hamiltonian methods. We shall pass then to a selection of methods based on mixed continuum–discrete representation of the solvent, to end up with the indication of some approaches based on a full discrete representation of the solvent.

8.1. CONTINUUM ASC METHODS.

An *ab initio* SCF–ASC procedure developed at Cornell by Scheraga's group (Grant *et al.* 1990) uses Connolly's algorithm (Connolly, 1983) to cover molecular surface with points at two different densities. The high–density covering is used to reduce local curvature effects in the self–polarization of apparent charge distribution on each tessera. The correction for this local curvature effect, see Section 6.1.2, has been first introduced in the original PCM version (Miertuš *et al.*, 1981) *via* an expansion in powers of the solid angle subtended by the tessera. The correction has a significant effect on the final numerical results of G_{el} and of its derivatives. Many other ASC versions introduce this correction; a slightly different formulation has been recently given by Purisima and Nilar (1995). Scheraga's method uses an iterative procedure to get the final wavefunction Ψ and the energy E (see eq.(36)), no calculations of G_{el} and of the other components of G are considered. The results, obtained with cavities using fixed Rashin radii (Rashin *et al.*, 1988), and the scheme proposed by Aguilar *et al.* (1989) to get radii dependent on the electronic distribution, typically converge within five iterations. This computational scheme has been later used in a classical formulation of the solvation problem we shall consider below.

Colvin and Melius' (1993) ASC method uses Zahuar's procedure (Zahuar and Morgan, 1985) to triangulate a contact surface based on a scaled van der Waals molecular surface. The electrostatic potential of the molecule *in vacuo* is used to determine effective atomic charges (the so-called potential derived (PD) charges; see Tomasi *et al.*, 1991, for a review). From the PD charges an approximation to the AS charges is derived, and introduced in the Fock matrix. The process is iteratively repeated (typically 8 iterations). The total solvation energy is determined by a single-point MP2 calculation. The electrostatic calculation is supplemented by G_{cav} values obtained with the help of Sinanoglu's formulas (Halicioglu *et al.*, 1969), see eq.(81). No G_{dis} contributions are considered. Some examples of applications may be found in Evleth *et al.* (1994) and Colvin *et al.* (1995).

In a recent paper Rashin (Rashin *et al.*, 1994) presents an ASC procedure applied to density functional theory (DFT), based on de Mon's program (Salahub, 1987), with expansion of DFT orbitals on various Gaussian basis sets and examining both local (LSD) and not local (NLSD) approximations. In the last years several DFT solvation procedures (Ruiz–López *et al.*, 1994; Contreras *et al.*, 1993; Contreras, 1994; Fortunelli and Tomasi, 1994; Tannor *et al.*, 1994; Adamo and Lelj, 1994; Chen *et al.*, 1994; Truong and Stefanovich, 1995a; Hall *et al.*, 1995) have been published using both ASC and other approaches, all examining the performances of the various options available for DFT calculations. The attention has been mainly addressed to the examination of solvation energy at equilibrium geometry (for which the use of NLSD methods is preferable, with good quality results). Presently, no extensive studies are available for DFT solvation description of chemical reactions.

The above quoted Rashin's paper uses the tessellation of the cavity surface already used in previous Rashin's works (Rashin and Namboodiri, 1988; Rashin, 1990); this tesselation is similar to that used by Grant (Grant *et al.*, 1990), but it has never been fully documented. The solution of the quantum problem is done iteratively as in the original PCM (Miertuš *et al.*, 1981), using Mulliken's solute atomic charges as the source of the ASCs. To compute ASCs the authors rightly consider the use of the full molecular charge distribution more natural and actractive, but they point out that the use of local multiple expansion is a practical alternative to avoid the problem of describing effects due to the components of the solute electronic cloud $\rho_{M,e}$ lying outside the cavity boundary. We have not mentioned, for brevity's sake, this problem in the previous Sections; its effect, which leads to an erroneous value about the sum of AS charges, not satisfying Gauss' theorem, was signalled in the first 1981 PCM version (Miertuš *et al.*, 1981). Since then, in PCM this error has been corrected through a re-normalization of the ASCs. The importance of this correction has been emphasized by the use of direct ASC methods (Cammi and Tomasi, 1995b), and in the calculation of derivatives G^α (Cammi and Tomasi, 1994) and of hypersusceptibility functions (Cammi *et al.*, 1995b). The reduction of the whole electronic charge distribution to atomic charges apparently eliminates the problem, but a re–normalization of ASCs is still convenient. In addition, the error introduced by the use of atomic charges may be larger than that due to the tails of $\rho_{M,e}$ outside the cavity (of the order of 30-40 % for Mulliken charges, less when potential–derived (PD) charges are used and perhaps even less when electric field derived charges (FD) are introduced as Luque and Orozco (1995) suggest). The use of atomic charges or local multipole expansions may be compulsory for classical methods not using a quantum–mechanical definition of $\rho_{M,e}$. For problems treated at the quantum–mechanical level, the user must decide when it is convenient to use a procedure which obtains local charges to compute $\{q\}$ from $\rho_{M,e}$.

In the discussion following Rashin's paper (Rashin *et al.*, 1994), Gao raises an important question to which we are giving here our answer. He asks if one has to use the Hamiltonian $(\hat{\mathcal{H}}^o_M + \hat{\mathcal{V}}_{el})$ or $(\hat{\mathcal{H}}^o_M + \frac{1}{2}\hat{\mathcal{V}}_{el})$ to get the variational wavefunction. The correct expression is the first (see eq.(56)), and one has to use the corresponding functional $E = \left\langle \Psi \left| \hat{\mathcal{H}}^o_M + \hat{\mathcal{V}}_{el} \right| \Psi \right\rangle / \langle \Psi | \Psi \rangle$ when an iterative procedure is used, because, at every step, $\hat{\mathcal{V}}_{el}$ is fixed. In direct procedures, on the contrary, the right functional to be used is $G = \left\langle \Psi \left| \hat{\mathcal{H}}^o_M + (1/2)\hat{\mathcal{V}}_{el} \right| \Psi \right\rangle / \langle \Psi | \Psi \rangle$, as $\hat{\mathcal{V}}_{el}$ linearly depends on $\Psi\Psi^*$. Both direct procedures, with G, and iterative procedures, with E, give exactly the same final results (Cammi and Tomasi, 1995b). If one passes to models including a non linear–dependence of $\hat{\mathcal{V}}_{el}$ on $\Psi\Psi^*$, the factor (1/2) before $\hat{\mathcal{V}}_{el}$ has to be changed.

The PCM computational scheme has been recoded by several other groups, thus making available different sources of this code. The versions we know are limited to some of the features we have summarized in Section 7, with some changes and implementations.

An *ab initio* version has been coded by Green (1991) at Hillier's group in Manchester, implemented in GAMESS (Guest and Kendrick, 1986) and used to study several chemical reactions, with comparison with the results obtained with other methods. We shall comment the related papers later. Gaussian94 program (Frisch *et al.*, 1994) contains three versions of the PCM. The first one, called PCM, is a revised and refined version of the original 1981 PCM implementation (Miertuš *et al.*, 1981). Tessellation is given in terms of meridians and parallels on each sphere, and AS charges renormalization (see above) is introduced. Different options on the number of tesserae on each sphere, and on atomic radii to be used in the cavity's construction (several sets are reported) are available. A second version, called IPCM, models the cavity on a isodensity surface (default density value 0.001 a.u.). The method of tessellation here required leads to a larger number of points on the cavity than the previous one. Both PCM and IPCM may be used at the HF, DFT, MP2, MP3, MP4(SQD), QCISD, CCD, CID, and CISD levels of the theory (even if for the evaluation of energy only). A third version, called SCIPCM, determines, in a self-consistent way, the isodensity surface (with different possible options), iteratively computed at the HF level.

Few tests on the performance of these G94-PCM versions are given in a recent paper (Cossi it et al., 1996b), which also presents the implementation of our direct PCM versions (both at the HF and DFT levels of QM theory) in the G94 program. Of course, these few tests will be rapidly outdated by abundant results coming from other groups. The availability of PCM in a so well–known and widely used set of programs such as Gaussian, opens the perspective of a wider use of this model (that we consider to be the most complete and flexible one among those presently available), with more and more sophisticated versions, and the possibility of further improvements in the approach itself. The use of self–consistent isodensity surfaces, here just outlined, is a clear indication that we can expect further progress.

There are several semiempirical formulations of the ASC technique, mostly based on the PCM. Using semiempirical formulas, one has to define the approximation used for the evaluation of the elementary matrix elements $\left\langle \chi_\mu \middle| \hat{\mathcal{V}}_{el} \middle| \chi_\nu \right\rangle$ of the effective Fock matrix $\widetilde{\mathbf{F}}$. All the authors we shall quote here have considered this point, and have adopted different expressions.

Miertuš and Frecer (Miertuš *et al.*, 1988; Frecer *et al.*, 1989; 1991) have elaborated a CNDO version of PCM, in which the number of tesserae on each sphere is strongly reduced. We would like to mention the elaboration of a set of approximations to compute G_{dis} and G_{rep}, based on discrete representations of first and second solvation shells (for more comments, see Tomasi and Persico, 1994).

Luque and Orozco (Orozco *et al.*, 1994; 1995; Luque *et al.*, 1994) have supplemented MNDO, PM3, and AM1 versions of PCM with a careful analysis of all factors involved in the calculation (i.e. scaling factors for semiempirical molecular electrostatic potentials, cavity radii for neutral, cationic and anionic solutes) using a variety of methods (Monte Carlo, QM/MM, and FEP simulations), to check the conclusions. Parametrization has been mainly determined for water solutions, but it has been recently extended to chloroform (Luque *et al.* (1995)) and CCl_4. In the remarkable number of publications of this group, only partially quoted here, there is a good wealth of information useful for further studies.

The Sakurai's group has presented the first complete formulation of the direct minimization procedure of G_{el} (Hoshi *et al.*, 1987; 1988). A characterizing feature of their formulation is the attention paid to the extension of the model to non–homogeneous solvents, described as regions, with different dielectric constants, divided by sharp boundaries. A general description and parametrization of the procedure for MNDO, AM1 and PM3 methods is given in a paper by Furuki *et al.* (1994).

Chudinov *et al.* (1992) present a PM3 version of PCM with a modified iterative sequence. Also the strategy for the definition of tesserae has been modified. A successive paper (Chudinov and Napolov, 1993) sketches a method for geometry optimization, focussing the attention on partial derivatives involving tesserae. Presently, this group, led by Basilevsky in Moscow, is mainly interested in dynamical aspects of chemical reactions, a subject treated in Section 5.

The direct minimization procedure has been also used by Wang and Ford (1992) in an MNDO and AM1 version of the program. These authors introduce an alternative partition of the spheres defining the cavity, which still starts from the pentakisdodecahedron. The program has been used to study S_N1, and S_N2 reactions as well as tautomeric equilibria (Ford and Wang, 1992).

Another semiempirical PCM formulation has been presented by Rauhut *et al.* (1993). They use a "marching–cube" algorithm to define tesserae (Lorensen and Cline, 1987). The flat mesh thus obtained is then projected on a van der Waals surface (factor λ=1.15). These authors point out that the full NDDO (Neglect of Diatomic Differential Overlap) expression of the $\left\langle \chi_\mu \middle| \hat{\mathcal{V}}_{el} \middle| \chi_\nu \right\rangle$ elements is convenient in order to have good results, when combined with the original PCM procedure for the surface charge renormalization.

The marching–cube algorithm has been used also by Kölle and Jug (1995) to define the tesserae of isodensity surfaces. The procedure is implemented in the semiempirical SINDO1 program (INDO with Slater–type orbitals, Li *et al.*, 1992). To compute AS charges the asymptotic density model ADM (Köster *et al.*, 1993) is used. This is an approximation to the calculation of molecular electrostatic potentials based on the cumulative atomic multipole moment procedure (CAMM, Sokalski *et al.*, 1992).

Fox *et al.* (1993) have published a further semiempirical method (INDO/S–CI) which is more addressed to the study of excited states. Zahuar's experience on triangulation of molecular surfaces is here exploited. In the method used here (Zahuar and Morgan, 1990) the triangulation of the closed surface is given in terms of a number of points (nodes) connected by curvilinear segments (edges), and followed by a ten–points Lagrangian interpolation. The method is used in the iterative formulation, and compared with a FEM (Finite Element Method) procedure (Zahuar and Morgan, 1988). The authors signal faster convergence than in the original PCM implementation and consider the FEM procedure more promising for large molecules. The procedure described in this paper implements a previous ASC solvation method used by Fox and Rösch (1992).

We define as "classical" a formulation of the molecular solvation which makes use of point charges or multipole expansions to describe the solute, without turning to a Schrödinger equation. On the other hand, there are methods which use point charges drawn from a quantum mechanical description of the solute to compute AS charges (we have already seen some examples). In our classification these latter still belong to the category of effective Hamiltonian quantum methods.

Classical descriptions have been used by Rashin in already quoted papers (Rashin *et al.*, 1988; 1990; 1994). An interesting output of these studies is the definition of cavity radii for neutral and charged atoms, which has been used in more recent literature.

The classical formulation of the ASC problem presented by Drummond (1988) has some aspects of interest. The supertensor formalism he introduces is similar, although not equal, to the BEM matrix formulation shown in eqs. (51–54). The set of linear equations there which was introduced there may be used in classical, or quantum iterative solutions of the problem (for the quantum use, see Grant *et al.*, 1990, and Coitiño *et al.*, 1995a), or, alternatively, in a direct calculation, *via* the inversion of the **D** matrix (see eqs. 52–53). Drummond's formulation makes easier the handling of the equivalent supermatrix he defines. This approach has not been tested in quantum calculations.

Claverie's papers represent a cornerstone in the elaboration of solvation models (Huron and Claverie, 1972; 1974). The evolution of this research led him, before his untimely death, to use a classical ASC approach (Langlet *et al.*, 1988) exploiting a lucid analysis of local multipole expansions (Vigné–Maeder and Claverie, 1988). Now this elaboration has been connected with SIBFA, a procedure which provides a detailed description of a molecule in terms of interacting fragments (Gresh *al.*, 1984; 1986), to give a versatile classical solvation program (Langlet *et al.*, 1995).

To end up this Section, we consider now an alternative ASC approach which strongly deviates from the basic set up previously discussed. The quantum system, the "solute", has been considered immersed in a continuum solvent distribution characterized, *prima facie*, by its dielectric response function. Klamt and Schüürman (1993) propose to replace the dielectric response function with the response function of a liquid electric conductor (the so-called COSMO model). In doing so, one has other electrostatic problems, with simpler boundary conditions. In fact, for a system of charges (the solute) in a screening conductor (where $\varepsilon = \infty$), the electrostatic potential on the cavity surface $\mathcal{S}$ is zero:

$$\Phi_{tot}(\vec{s}) = \sum_{\alpha} \frac{Z_\alpha}{|\vec{s} - \vec{R}_\alpha|} - \int_V \frac{\rho_{M,e}(\vec{r})}{|\vec{s} - \vec{r}|} d^3r + \int_S \frac{\sigma(\vec{r})}{|\vec{s} - \vec{r}|} d^2r = 0 \qquad (98)$$

where $\vec{s}$ is on $\mathcal{S}$; for the definition of all other quantities, see eqs.(40)–(46).

To obtain the surface charges $\{q\}$ for a conductor, Klamt and Schüürman introduce an empirical scaling factor $f(\varepsilon) = (\varepsilon - 1)/(\varepsilon + 1/2)$. This method, which presents other features deserving attention, has been implemented in the MOPAC package (Stewart, 1990). The same idea has been exploited by Truong and Stefanovich (1995b, 1995c) at the *ab initio* level; it also includes dispersion, repulsion, and cavitation contributions. The method, called GCOSMO, uses GEPOL93 for the cavity, and a scaling factor $f(\varepsilon) = (\varepsilon - 1)/\varepsilon$. Both COSMO and GCOSMO give expressions for the analytical derivatives of G_{el}.

8.2. MULTIPOLE EXPANSION METHODS (MPE).

The expansion of the electrostatic potential into spherical harmonics is at the basis of the first quantum–continuum solvation methods (Rinaldi and Rivail, 1973; Tapia and Goschinski, 1975; Hylton McCreery *et al.*, 1976). The starting points are the seminal Kirkwood's and Onsager's papers (Kirkwood 1934; Onsager 1936): the first one introducing the concept of cavity in the dielectric, and of the multipole expansion of the electrostatic potential in that spherical cavity, the second one the definition of the solvent reaction field and of its effect on a point dipole in a spherical cavity. The choice of this specific geometrical shape is not accidental, since multipole expansions work at their best for spherical cavities (and, with a little additional effort, for other regular shapes, such as ellipsoids or cylinders).

The first implementations of the MPE quantum–continuum solvation methods here introduced have been limited to spherical cavities. Tapia and Goschinski (1975) use a spherical cavity, reducing the expansion of the reaction potential to the dipole only (the quantum analog of the Onsager model). Rivail's group also started with a spherical cavity (Rinaldi and Rivail, 1973; Rivail and Rinaldi, 1976) using an expansion in terms of spherical harmonics extended up to the total angular moment l=7. The method has been later extended to ellipsoidal cavities (Ruiz–López and Rinaldi, 1983), and to *ab initio* SCF methods (Rinaldi *et al.*, 1983).

The SCRF set of procedures elaborated during the years by Rivail's group in Nancy is the most complete one to be found among MPEs. We shall use here the notation introduced by that group. By expressing the Schrödinger equation at the SCF level (and including electrostatic terms only: $\hat{\mathcal{V}}_{int} \rightarrow \hat{\mathcal{V}}_{el}$), they write the Fock matrix elements as:

$$F_{\mu\nu} = F_{\mu\nu}^{(0)} + \sum_{l,l'} \sum_{m,m'} M_l^m f_{ll'}^{mm'} \left\langle \chi_\mu \left| M_{l'}^{m'} \right| \chi_\nu \right\rangle \tag{99}$$

where M_l^m is a component of the multipole expansion of the solute charge distribution, and $\left\langle \chi_\mu \left| M_{l'}^{m'} \right| \chi_\nu \right\rangle$ a one–electron integral over the basis set $\{\chi\}$ involving the operator related to the m'th component of the multipole of rank l'. The factors $f_{ll'}^{mm'}$ depend only on the geometry of the cavity and on the dielectric constant ε (as the **D** matrix in BEM–PCM method, eqs. 52–53). By introducing the derivatives of the electric field, or 'reaction field', components, R_l^m:

$$R_l^m = \sum_{l'} \sum_{m'} M_{l'}^{m'} f_{ll'}^{mm'} \tag{100}$$

we can rewrite the fock matrix in a form which is more easily comparable with eq.(61)

$$\mathbf{F} = \mathbf{h} + \mathbf{G}(\mathbf{P}) + \sum_{l} \sum_{m} R_l^m \mathbf{M}_l^m(\mathbf{P}) \tag{101}$$

where $\mathbf{P} = \mathbf{CnC}^\dagger$ is the one–electron density matrix. The solution of the corresponding Hartree–Fock equation:

$$\mathbf{FC} = \varepsilon\ \mathbf{SC} \tag{102}$$

is obtained by iterative steps.

One of the advantages of this formulation is that the $f_{ll'}^{mm'}$ factors have an analytical form for regular cavities, hence it is relatively easy to get energy derivatives at fixed cavity (Rinaldi *et al.*, 1992). The Nancy's SCRF procedure (available as QCPE program, Rinaldi and Pappalardo, 1992) has been recently supplemented by many additional options, i.e. calculation of G_{dis} and G_{cav}, calculation of G_{el} at the MP2, GVB and MCSCF levels of the quantum theory (Chipot *et al.*, 1992; Rinaldi and Rivail, 1995).

When the solute has a regular shape, so as to be reasonably well inserted in an ellipsoid cavity, the method works efficiently and fast. The problems arise when the solute is irregular, because the expansion method does not converge, and truncated expansions may give poor results. Extensions to multicenter expansions and to cavities of general shape have been now introduced in Rivail's SCRF procedure (Dillet *et al.*, 1993). This new feature will surely improve convergency. The analytical G derivatives are obtained with a new code (Dillet *et al.*, 1995), which is based on the solution of a set of linear equations and gives the derivatives of $f_{ll'}^{mm'}$ factors at a low computational cost. Another problem regards the description of solvent–separated solutes. The approach of reactants in bimolecular chemical reactions is ruled by solvent–mediated interaction forces. Their description in the MPE methods is quite difficult.

Another MPE procedure, well implemented for the *ab initio* description of the solute, has been elaborated by Mikkelsen *et al.* (1988). The method is limited to spherical cavities and it is used not for chemical reactions, but for the study of solvent effects on a number of properties of solutes at fixed geometry (electronic spectra, photoionization, hyperpolarizabilities) with results of remarkable interest (Ågren *et al.*, 1993; Mikkelsen *et al.*, 1994).

A solvation model, in which the reaction field is reduced to a dipolar term in a sphere (i.e. the Onsager model), was introduced in Gaussian 92 (Frisch *et al.*, 1992). This method (Wong *et al.*, 1991) is accompanied by a clear formulation of the formal quantum–mechanical problems, including second derivatives (Wong and Wiberg, 1992), and calculations beyond the Hartree–Fock approximation. It is still present in the last version of Gaussian program (Frisch *et al.*, 1994), and it is easy to use. However, users are warned against the serious limitations due to the fixed spherical cavity and to the reduction of the reaction field to the dipole term only.

Karelson–Zerner's formalism was originally expressed on the basis of the Onsager model implemented at the semiempirical level (Szafran *et al.*, 1993). It has been used for several applications, mainly regarding solvent effects on electronic spectra and tautomeric equilibria. The model now evolves toward more faithful descriptions of the reaction potential *via* a partition of the solute charge distribution into fragments (Karelson *et al.*, 1993). Each fragment is embedded in a separate spherical cavity. For each sphere charge and dipole terms are considered to describe the reaction field and the mutual interactions among terms related to different spheres are included in the procedure.

8.3. GENERALIZED BORN'S MODEL AND AMSOL.

Born's model (Born, 1920) gives the solvation free energy of a spherical ion as:

$$\Delta G_{el}(\text{Born}) = -\frac{\varepsilon - 1}{2\varepsilon}\frac{q^2}{a} \tag{103}$$

The ion is described by a conducting sphere with charge q and radius $\overset{\circ}{a}$, or by a point charge at the center of a void spherical cavity ($\varepsilon = 1$) with radius a (the two electrostatic problems are completely equivalent).

Extensions of this formula to an assembly of point charges in a void cavity may lead to expressions of the type:

$$\Delta G_{el}(\text{gen.Born}) = -\frac{\varepsilon - 1}{2\varepsilon}. \sum_k \sum_{k'} q_k q_{k'} \gamma_{kk'} \tag{104}$$

where k labels the atomic center, and $\gamma_{kk'}$ is an integral of Coulomb type.

We would like to stress here a terminological issue. The description of the solute charge distribution in terms of point charges is the lowest level of a local multiple expansion of $\rho_M = \rho_{M,e} + \rho_{M,n}$. We have considered this description several times in the previous Sections. Generally, this description is accompanied by a definition of the reaction potential Φ_{el} in terms of ASC, MPE approaches (or in terms of other approaches, as Image Charge, IM, or Finite Difference, FD, we have not considered yet). These combined descriptions of ρ_M and Φ_{el} have been sometimes named "Extended" (or "Modified") Born expression, and also "Generalized Born". Actually, they have little in common with the method resorting to eq.(104) to which the denomination "Generalized Born" should be reserved, following Cramer and Truhlar's choice.

A very interesting semiempirical expression for $\gamma_{kk'}$ of eq.(104) has been suggested by Still *et al.* (1990) for a cavity of molecular shape (interlocking spheres). Their expression contains several parameters to be determined, describing, *inter alia*, the radii a_k, and the portion of the 'exposed area' for each sphere, when the others are placed at the desired positions.

This formalism has been adopted, and modified, by Cramer and Truhlar (Cramer and Truhlar, 1991, 1992) in a procedure which belongs to the category of continuum effective Hamiltonian quantum methods (for reviews see Cramer and Truhlar, 1994, 1995a, 1995b).

The Schrödinger equation is expressed at the Hartree–Fock level (only semiempirical calculations, with AM1, Dewar *et al.* 1985, and PM3, Stewart 1989, parametrization, have been considered), and the solution is found by extremizing the energy functional G (which is the sum of the internal electronic and internuclear Coulombic energy of the solute and of the solvent's free energy). The elements of the Fock matrix are defined as follows:

$$F_{\mu\nu} = F_{\mu\nu}^{(0)'} + \frac{\varepsilon - 1}{\varepsilon} \sum_k \sum_{k'} \frac{q_k}{\gamma_{kk'}} \frac{\partial q_{k'}}{\partial P'_{\mu\nu}} - \frac{1}{2}\frac{\varepsilon - 1}{\varepsilon} \sum_k \sum_{k'} q_k q_{k'} \frac{\partial \gamma_{kk'}}{\partial P'_{\mu\nu}} \tag{105}$$

The $\gamma_{kk'}$ terms are the Coulomb functions introduced in eq.(104), $P'_{\mu\nu}$ are the relaxed (i.e. polarized by the solvent in a self–consistent manner) elements of the density matrix $\mathbf{P} = \mathbf{CnC}^\dagger$, and $F_{\mu\nu}^{(0)'}$ are the elements of the *in vacuo* Fock matrix, in the form used in the semiempirical approach, but referred to the relaxed $\mathbf{P}'$ matrix. The charges q_k are drawn from $\mathbf{P}'$, according to the Mulliken's rules, or according to a new definition given by the authors (Storer *et al.*, 1995). This is the key point which introduces a non linear–term in the Hamiltonian.

The electrostatic term of the free energy, called G_{ENP}, is given by:

$$G_{ENP} = \frac{1}{2}\sum_{k}\sum_{\mu}\sum_{\nu} P'_{\mu\nu}(h_{\mu\nu} + F^{(0)'}_{\mu\nu}) + \sum_{i}\sum_{j>i}\frac{Z_i Z_j}{r_{ij}} - \frac{\varepsilon - 1}{2\varepsilon}\sum_{k}\sum_{k'} q_k q_{k'} \gamma_{kk'} \tag{106}$$

where $h_{\mu\nu}$ are the elements of the one–electron component $\mathbf{h}$ of the Fock matrix.

The electrostatic term G_{ENP} is accompanied by a second term G_{CDS} which, as the acronym indicates, collects cavitation, dispersion, and solvent structure contributions, the latter regarding dielectric saturation and other reorganization effects, when present. G_{CDS} is modeled in terms of an interfacial atomic surface tension term, σ_k, and of the parameter A_k giving the solvent–accessible surface area for atom k:

$$G_{CDS} = \sum_{k} \sigma_k A_k(\beta_k, \{\beta_{k'}\}) \tag{107}$$

where the radius β_k is the sum of the van der Waals radius of atom k and of the solvent radius.

The method has several parameters with non–assigned numerical values, i.e. those cointained in $\gamma_{kk'}$, and σ_k. They are simultaneously optimized to fit a large set of experimental solvation free energies, ΔG_{sol}. This fitting has been done for water in the first set of papers, and now extended to n–hexadecane as solvent (Giesen *et al.*, 1995).

The method, called AMSOL and available at the QCPE (Cramer *et al.*, 1993), manages to give reasonable estimates of ΔG_{sol} for compounds remarkably different from those used in the calibration of the parameters. This indicates the strength of the parametrization employed, as a recent report clearly confirms (Liotard *et al.*, 1995). Moreover this method is fast, easy to use, and provided with useful options, among which we quote geometry optimization. The main defects are due to the semiempirical nature of the quantum procedure. In some cases semiempirical methods give, nuclear geometries remarkably different from experiment and good *ab initio* calculations. These defects are more evident, when non–covalent interactions (e.g. hydrogen bonds) play an important role, or near the Transition State in chemical reactions. It seems to us that AMSOL does not introduce any further errors in the geometry, however it is not able to eliminate AM1 and PM3 errors if any.

Another limit of semiempirical methods is due to the fact that they do not provide a series of results of increasing accuracy, as *ab initio* methods do, from which the researcher often derives information about the reliability of the calculations. One possible improvement of the method is to resort to *ab initio* calculation *in vacuo*, and then to compute ΔG_{sol} by using AMSOL. The free energy hypersurface will be given by:

$$G(\mathbf{R}) = E^{o}_{ab\ in.}(\mathbf{R}) + \Delta G_{AMSOL}(\mathbf{R}) \tag{108}$$

This partition of $G(\mathbf{R})$ is frequently used (see, for example, Section 3). The use of AMSOL has the merit of introducing, even if at a low computational level, solvent–induced polarization at an SCF level.

8.4. FINITE DIFFERENCE METHODS (FDM).

Many continuum solvation methods prefer to attack the electrostatic problem by resorting to grid integration of the Poisson equation. The use of 3D grids makes it convenient to extend the model we have considered till now, characterized by a constant value of ε, and by the corresponding Poisson and Laplace equations, i.e.

$$\nabla^2\Phi(\vec{r}) = -4\pi\rho_M(\vec{r}) \quad \text{inside the cavity} \tag{109}$$

$$\nabla^2\Phi(\vec{r}) = 0 \quad \text{outside the cavity} \tag{110}$$

(Φ is here the total electrostatic potential), to the following general expression

$$\vec{\nabla}\cdot[\varepsilon(\vec{r})\vec{\nabla}\Phi(\vec{r})] = -4\pi\rho(\vec{r}) \tag{111}$$

The dependence of ε on position allows to describe both local saturation effects, and, in salt solution, the dependence of ε on concentration. In numerical grid methods (there are families of methods, called Finite Elements Methods, FEM, and Finite Differences Methods, FDM – see Tomasi and Persico (1994) for a short explanation of differences) there is nothing but a little increase in complexity in treating eq.(111) instead of eqs.(109–110). The treatment of the latter model by means of a MPE method is more difficult, and a specialized PCM version (Cossi *et al.*, 1994) is necessary to exploit it with a BEM–derived method (see Juffer *et al.*, 1991, for another proposal based on the BEM technique).

FD methods are widely used in combination with classical point charge description of the solvent (FEM is less used; see You and Harvey, 1993, for a recent implementation of this approach). The first formulation of the method is due to Warwicker and Watson (1982) (see Davis and McCammon (1990) for a review). The use of 3D numerical grids makes the calculations generally computer intensive, and, in addition, the evaluation of ΔG_{el} is rather inaccurate.

Many shortcomings have been eliminated. A double calculation, with and without solvent, has been introduced by Gilson and Honig (1988) to eliminate the solute self-energy and part of the spurious contributions to the Coulombic interaction. A more rapid procedure, avoiding the second calculation, has been proposed by Luty *et al.* (1992). Acceleration of the convergence has been reached by adopting mathematical methods that are more efficient than the Gauss–Seidel iteration scheme initially used (Davis and McCammon, 1989; Nicholls and Honig, 1991). The number of grid points has been reduced by introducing a focussing technique (Gilson *et al.*, 1988) according to which a coarse grid is progressively refined in the vicinity of the solute.

The reduction of the solute charge distribution to a set of fixed point charges makes it possible the application of the FD method to the study of large molecules, such as polypeptides or proteins.

There are several solvation programs using FDM approaches. The program of larger use is DelPhi (Nicholls and Honig, 1991; Honig *et al.*, 1993), of which there are different versions. We also mention here the program due to Bashford (Bashford and Karplus, 1990; Bashford and Gewert, 1992).

The formulations of the FD procedures we have sketched here are not suited to study chemical mechanisms; they may give reaction energy differences, as pK_a values and tautomeric equilibria, or information about processes not involving bond forming and breaking, as conformational changes in large molecules.

Both FD programs we quoted above (DelPhi and Bashford's approach) introduce a dielectric constant larger than ε=1 for the region of space occupied by the solute, in order to mimic solute polarization. The chosen values range around ε=2 and ε=4. The estimate of the electronic polarizability within a molecule leads to ε=2, a larger value indicates that there is an attempt to mimic also orientational polarizability. Another option is to assign different values of ε to different regions of the molecule, exhibiting different electronic or orientational polarizabilities (the local dielectric constant, LDC, Sharp *et al.*, 1992). However, these refinements make the method more cumbersome and more subjected to arbitrary modelization.

At present, there is a renewed interest in grid calculations in the quantum description of molecular systems (*in vacuo*) that may be exploited for FD calculations of solvent effects. The DelPhi program is now extended (Tannor *et al.*, 1994) to the use of a high-level *ab initio* quantum mechanical procedure (Friesner, 1991). This new quantum program, called PS–GVB (Ringnalda *et al.*, 1994), uses a synthesis of grid techniques with standard orbital methods, with a noticeable speed-up with respect to current programs. In this case DelPhi uses ε=1 for the internal space. The authors use an iterative solution of the non-linear problem, with the effective potential represented by a few hundred charge points, passed from DelPhi to PS–GVB. The calculations are limited to G_{el}, the other terms being introduced with an empirical parameter. The method could be used for chemical reactions, but we only know test calculations on solvation energies.

Bashford's program has been coupled to a Density Functional quantum program (Chen *et al.*, 1994). In this computational set up, use is made of the Amsterdam Density Functional program (ADF, Te Velde *et al.*, 1992). The reaction potential is defined in terms of MEP fitted atomic charges (CHELPG program, Breneman and Wiberg, 1990). This choice, which is considered as provisional, is due to the acknowledgement that the use of the complete QM charge distribution needs to take into account the electron density distribution lying outside the molecular volume, a problem easy to solve for the ASC methods, but which gives problems for FD methods. The dielectric constant is set equal to 1 for the molecular space, while the non–linear coupled QM and electrostatic problem is solved iteratively (5–7 cycles). The method uses a large number of grid points (24000–50000) a few hundred of which are used for the reaction potential. This method has been applied until now to test calculations of the solvation energy.

These two examples show that there is an evolution of FD methods from classical to quantum descrition of the solute. Therefore the new approaches we have resumed belong to the class of effective Hamiltonian methods, with continuous description of the solvent. It is not possible yet to give a definitive appreciation of these new proposals, and we shall confine ourselves to express some provisional comments. The quality of the results is comparable to that of the best versions of the ASC and MPE approaches, with a lower computational efficiency. Grid methods require the evaluation of the necessary electrostatic property of a larger number of points than ASC methods, and this may represent a decisive factor. However, the methods we are considering here are at the first stages of their elaboration, and their efficiency can be surely improved. Other features of some relevance in the study of chemical reactions, as the stability of the results with respect to the introduction of some solvent molecules in the "solute" and the calculation of analytical derivatives G^{α}, have not been considered yet.

8.5. ON THE DEFINITION OF THE CAVITY.

In the previous Sections, we have paid attention to the definition of the cavity in the different computational procedures we have examined. A proper definition of the cavity plays an important role in assessing the quality of continuum solvation calculations. It is convenient to collect under a specific heading some remarks on this point.

From an operational point of view the best cavity is that giving a better agreement between computed and experimental values. This pragmatical approach has been attempted but there are several difficulties forbidding a general use of it. The experimental values one can directly use are limited: they mostly refer to solvation energies, ΔG_{sol}, of small – or medium – size molecules at their equilibrium geometries. Other experimental values must be used indirectly, i.e. on the basis of their interpretation according to some models. In addition, the calculation of ΔG_{sol} is partitioned into different terms (see Section 2) computed at different levels of accuracy, and this decreases the level of confidence in the results coming out from minimization of the deviation of computed ΔG_{sol} values with respect to the experiment.

One has thus to resort to other approaches based both on modellistic considerations about naive pictures and on *ad hoc* calculations of other properties. Modellistic considerations agree with numerical evidence in discarding cavities having a regular shape. They may be used for a first guess, being their use cheaper in some methods, but they often lead to erroneous quantitative predictions. Cavities modeled on the solute shape are safer, and lead to more satisfactory results. To obtain these cavities it is convenient to start from a shape function (Pomelli and Tomasi, 1995) expressed as the union of hard sphere–shape functions $f_{hs}(A)$ centered on the N atoms of the molecule M:

$$f_{hs}(M) = \bigvee_{A=1}^{N} f_{hs}(A) \qquad A \in M \tag{112}$$

where $\bigvee$ is the logical operator "or", which runs on all the atoms, and hence on the spheres, and

$$f_{hs}(A) = f_{hs}(\vec{r}, \vec{R}_A, \lambda(A), R_{vdW}(A)) \tag{113}$$

is the atomic shape function, related to the van der Waals radius of atom A, $R_{vdW}(A)$, modified by a suitable numerical factor λ, and depending on the $\vec{R}_A$ position of the nucleus. When the R_{vdW} values have been selected the only parameters left free are the factors λ.

With the definition (112) one immediately arrives at the molecular (or van der Waals) cavity, characterized by the volume $\mathcal{V}_W$ and the surface $\mathcal{S}_W$. However, modellistic considerations suggest the use of other cavities, that may be derived from the van der Waals one by introducing further parameters related to the solvent molecules.

The first one is the cavity giving the solvent accessible surface $\mathcal{S}_{SA}$. This surface is defined by introducing a sphere representing the solvent (with radius R_s giving the correct value of the solvent molecular volume) and making it roll on the $\mathcal{S}_W$ surface. $\mathcal{S}_{SA}$ is defined by the close surface described by the center of the solvent sphere in its movement around the solute. The resulting sphere corresponds to an increment of R_s to the hard sphere shape functions, $\mathcal{S}_W$. The surface $\mathcal{S}_{SA}$ may be used to compute G_{cav}, and also G_{dis}, when atom–atom dispersion functions are used. For the latter property we prefer to introduce a set of $\mathcal{S}_{SA}$ surfaces (Floris and Tomasi, 1989), the meaning of which will be made clear by the following example. Let us suppose that the solvent is methane: an atom–atom calculation of G_{dis} will be more realistic if the contributions due to the C atoms and to the H atoms are computed with different increments of the $f_{hs}(A)$ functions. Modellistic reasons suggest an increment equal to R_s for C, and to $R_s - R_{\mathrm{C-H}}$ for H. In fact, the minimal distance of approach of H atoms will be always smaller than that for C atoms only for an amount equal to the C–H bond length.

$\mathcal{S}_{SA}$ is not appropriate for the calculation of G_{el}. For this quantity, it is convenient to resort to the solvent excluding surface $\mathcal{S}_{SE}$ defined as the surface limiting the portion of space not accessible to the solvent molecules. This cavity may be defined in terms of f_{hs}, supplemented by an obstruction function f_{obs} (Pomelli and Tomasi, 1995), such as:

$$f_{hs}^{SE} = f_{hs} \bigvee f_{obs} \tag{114}$$

To evaluate f_{hs}^{SE} the van der Waals spheres are not sufficient. The analytical definition of f_{obs} is often replaced by approximate expressions, a set of additional spheres as in GEPOL (Pascual-Ahuir *et al.* 1987), by analytical functions as in Connolly's algorithm (Connolly, 1983) and in DefPol (Pomelli and Tomasi, 1995).

The definitions of $\mathcal{S}_{SA}$ and $\mathcal{S}_{SE}$ we have given are not completely satisfactory, as they assume that the solvent molecule can be represented by a sphere. This is an assumption which is not realistic for solvents with a more complex shape. Work is in progress to generalize both definitions to solvent of different shape.

These elaborations have been based on modellistic considerations. To reach a final definition of cavities we have to determine the $\lambda(A)$ factors introduced with eq.(113). Accurate sets of $\lambda(A)$ values should depend, to some extent, on the nature of chemical groups which atom A belongs to, as well as on the nature of the solvent (it must be recalled that the cavity is not an intrinsic property, but it measures an interaction). For example, it is not surprising to find that the recommended value for the H atoms is a bit smaller when it is bound to a heteroatom than when it is bound to an aliphatic carbon (Bachs *et al.*, 1994), or that the value is different when the group is formally neutral or charged ($-NH_2$ vs $-NH_3^+$ for instance, Bonaccorsi *et al.*, 1984b; Orozco and Luque 1994). For the same reasons the recommended λ value for a given atom in a given molecule may be found to be dependent on the solvent (Bonaccorsi *et al.*, 1990; Cossi *et al.*, 1994).

The best strategy to be followed in order to get accurate sets of λ values has not been defined, so at present more or less complex statistical elaborations of some data are used. Among the numerical data that have been used we mention: solvation and solvent transfer energies, intrinsic solute properties (electron isodensity surfaces, isopotential electronic surfaces, multipole expansions of local charge distribution), isoenergy surfaces for the interaction with selected probes (water, helium atoms), Monte Carlo simulations with solutes of various nature. All these sets of data deserve comments, that are here severely limited not to unduly extend this Section.

The use of isodensity to define $f_{hs}(M)$ (Frisch *et al.*, Gaussian94; Kölle and Jug, 1995) does not reflect the different strength of the interaction, and the effect this has on $\lambda(A)$ for different atoms. We suspect that, by using cavities based on isodensity surfaces, the model could no longer be able to properly treat solute–solvent hydrogen bonds. Analogous remarks apply for the use of the electronic component of the molecular potential to define $f_{hs}(M)$ (Rivail *et al.*, 1985).

In an interesting paper, Amovilli and Mc Weeny (1991) explore the use of the interactions to define the molecular surface. The corresponding cavity is of direct use for solvents without permanent dipole and low polarizability, but it is not safe enough for solvents such as water.

Monte Carlo calculations give atom–atom correlation functions, g_{AS}, from which important elements on the relative values of the λ values can be obtained, but in standard MC calculations electrostatic, dispersion and repulsion contributions are mixed together, and angularly averaged $g_{AS}(r_{AS})$ functions are produced. Some important corrections must be introduced before using these data in the calibration of $\lambda(A)$ factors (Floris *et al.*, 1993).

The examples we have given here will suffice to indicate the problems connected with accurate definitions of the λ parameters.

To end up this Section it is convenient to spend a few more lines to firmly dispel any impression that the problem is still unresolved, and that the cavities in use are not reliable. Fortunately, the situation is different: our early proposal of using an universal factor $\lambda = 1.2$ for all atoms in aqueous solutions gives errors on the average below 1 kcal/mol. Further refinements, as those we have mentioned, reduce the error to lower bounds. Passing to other solvents (which are less sensitive than water to the specific $\lambda(A)$ values), λ ranges in the interval 1.15–1.5 (with few exceptions). The recommended values reported in the literature can be used with confidence.

We have based this discussion mostly on material related to the PCM procedure: problems, tentative solutions, and also conclusions are similar for the other methods using cavities modelled on the solute molecular structure.

9. Discrete solvation methods.

The process of elaborations of solution models presents an important variant at its very beginning. We have thus far considered models based on a continuous distribution of the solvent, uniform or not uniform, in some cases supplemented by a molecular description of the first solvation shell or of a part of it. We have now to consider models in which the whole solvent, or a large part of it, is represented by discrete units, i.e. the solvent molecules.

Some authors indicate the models belonging to the first category as 'macroscopic models' and those of the second one as 'microscopic models'. This is a misleading definition, as in both sets the main portion, the solute, is treated at the detailed atomic, or subatomic, level necessary for chemical descriptions, which for convenience we may call microscopic. Macroscopic models for solutions are something else, of limited usefulness in order to study the details of chemical processes, and they are not treated in this chapter.

The difference between continuum and discrete models stays, as we have said, in the description of the secundary, but larger, portion of the material model, the solvent. This is a delicate point, which deserves some comments to supplement what said in Section 1.

Continuous models have been considered with some suspicion in the seventies (and also later: all the suspicions are hard to be dispelled, even when not motivated and when evidence is against them) by theoretical chemists, confident to be able to describe the properties of a molecule, or of an assembly of molecules, with QM methods. Continuum models, they argued, are good to grasp some essential features of the problem in a "physical" way, as the models of Kirkwood, Onsager, Marcus gave, but not precise enough to reach the "chemical" level of accuracy for which a complete molecular description of the system is necessary. It is not a valid objection, they continued (we are here collecting opinions and sentences scattered in different papers of various authors as belonging to the whole category of chemists using QM methods), to remark that the present level of computational methods and facilities is not sufficient to treat the solution at the desired degree of accuracy. It is better to keep the molecular level of description, to do what is possible with the available computational resources and to work to improve them; to pass to a continuum description of the solvent will not permit to improve methods.

These comments are partially correct. In fact, improvement of continuum models passed through the analysis of results of discrete models, not regarding the whole solution but by far smaller solvation aggregates. The calculations of small molecular clusters, including one or more solvent molecules (generally water) with QM methods, often of low level but treating all the molecules on the same footing, started in the late sixties. This approach, later called the "supermolecular approach" (the name is due, we think, to Pullman's group) gave, and still gives, important information about the effect of non-covalent interactions between solute and solvent molecules, and by extrapolation, about some solvent effects on solute properties. It has been also extended to study some chemical reactions: we quote here as example a study on the $CH_3F + F^-$ reaction in water at the CNDO level (Cremaschi *et al.*, 1972), followed by other similar studies of Simonetta's group, but other examples could be done.

It came out immediately clear that the supermolecule approach cannot represent the method to be used in extensive studies of solvent effects. The computational costs increase in the *ab initio* versions with more than the fourth power of the number of basis set functions, at a given nuclear geometry of the supermolecule. Even more important it has been the recognition that, when the size of the solvation cluster exceeds some very low limits, the number of different nuclear conformations at an equivalent energy increases exponentially: computational costs increase in parallel, and the introduction of thermal averages on these conformations becomes necessary. These facts, and some attempts to overcome them, are well summarized in a Clementi's monograph (Clementi, 1976). The problem of multiple equivalent minima still plagues some discrete solvation models.

An important step for the evolution of models has been given by the analysis of the intermolecular interactions in dimers, trimers, and larger clusters, involving one, and two, and more solvent molecules, respectively. Decompositions of the interaction energy, ΔE, have been prepared by several groups (Bonaccorsi *et al.*, 1969; Kollman and Allen, 1970; Dreyfus and Pullman, 1970; Morokuma, 1971) but the most complete was proposed some years later by Kitaura and Morokuma (1976). This method, still in use, adopts elements given by the quoted methods proposed by Bonaccorsi, Dreyfus, and Morokuma, adds a definition of a further term, and gives both a satisfying formal expression and an efficient implementation of the procedure.

From the examination of ΔE decompositions performed in our group in Pisa (of which we quote the first papers: Bonaccorsi *et al.*, 1971; Alagona *et al.*, 1972, 1973) we learned that solute-water interactions can be reduced to the interaction terms of electrostatic origin, i.e. those described by a simple Hartree product wavefunction $\Psi_{AS} = \Psi_A \cdot \Psi_S$, that the validity of this approximation increases when the number of water molecules increases, and that interaction energy, as well as solvent effects on the solute electronic distribution, can be faithfully described by "projecting" solvent electrostatic effectors on a surface near to, but a bit larger than the van der

Waals envelope of the solvent. This projection can be expressed in terms of a response function, whose kernel contains a damping factor (the "dielectric constant") very near to the optical dielectric constant of water, ϵ_{opt}, when the water molecules are held fixed, or rapidly increasing towards the static dielectric constant, ϵ_0, when water molecular motions are allowed and their number in the cluster increases. This is the origin of our PCM model (more details can be found in Tomasi, 1982). Surely, similar considerations spurred Rivail and coworkers to elaborate their SCRF method (Rivail and Rinaldi, 1976). An additional contribution to the formulation of today continuum models came from the nice analysis given by Kołos (Kołos, 1979; Clementi *et al.*, 1980) of the importance of dispersion contributions.

This said for the role played by discrete models in the supermolecule approach in suggesting and supporting new versions of the continuum model, we may pass to consider other evolutions of models based on this approach.

The interaction energy hypersurface for dimers (essentially described in a space with dimensionality ≤ 6, being the internal monomer's geometry kept fixed) was used to get simplified analytical expression of $\Delta E(R_{AB})$. These simpler expressions were used either to reduce computational times in discrete models similar to the supermolecule approach, or in computer simulations of the liquid.

An example of the first approach can be found in a clever paper by Noell and Morokuma (1976) (others will be discussed later), and one application of the second approach (probably the first application) in a paper by Popkie *et al.* (1973). The first approach belongs to the family of effective Hamiltonians (see Section 1) and introduces more chemistry in the description of the solute. However, it presents the same inconveniences of the supermolecule approach, in spite of the considerable reductions of computational times. In this Section we shall examine other more effective versions of this approach. The second approach has been largerly exploited in the following years. The Monte Carlo simulations of pure water given by Clementi and coworkers in the above quoted paper (Popkie *et al.*, 1973) was followed by many other applications to solutions. The activities of Clementi's group in this field during these initial years are well summarized in a monograph (Clementi, 1980). In the present Section we shall consider further extensions, more addressed to the study of chemical reactions.

Popkie's potential (Popkie *et al.*, 1973) used in that Monte Carlo simulation has some advantages with respect to the empirical potentials used in the few preceding Monte Carlo (Barker and Watts, 1969) and molecular dynamics (Rahman and Stillinger, 1971) simulations of pure water, since the use of a QM definition of the $\Delta E(R_{AB})$ hypersurface allows one to improve the potential exploiting progresses in the QM calculations of variational energies. For many years, this approach has been limited to the use of two-body potentials, with monomers at fixed geometries, and without relaxation of their electronic distributions; now the number of simulations with explicit, or effective, three-body potentials, or with non-fixed internal geometry is increasing.

We have considered necessary to add this short introduction to the Section devoted to discrete models in order to compare motivations and justifications for both families of methods, and to stress their respective limits and problems. The starting point is the same in both families, the problems are different. We shall see how people working with discrete models have been able to overcome many of the problems they had to face.

9.1. LANGEVIN DIPOLE AND SURFACE CONSTRAINED MODELS.

The Langevin dipole model (LD) developed by Warshel (Warshel and Levitt, 1976) can be considered as an intermediate step between discrete and continuum models. Solvent (actually water) polarization is described by introducing a grid of polarizable point dipoles, responding to other electic fields according to Langevin's formula:

$$\mu_k = \mu_0 \{coth(x_k) - 1/x_k\} \vec{E}_k / |E_k| \tag{115}$$

where in $x_k = C\mu_0|E_k|/kT$ there is a factor, C, which reflects the resistance of the water molecules to orientation, essentially due to hydrogen bonds between themselves. The electric field $\vec{E}_k$ is determined with an iterative procedure.

This formulation gives a practical solution to the problem we signaled before, namely the quite large degeneracy of solvent molecule conformations. Being the solvent molecules held here fixed at the nodes of a 3D grid, and being their orientation limited to the mean orientation given by Langevin formula, the conformational problem is easier to solve (on this problem see also Warshel and Russel, 1984; Russel and Warshel, 1985; Edelstein *et al.*, 1991). The solvent reaction potential obtained in such a way is inserted in an effective Hamiltonian which deserves further comments. The whole system, solute plus solvent, is divided into three parts. One regards the solvent, which is described as said above. The solute, supposed to be large (the example shown in that paper is the lysozyme/substrate), is partitioned in two parts, one treated with classical potential of MM type, the other with a semiempirical QM formulation expressed in terms of hybrid orbitals (Warshel and Karplus, 1972). Couplings between QM and MM parts are included, as well as between solute and solvent via the reaction field.

Remark the year, 1976, of publication of this paper. It is one of the first proposals to use an effective QM Hamiltonian to treat solvation effects. The partitioning between solvent and solute is treated according to the views provided by the analysis of ΔE from supermolecule calculation. The procedure contains also an attempt of introducing combined MM/QM description of large solutes, a theme treated later by other authors, see e.g. the work done by van Duijnen (van Duijnen *et al.*, 1992; de Vries *et al.*, 1995; van Duijnen and de Vries, 1995), which will be described more extensively in another chapter of this book.

Of course this LD method was open to ameliorations. For example, the description of the reaction field does not pay much attention to the boundary conditions. The description given here resembles that of FD methods: now, we know that in FD methods a finer grid is necessary near the solute surface, and the same presumably holds for the LD methods too. Actually, this refinement has been introduced later.

Warshel has been quite active in improving his method, and it is convenient to summarize here 20 years of activity in this field, in order to have another overview of the evolution of methods, as we have had for the continuum ones, and to re-visit concepts and approximations the reader can find in other pages of this chapter, where they are introduced by following another line of exposition.

In a following methodological paper Warshel (1978) drastically simplifies the solute description (here addressed to describe ionic dissociation), paying more attention to the solvent, which is partitioned into three regions. Region I contains molecules described as compressible soft spheres, with a point dipole at their centre, while Region II is limited to a single layer of fixed solvent molecules, constrained to stay on the surface of a sphere which contains the solute and the molecules of Region I. Region III is a continuous dielectric encircling inner regions. This definition of the model, called SCSSD (surface constrained soft sphere dipoles) is an attempt to alleviate boundary constraints of the LD method. Here, there is no more a grid of Langevin dipoles, but soft spheres. This change requires the introduction of Boltzmann thermal averages to get enthalpy and free energy of solvation. The thermal averages, which could be evaluated with Monte Carlo simulations (a costly and rather unprecise procedure at that time), are simply replaced by the minimum energy values. The paper also considers adiabatic surfaces in the solvent coordinate space, a subject treated more in detail in subsequent papers.

For the description of chemical reactions which involve bond breaking/forming processes, Warshel introduced in 1980 (Warshel and Weiss, 1980) a valence-bond formulation of the solute undergoing reaction. The solute is described by the minimal number of VB structures, let us say for simplicity the two structures $\Psi_1 = X - Y$ and $\Psi_2 = X^-Y^+$, where X is more electronegative than Y (the approach can be extended to a larger number of VB structures, for example the three structures of the proton-transfer process, $AH + B \rightarrow A^- + BH^+$, of the same paper, but the number of structures is always kept at the minimum). One has thus to solve a 2x2 secular problem, once known the values of the H_{11}, H_{22} and H_{12} matrix elements. These are computed *in vacuo* using a very simple empirical procedure (thence the name of empirical valence-bond theory, EVB, used by Warshel). In solution the diagonal elements H_{11}^o and H_{22}^o are corrected taking into account solvent effects via a modified SCSSD procedure, while the H_{12}^o element is left unchanged with respect to the *in vacuo* value.

The changes in SCSSD regard the replacement of the permanent dipole moment of each soft solvent sphere by a couple of dipoles, permanent and induced. The permanent dipoles are kept fixed at the orientations minimizing the energy for the solute ground state, while induced dipoles are allowed to be different for each VB structure. A nested iterative procedure is adopted to fix both dipole distributions, and related contribution to the various H_{ii} elements.

This method deserves further comments. It is a method addressed to describe the main features of a chemical reaction in solution, not paying much attention to fine details, but aiming at simplicity and versatility. It introduces in the discrete solvent model several concepts of capital importance , used, until that time, only for reaction *in vacuo* or for reactions in solution described at the basic continuum level.

The introduction of VB structures (the empirical nature of EVB does not matter here) makes easier the description of the process in terms of diabatic potential surfaces. The latter have been used to describe reactions *in vacuo* since long time, more than 30 years before the formal definition given by Smith (1969), and continue to represent an important tool for an accurate study of many chemical processes *in vacuo* (see e.g. the reviews given by Sidis (1992) and by Pacher *et al.* (1993)), and for qualitative and semiquantitative characterizations of organic reactions (see e.g. Shaik and Hiberty, 1991). Their use in solution has been spurred by Marcus' model on electron transfer (ET) reactions (Marcus, 1956), and extended by the Russian school in the sixties to other chemical processes (Levich and Dogonadze, 1959; Levich, 1966; Dogonadze, 1971; German *et al.*, 1971, Albery, 1980).

Marcus' model and diabatic states make more immediate (even if not strictly necessary) the introduction of the dynamical solvent coordinate, of which in Section 5.2 we have given a definition based on parameters of the continuum model, but other definitions are possible. Actually, the Russian school used this concept without giving formal definitions, at the best of our knowledge (several papers have been published in relatively minor Russian journals, with limited circulation in western countries in those years), and basing it on a description of the solvent as a continuum or a set of oscillators. Warshel's definition, given in a subsequent paper (Warshel, 1982) is probably the first resorting to the discrete molecular solvent model. The coordinate in question is defined in terms of the orientation assumed by the polarizable dipoles in the SCSSD model.

A refined description of the solvent coordinate, Q, is given in another paper (Hwang *et al.*, 1988), in which the EVB approach is used and solute–solvent interactions are treated with a different model, called SCAAS (surface constrained all atom solvent, Warshel and King, 1985; see also King and Warshel, 1989). In this model the space is divided into three regions by two concentric spheres. The outer region is treated as a continuum dielectric, while water molecules of the two internal regions are treated explicitily, each one characterized by two coordinates, distance r from the centre, and orientation θ. The surface of the sphere which surrounds the two internal regions is constrained by additional forces that simulate the effect of the bulk solvent. The solvent coordinate Q is coupled to a spatial coordinate R to study energetics and dynamics of a S_N2 reaction, making use of simulations via trajectories to evaluate the time-dependence of both coordinates $< Q(t) >$ and $< R(t) >$.

LD, SCSSD, SCAAS models we have briefly presented according to their original formulations can be, and actually have been improved. We quote here the reduction of 'overpolarization' given by the original LD model (Warshel and Russel, 1984; Xu *et al.*, 1992), the introduction of statistical averages in the LD model (Lee *et al.*, 1993), and of a local field (LF) model to speed up simulations (Lee and Warshel, 1992); but several other improvements can be found in the literature. More details can be found in several review papers (Warshel and Åqvist, 1991; Warshel and Parson, 1991; Åqvist and Warshel, 1993). All the quoted models have been used, with MO or EVB descriptions of the solute (the QM description has to be selected according to the problem), for a variety of applications, among which we quote a sizeable number of reactions, proton transfer, hydride transfer, S_N2 reactions, and others, studied under their static and dynamic aspects (for the latter mostly with EVB descriptions). We quote here the most recent papers we know regarding chemical reactions: Kong and Warshel, 1995, Muller and Warshel, 1995. A large portion of these studies regards proteins and enzymes and these will be considered in another chapter of this book.

This relatively well detailed description of Warshel's activity (many points have been omitted for the sake of brevity) should permit to get an appreciation of some basic features of the discrete solvent models leading to effective Hamiltonians. A partitioning of the molecular assembly under the form of a Hartree product, $\Psi_A \cdot \Psi_S$, is almost compulsory. The description of solvent molecules (and of the interactions they have with the solute) is reduced to simple terms. The description of the energetic properties of the system requires the use of simulations (even if simpler descriptions can give the essential aspects of some phenomena, as the successful applications of the LD model in the original formulation shows). To reduce computational times for these simulations one has to introduce models and algorithms able to keep the number of solvent molecules within reasonably small values and here the use of continuum distributions plays an important role (the SCAAS model is an example of the ingenuity necessary to reach this goal). The description of boundary conditions represents a problem, different with respect to those found in continuum models, but still delicate in many aspects. The boundary conditions are even more delicate when one considers large solutes: we have shortly resumed the solution to this problem suggested by Warshel in 1976 and then refined in the following years; this theme will be considered again in the next Subsection.

We have also learned that the use of diabatic descriptions is very instrumental to describe the dynamics of reactions. This theme has been treated more in detail in another Section. The EVB description introduced by Warshel is very effective in describing the essential aspects of the phenomenon. Here, researcher's good physical sense and experience play an essential role in selecting the VB structures to be used, and the level of approximation in the description of Hamiltonian matrix elements. The chemical approach to the problems of reactions (we introduce again a difference between physical and chemical approach to a problem of molecular sciences) requires more details and more precision: the definition of VB methods for accurate description inserted in foolproof computational packages to be used by non-specialists, still constitutes a serious challenge.

9.2. COMBINED QUANTUM MECHANICS AND MOLECULAR MECHANICS METHODS.

As remarked in the previous subsection, discrete solvent models require the use of statistical procedures to properly describe solvent effects. In exposing the evolution of Warshel's methodologies in Subsection 9.1 several times we indicated the use of computer simulations, and the devices introduced in the model to reduce the number of discrete components, which are objects of the simulation, and hence the computational times.

Computer simulations, Monte Carlo or molecular dynamics, in fact appear to be the actual most effective way of introducing statistical averages (if one decides not to pass to continuous distributions), in spite of their computational cost. Some concepts, such as the quasi-structure model introduced by Yomosa (1978), have not evolved into algorithms of practical use. The numerous versions of methods based on virial expansion, on integral equation description of correlation functions, on the application of perturbation theory to simple reference systems (the basic aspects of these methods have been masterly described by Barker and Henderson (1976) in a well-known review, which readers are referred to) all consider the physical aspects of the problem, relegating chemistry in the background. The considerable development of these methods since Barker and Henderson's review little has added thus far to the chemical picture of solutions, and of chemical reactions in liquid media. Progresses in this direction can be expected, however. One example is just given by the recent proposal of inserting the reference interaction site model (RISM, Chandler and Andersen, 1972) in an effective Hamiltonian to study chemical reactions in solution (the RISM-SCF model, Ten-no *et al.*, 1993). RISM is an integral equation method, less expensive than simulations, already used to test some aspects of the continuum PCM model (e.g. Floris *et al.*, 1993) and to study some reactions (Chiles and Rossky, 1984; Jorgensen *et al.*, 1987; Kuharski and Chandler, 1987; Yu and Karplus, 1990). The RISM-SCF method will be considered in the following pages.

In recent years there has been a renewed interest in methods combining QM calculations with simulation procedures. The subject deserves a scrutiny more detailed than that we gave in introducing Warshel's LD method (Warshel and Levitt, 1976). In these methods Molecular Mechanics-type (MM) potential functions are used to describe both solute–solvent and solvent–solvent interactions, while interactions within the solute are described at the QM level.

The effective Hamiltonian assumes the form:

$$\hat{\mathcal{H}}^{eff} = \hat{\mathcal{H}}^{o}_{M} + \hat{\mathcal{H}}_{MS} + \hat{\mathcal{H}}_{SS} \tag{116}$$

where M stands for the solute, and S is a collective index denoting the solvent molecules included in the model. The number of solvent molecules explicitly considered in this Hamiltonian must be finite (although it can be large), and consequently periodic boundary conditions, or continuum descriptions, must be included in numerical calculations to represent the whole solution.

There are some important differences with respect to the effective Hamiltonian used in continuum methods, that we report here for the reader's convenience (see eq.(1)):

$$\hat{\mathcal{H}}^{eff} = \hat{\mathcal{H}}^{o}_{M} + \hat{\mathcal{H}}_{MS} + \hat{\mathcal{V}}_{int} \tag{117}$$

Both $\hat{\mathcal{V}}_{int}$ and $\hat{\mathcal{H}}_{MS}$ describe solute–solvent interactions. The first one uses averaged solvent distributions and the corresponding response functions (in almost all applications the distribution is considered uniform, as already indicated). On the contrary, $\hat{\mathcal{H}}_{MS}$ uses, istantaneous solvent distributions.

The necessary averaging is provided by classical computer simulations, either Monte Carlo or molecular dynamics, where the set of interaction potentials also includes M–S terms. The latter are modified on the basis of the solute electronic distribution $\Psi\Psi^*$ according to the rules assumed in the definition of the MM potentials.

At a given solvent configuration sampling, energy is given by the expectation value of the effective Hamiltonian:

$$E_{tot} = \left\langle \Psi \left| \hat{\mathcal{H}}^{eff} \right| \Psi \right\rangle = \left\langle \Psi \left| \hat{\mathcal{H}}^{o}_{M} + \hat{\mathcal{H}}_{MS} \right| \Psi \right\rangle + E_{SS} \tag{118}$$

The energy of the MM part (the "solvent") has been taken outside the expectation value integral, as it does not depend on the electron coordinates of the QM part. The difference between E_{tot} of eq.(117) and the corresponding E of eq.(57) of continuum models is worth noting. E_{tot} is an energy, while E is related to a free energy; E_{tot} depends on a specific configuration of the solvent molecules, while E is related to an averaged distribution.

The explicit expression of $\hat{\mathcal{H}}_{MS}$ depends upon the choice made for MM interaction potential. The expression generally used is reported in the following equation:

$$\hat{\mathcal{H}}_{MS} = -\sum_{s}\sum_{i} \frac{q_s}{r_{si}} + \sum_{s}\sum_{\alpha} \frac{q_s Z_\alpha}{R_{s\alpha}} + \sum_{s}\sum_{\alpha} 4\epsilon_{s\alpha}\left[\left(\frac{\sigma_{s\alpha}}{R_{s\alpha}}\right)^{12} - \left(\frac{\sigma_{s\alpha}}{R_{s\alpha}}\right)^{6}\right] \tag{119}$$

The index s refers to the atoms of the solvent molecules S, whereas i and α to the electrons and nuclei of the QM solute. In this expression the interaction is reduced to coulombic terms involving net charges on the solvent atoms (q_s), and to dispersion–repulsion terms (van der Waals contributions) expressed in terms of two sets of parameters, $\{\epsilon_{s\alpha}\}$ and $\{\sigma_{s\alpha}\}$ coefficients, which do not change during the calculation.

The polarization of the solvent molecules is here missing. This contribution may be recovered by adding a further term to $\hat{\mathcal{H}}_{MS}$:

$$\hat{\mathcal{H}}^{pol}_{MS} = -\frac{1}{2}\sum_{s}\sum_{i} \frac{\vec{\mu}^{ind}_{s}}{r^3_{si}} \cdot \vec{r}_{si} + \frac{1}{2}\sum_{s}\sum_{\alpha} Z_\alpha \frac{\vec{\mu}^{ind}_{s}}{R^3_{s\alpha}} \cdot \vec{R}_{s\alpha} \tag{120}$$

In this simplified expression the electronic solvent molecule polarization is given in terms of the induced dipoles $\vec{\mu}^{ind}$, which may be calculated by

$$\vec{\mu}^{ind}_{s} = \alpha_s \vec{E}^{T}_{s} \tag{121}$$

The atomic polarizability α_s may be taken as a fixed parameter, but the total electric field on atom s, $\vec{E}^{T}_{s}$, must be computed at every step. This is a quite computer demanding calculation, hence the solvent polarization contribution is often discarded.

We may obtain a rough estimate of the effect of this contribution by comparing this method with the continuum one. After a proper average (not yet introduced at this stage of the calculation) the QM/MM method without $\hat{\mathcal{H}}_{MS}^{pol}$ gives a description of the solute–solvent interaction corresponding to the sum of G_{dis} and G_{rep} plus a portion of G_{el}. The portion of G_{el} included here corresponds to the orientational contribution to the polarization vector $\vec{P}$ (or, equivalently, to the orientational contribution to the electric susceptibility of the medium). The missing part corresponds to the electronic component of $\vec{P}$. The partition of $\vec{P}$ into $\vec{P}_{or}+\vec{P}_{el}$ corresponds to the Pekar partition $\vec{P} = \vec{P}_{slow} + \vec{P}_{fast}$ we have introduced in Section 4.2 (see eq.(18)). The contribution to $\vec{P}$ due to $\vec{P}_{fast}$ is far from being negligible, as the effects on solute energy and electron charge distribution are concerned. However, it must be remarked that the combined QM/MM approach is often applied to low level of the quantum theory (semiempirical wavefunctions), hence the errors due to neglecting solvent polarization are smaller than the ones due to the QM theory level.

The above outlined computational methodology is coupled to a computer simulation (molecular dynamics or Monte Carlo) to get energetic quantities with a thermodynamically defined status, in particular free energies. We shall not give more details on the method, referring the reader to few selected papers for a better appreciation of the combined QM/MM solvation methods (Field *et al.*, 1990; Waszkowycz *et al.*, 1991; Bash *et al.*, 1991, 1993; Luzhkov and Warshel, 1992; Gao, 1992, 1993, 1995; Lin *et al.*, 1995; Stanton *et al.*, 1995; Muller and Warshel, 1995) and for the large literature regarding the calculation of free energies from simulations (Kollmann, 1993; van Gunsteren *et al.*, 1993).

A variant of the combined QM/MM approach introduces a hybrid description of the solute. The main motivation for the introduction of this additional approximation lies in computational costs. Combined QM/MM calculations are quite costly, even when all the possible simplifications are introduced in the QM part and in the MM interaction potentials. On the other hand, QM formulation is more reliable than an empirical potential function to describe chemical reactions which involve bond–formation and disruption processes. To temperate contrasting factors, i.e. the need for a QM description and the computational costs, one may resort to the well established fact that, in chemical reactions, the quantum bond–breaking and bond–forming processes are limited to a restricted portion of the molecular system, with the remainder playing an auxiliary role. Hence, it may be convenient to resort to hybrid descriptions, where the active part of the molecule is described at the QM level and the remainder *via* MM potentials.

The problem of "cutting" some covalent bonds to defined subsystems to be handled separately is quite old, and, for its relevance in the whole field of quantum chemistry, it has been the object of a deluge of papers. A selected bibliography can be found in a review which also considers solvation problems (Tomasi *et al.*, 1991).

The specific topic of interest here, i.e. the search for methodologies which are able to describe with the necessary accuracy both conformational and reaction energy surfaces in hybrid QM/MM models for relatively large molecules, has given rise to a considerable number of papers. Some have been already quoted; we report here a further selection: Singh and Kollman (1986), Eckert *et al.* (1990), Kawamura–Kuribayashi (1992), Ferenczy *et al.* (1992), Thery *et al.* (1994), Maseras and Morokuma (1995), to which all papers referred to the above quoted combined QM/MM methods must be added.

All the methods we have thus far considered use Monte Carlo or molecular dynamics simulations. An alternative approach to get thermodynamic quantities in an effective Hamiltonian+discrete model is based on the use of the integral equation RISM method. The RISM-SCF method proposed by Ten-no and coworkers (Ten-no *et al.*, 1993, 1994; Kawata *et al.*, 1995) combines a QM description of the solute with a RISM description of the whole system in a way which deserves attention.

In the RISM methods (there are several variants) attention is paid to site–site interactions from which site–site correlation functions are derived. There are no formal limits on the number and quality of such sites. In the RISM-SCF model a difference is introduced between sites belonging to the solute and to the solvent. All the site–site potentials are expressed as a sum of a short–range term and of a Coulombic long–range term:

$$u_{\alpha\gamma}(r) = u^*_{\alpha\gamma}(r) + q_\alpha q_\gamma / r \tag{122}$$

Let us call λ the sites which belong to the solute. The q_γ charges are obtained from the MOs given by the solution of a Hartree–Fock equation including an electrostatic potential $\mathcal{V}_s$ having as sources the charges of all the solvent sites (it is expressed in terms of the correlation functions $\mathbf{h}_{\lambda\alpha}$ where α runs over all solvent sites). Solvent charges are fixed, solute charges are obtained with an iterative procedure. One starts from a first guess of q_λ values, obtains with RISM procedures the solvent–solvent and then solute–solvent correlation functions. With the last functions $\mathcal{V}_s$ is fixed, the H–F equations solved and new q_λ charges are obtained under the constraint that q_λ charges have to reproduce the electrostatic molecular potential (a 3D grid is used here). With these new q_λ charges one computes again the solute–solvent site–site correlation function. When convergence is achieved one has information on the solute's properties from the QM wavefunction, and on the solvation free energy from the correlation functions, using for this last point one of the several available approximate closure relations.

The method is innovative and opens new perspectives. By comparing it to the other simulation–based QM/MM methods, one immediately recognizes as advantage the much lower computational cost, which makes less stringent the necessity of keeping low the number of solvent molecules, and as disadvantage the need of using closure relations (a tricky subject, which has given some bad surprises in the past). Under other aspects, the two approaches are similar. Both use interaction potentials of empirical origin (the parameters used in $u^*_{\alpha\gamma}$ of eq.(122) are the same as those used in some simulations), and solvent polarization effects are missing.

The proposed version of the RISM-SCF method presents several computational points, namely the number of interaction sites both in the solute and in the solvent, the definition of $\mathcal{V}_s$, the selection of the closure relation, etc., requiring further tests, which, we hope, will be done in the next future.

9.3. METHODS NOT USING EFFECTIVE HAMILTONIANS.

A possible further step in the simplification of the discrete model could be given by a full MM description of the solute. This means to abandon the effective Hamiltonian approach. At present, it is well established that MM methods, when properly parametrized, can give reliable descriptions of the conformational energy hypersurface *in vacuo*, $E^o(\mathbf{R})$. The same applies for the solvation energy on the conformational surface, $\Delta G(\mathbf{R})$: see, e.g. Sharp (1991), Zauhar (1991), Montagnani *et al.* (1992), and Varnek *et al.* (1995). MM energies are less reliable when the portion of the energy hypersurface under examination corresponds to reaction transition states (however, there are progresses in this field, see Eksterowicz and Hank, 1993). In MM, the description of solute polarization effects is given by indirect means, through the introduction of an effective screening constant in the non–bonded coulombic contribution (see, e.g. Weiner *et al.* (1986), Lavery *et al.* (1986), Fritsch and Westhof (1991)). The method is approximate and not tested for geometries corresponding to bond–breaking and forming situations. The suggestion to potential users is to wait until the approach will be better parametrized and tested.

Another approach using a discrete description of the solvent, and not based on effective Hamiltonians is giving quite interesting results for chemical reactions in solution (Jorgensen, 1989b). It is based on the *ab initio* determination of the reaction coordinate (RC) *in vacuo*. A number of points on this coordinate is selected, and the corresponding wavefunctions are determined (often of good quality). From these wavefunctions the atomic charges q_i of the solute are determined and then inserted in the coulombic term of the MM solute–solvent interaction energy: $\sum_i \sum_s \frac{q_i q_s}{r_{is}}$ (see, for a comparison, eq.(119) which reports the quantum mechanical operator for this classical term). A computer simulation (generally Monte Carlo), followed by the determination of the free energy change, is given for each point, thus obtaining $\Delta G_{sol}(\mathbf{R})$ (electrostatic as well as dispersion–repulsion terms are included) which can supplement the already calculated $E^o(\mathbf{R})$ values along the reaction coordinate (RC). Several recent Jorgensen's papers give some nice examples of the application of this procedure to 'true' chemical reactions (see Jorgensen *et al.*, 1994, for an overwiew). The information gained in such an approach allows to obtain quite interesting insights on the solvent effects on the reaction mechanism.

The procedure has some limitations which we shall briefly examine. The description of the reaction coordinate is performed on the energy hypersurface of the isolated molecule. Hence, the solvent effects on the solute geometry are completely discarded. Moreover, the Monte Carlo simulations use classical expressions of the solute–solvent interaction potential. This means that the solvent effects on the solute electronic charge distribution are neglected, and that it is not possible to insert any solvent molecules in the QM description of bond–breaking and forming processes, unless by starting the calculations from the very beginning. The first limitation has a counterpart in those continuum solvation calculations which exploit geometries determined *in vacuo*. In the latter case such an approximation is adopted as to reduce computational costs, being the determination of TS and RC more costly in solution than *in vacuo*. This is also the reason why SCF geometries are so widely used to study reaction mechanisms *in vacuo* with correlated wavefunctions. In a number of cases this approximation works quite well, in others it does not. At present, there is more experience in electron correlation effects than in solvent effects. Now there is a tendency to use geometries determined with some correlation contributions, and to refine the description at the electronic level. However, solvent effects on the geometry could be larger than electron correlation effects.

The lack of solvent polarization effects on the solute electronic charge distribution should be more important at the points of the energy surface where there are loose electrons, generally around the TS point. Surprisingly, the available information is scarce. It would be easy to collect information on this point by comparing values of ΔG_{el} at the TS of selected reactions obtained from the two different approximations 1 and 3 of Section eq.(6.6), i.e. with the fully relaxed density function ρ_M, and with the rigid function ρ_M^o computed *in vacuo*.

In Section 4.2 we stressed the importance of including active solvent molecules in the QM description of the solute. Of course, it is possible to include these molecules in the *ab initio* calculations preceding Monte Carlo simulations. However, there are now some examples (Coitiño *et al.*, 1995c) indicating that in some cases the examination of small clusters *in vacuo* must be somewhat extended to larger clusters in order to identify active solvent molecules.

We have devoted considerable space to discuss the limitations of the Monte Carlo approach based on an *ab initio* reaction coordinate, as this approach seems to us to be quite important as a source of benchmarks for other approaches. The limitations we have highlighted can be easily removed, for example, by using *ab initio* calculations with the continuum approach as a starting point.

9.4. AN ATTEMPT OF CONCLUSION.

To fulfill our promise to give beginners a guide in the jungle of solvation methods, we are compelled to resume here, in a short space, our feelings and suggestions.

An important factor to be considered is the computational cost. Continuum methods are noticeably less expensive than simulation methods based on discrete models. On the other hand, simple properties, as the solvation energy ΔG_{sol} of small and medium–size solutes are computed equally well with both continuum and discrete methods, reaching chemical accuracy (Orozco *et al.*, 1992; Tomasi, 1994; Cramer and Truhlar, 1995a). There is large numerical evidence ensuring that the same conclusion holds for other continuum and discrete methods as well. The evaluation of ΔG_{sol} at TS has given almost identical results in several cases, but here numerical evidence is not sufficient to draw definitive conclusions.

Accuracy and predictive power are even more important factors. We shall discard here the cases where a rough guess of solvent effects is sufficient. To have a good description of $G(\mathbf{R})$ over the set of geometries $\{\mathbf{R}\}$ necessary to study a reaction we need to examine both components of $G(\mathbf{R}) = E^o(\mathbf{R}) + \Delta G_{sol}(\mathbf{R})$. A good precision in both components is necessary. Good representations of $E^o(\mathbf{R})$ may be wasted by poor descriptions of $\Delta G_{sol}(\mathbf{R})$, and conversely a good evaluation of $\Delta G_{sol}(\mathbf{R})$ is of little use if it is not accompanied by an adequate description of $E^o(\mathbf{R})$.

The characterization of the geometric aspects of a reaction mechanism (definition of the RC, geometry of TS, intermediary complexes, etc.) performed on the basis of *in vacuo* calculations alone is a risky strategy. It is by far safer to rely on methods giving $G(\mathbf{R})$ directly. This is why we consider it advisable to use accurate continuum methods or combined QM/MM simulations, where the QM part is treated at a sufficient degree of accuracy, as the hybrid QM/MM description of the solute can give.

Attention must be paid to the occurrence of 'active' solvent molecules. They must be included in the QM part. Classical MM interaction potentials are notable to describe these effects. The identification of the true number of active solvent molecules must be given by trial: the program one uses must be able to do so.

The combination of all the points reported above seems to indicate versatile and efficient *ab initio* procedures as the best choice. However, there are other considerations to be added. Both continuum and discrete approaches suffer from limitations due to the separation of the whole liquid system into two parts, i.e. the primary part, or solute, and the secondary larger part, the solvent. These limitations cannot be eliminated until more holistic methods will be fully developed. We have already discussed some problems related to the shape of the cavity, which is the key point of this separation in continuum methods. We would like to remark that discrete methods suffer from similar problems of definition: a tiny change in the non–boded interaction parameters in the solute–solvent interaction potential corresponds to a not so small change in the cavity shape.

In doubtful cases, a comparison of the descriptions obtained with the two approaches could be decisive. This is why we consider computer simulations based on accurate potentials drawn from *ab initio* calculations as an important instrument to be mastered and used.

Our final suggestions may be thus expressed: for an accurate study, use first continuum models to gain a first insight on the phenomenon. These methods are less expensive and more foolproof. Once one has made a judicious choice of the relevant parameters, namely basis set and procedure for the QM portion, method used to compute $\hat{\mathcal{V}}_{int}$, and cavity, the final outcome of the calculations is unequivocally defined. On the contrary, computer simulations are more costly and delicate. A judicious choice of the MM parameters is less immediate, and the quality of the results depends on the completeness one has reached in spanning the configuration space.

When the first approximate description is gained, one can pass to further checks and refinements. The strategy suggested here is flexible enough to allow the use of all (or almost all) the methods we have examined. The first order description may be of low accuracy but cheap, or more accurate if we hope to find a satisfactory description at this first stage, and hence requiring only minor refinements. The final choice is thus left to the potential users; it will depend on personal tastes, on the size and characteristics of the system under examination, on previous experience.

Unfortunately, previous experience is for the moment quite limited, even if it is growing fast. The following section is dedicated to the presentation of some elements of this available experience, and it will end with further suggestions about the strategy to be used in the study of chemical reactions.

10. Computational results: a test of methods and a source of suggestion about their use.

A critical examination of the results of a set of calculations on solvent effects allows to learn something more about the underlying models and the computational procedures. With this limited perspective (we shall not attempt to draw any conclusion about specific chemical systems) we shall try to derive some methodological hints for future researches from the considerable body of existing calculations.

10.1. TAUTOMERIC EQUILIBRIA.

It is convenient to start with some simple chemical phenomena as reaction equilibria. The attention will be focussed on the balance between the quality of the solvation and internal energy differences. The studies on tautomeric equilibria are quite appropriate for this purpose. The problem is simple, being reduced to the comparison of the energies of some molecular structures, all of them at an equilibrium geometry. The numerical evidence is large, covering all the methods we have examined and with applications of different procedures to the same chemical system. In some cases experimental data for a comparison are available. After saying this, we would like to stress that a critical examination is an exacting task, and that its exposition requires considerable space. Fortunately, we can save this space by resorting to the careful analysis given by Cramer and Truhlar in their last review (Cramer and Truhlar, 1995b) in order to better document the conclusions we are drawing here. Among the various tautomeric systems we may limit our attention to heterocycles. Tautomeric equilibria in heterocycles are quite sensitive to the medium (Reichart, 1990), and presumably calculations of equilibria will be sensitive to fine details of the procedure. Cramer and Truhlar examine results regarding several classes of heterocycles (more than seventy different tautomers in a number of different solvents) which are reported in about forty papers, all using continuum models. To these papers we could add an almost equal number of computer simulation results.

Among these, the papers which compare the performances of different methods are particulary interesting. We quote here those from Hillier's group (Gould *et al.*, 1992; Parchment *et al.*, 1992; Woodcock *et al.*, 1992; Burton *et al.*, 1993) and from Luque and Orozco's group (Orozco and Luque, 1995), as they better allow a balanced appraisal over a number of methods for a given system.

To draw some general conclusions from a single paper or from a single solute is quite risky. As a matter of fact, in the set of papers we are examining there are various contrasting conclusions, but in the long run, i.e. examining all the body of computational evidence, sounder general conclusions seem to us fairly well established. A good computation of the electronic wavefunction, given with a good basis set and including correlation, needs to be coupled with a good description of the cavity for the continuum methods, or with well done simulations, in which a relatively large number of solvent molecules are considered, and accurate potentials are used. Several good results found in calculations not completely fulfilling these criteria seem due to fortuitous cancellations of errors, while others seem to indicate that some classes of tautomeric equilibria are less sensitive to the description of the electron distribution of the solvent stabilization.

In a remarkable number of cases a complete agreement among the various computational methods and the experiment has not been reached. This may be due to several reasons among which we would like to point out changes (or equilibria) of conformations in the O–H and N–H groups, which have not been taken into account in the reviewed papers, as well as the importance of using more refined $\lambda(A)$ factors in the definition of the cavity (see Section 8.5), as some unpublished checks made by Luque seem to indicate. We would like to stress that, generally speaking, there is a good agreement between continuum methods and free energy calculations *via* computer simulations. There is not any significant evidence which suggests the introduction of "active" solvent molecules in the compounds so far examined.

We shall not extend our analysis to other chemical equilibria, for which there is a considerable body of results, although not so abundant as for heterocyclic tautomery. The conclusions would be similar, even for those cases which involve charged species and consequently larger effects on the solvent organization.

10.2. THE ACTIVE ROLE OF SOLVENT MOLECULES IN REACTION MECHANISMS.

As a second point in our examination of numerical results, we shall consider the active role of solvent molecules (in particular water) in reaction mechanisms. This problem is more complex than tautomeric equilibria considered in the previous subsection, and its analysis would require longer discussions. For this reason we shall confine ourselves to show examples of two basic patterns of active intervention of additional water molecules. The reader is warned that the reactive role of the solvent molecules is not limited to these two basic mechanisms. Both mechanisms have been considered in a recent report by Rivail *et al.* (1994) which we take as a starting point for our analysis. In this report Rivail *et al.* compare two different reactions, the hydrolysis of formamide and the ionic dissociation of HCl in water. We shall examine the two cases separately.

The study of formamide hydrolysis performed at Nancy (Rivail *et al.*, 1994; Antonczack *et al.*, 1994) can be here accompanied by studies on other reactions (hydration of oxoaldehydes and aldol condensation) performed at Pisa (Coitiño *et al.*, 1994, 1995b, 1995c). In all cases the calculations have been performed at a good level, MP2 over an extended basis set, taking into account solvation effects in all their contributions and with cavities of appropriate shape. To simplify the discussion we shall confine ourselves to neutral reactions; the quoted papers also consider proton assisted reactions. The nominal reactions are reported below:

$$\begin{aligned} \mathrm{H_2N-CHO + H_2O} &\longrightarrow \mathrm{NH_3 + HCOOH} \\ \mathrm{OCH-CHO + H_2O} &\longrightarrow \mathrm{OCH-CH(OH)_2} \\ \mathrm{CH_3CHO + CH_2{=}CHOH} &\longrightarrow \mathrm{CH_3CHOH-CH_2-CHO} \end{aligned}$$

In all cases the introduction of a water molecule in the QM solute improves the agreement with the experimental data. The mechanisms collected here are remarkably different but they all share the feature of inducing the breaking of a O–H bond in the nominal reaction. It is well known, since a long time (see e.g. Alagona *et al.*, 1978), that the interaction with a water molecule makes the H–X bond break easier. This is a quantum electronic effect which can be rationalized by using common chemical concepts, as electronegativity differences, and measured with standard tools for the analysis of electronic effects, such as the position of the localized orbital charge centroids, the MEP value at selected points, the population analysis, etc. In the formamide hydrolysis and glyoxal hydration the additional water molecule reduces steric constraints on the TS geometry, but the examination of the aldol condensation shows that this is not the unique factor which reduces the energy barrier. It is often said that a six–membered ring, held together by hydrogen bonds, exhibits a particular stability. This may be the case for the formamide hydrolysis and for the glyoxal hydration, for which the lowering of the barrier due to the inclusion of a water molecule in the model, with consequent formation of a six–membered ring TS, has been computed to be 15.3 and 29.8 kcal/mol, respectively (these values are measured from the separate reagents energy *in vacuo*). In the case of aldol condensation the addition of the water molecule leads to an eight–atom ring, with an energy gain of 15.9 kcal/mol with respect to the six–atom ring found in the model without the additional H_2O molecule. It is evident that the electronic factors which stabilize larger cycles are still active at this level of complexity of the cyclic TS structure.

Bulk solvent alone is not able to describe these effects which are to be ascribed to the quantum behaviour of a single water molecule. Bulk solvent, in its continuum representation, indeed modifies the picture of the reaction mechanism we have resumed here. Its effect on the reaction energy profile leads to an increase of the energy barrier. According to the calculations, this increase amounts to 22.5, 9.9, and 5.8 kcal/mol for the three reactions, respectively (these values are obtained as differences of the energy barriers with respect to separate reagents, computed for the model with an additional H_2O molecule, both *in vacuo* and in the continuum medium). The energy difference between TS and the separate reagents is not the best estimate of the reaction barrier, since in all cases under examination there is the formation of a preliminary complex, sufficiently stable to allow thermal re-equilibration before the barrier crossing. The initial complexes are less stable in the bulk solvent than *in vacuo*, and the shift of the reference energy leads to a decrease of the barrier. The numerical data can be found in the reference papers.

In addition, we remark that evaluation of reaction barriers cannot be limited to E or G values, but requires an appreciation of the full free energy, which includes entropic contributions due to internal molecular motions. Numerical estimates may be found, once again, in some of the already quoted papers.

The studies we have briefly resumed allow to assess a couple of important points, which we shall resume here. A good computational level in the description of both the solvent electronic structure and solvation effects is not sufficient to ensure a good description of the reaction. Attention must be paid to the occurrence of specific quantum–type solvation effects which are hard to be described with continuum models as well as with classical MM interactions. The energetic characterization of the reaction must be performed with the due care; for example, in the reactions we have considered, differential entropic contributions are not negligible. We shall add another point. The inclusion of an additional water molecule in the "solute" has been suggested here by general considerations. The validity of this enlarged model has been first tested *in vacuo* and then confirmed by calculations with the whole solvent. Extension of the same approach to another additional water molecule shows that there are no other substantial improvements in the description. One active water molecule is sufficient here, but this is not the general case. An instructive example may be found in the same class of reactions.

Experimental studies on the glyoxal hydration in solutions at low water content suggest an apparent third–order dependence of the reaction rate on water concentration (Eigen, 1967). The search for a third active water molecule, performed with the strategy we have outlined, has failed; *in vacuo* the third water molecule better solvates the hydronium ion and has no active role in the reaction. This apparently confirms other views on the reaction mechanism (Sörensen and Jencks, 1987). Pursuing further the search, it has been found that a fourth water molecule plays an active role, *via* a mechanism which involves a cyclic structure with a H_2O–H_2O–H_3O^+ chain (Coitiño *et al.*, 1995c). More recent calculations, still unpublished, indicate that this structure may be reached without the need of introducing four water molecules in the QM model, but by directly using the PCM procedure without exploiting any previous *in vacuo* description. This example shows that the application of solvation methods to more complex cases leads to an evolution of computational strategies.

We shall now proceed to the second example of active participation of solvent molecules in reaction mechanisms which has been considered by Rivail *et al.* (1994), namely the dissociation of HCl. In this study (Rivail *et al.*, 1994; Chipot *et al.*, 1994) the dissociation process has been examined with the aid of two solute models, i.e. $HCl{\cdot}H_2O$ and $HCl{\cdot}(H_2O)_2$, both considered *in vacuo* as well as in continuum solvents with different dielectric constants. Once again, electronic wavefunctions of good quality have been used (SCF and MP2/6–31+G** calculations have been performed). Bulk solvation effects have been described with the SCRF formalism, using an ellipsoid cavity.

In vacuo and in low–polarity solvents the monohydrated complex $HCl{\cdot}H_2O$ exhibits a very flat potential energy surface along the hydrogen bond coordinate, with a single minimum corresponding to a neutral hydrogen–bonded complex. In bulk water a second minimum is observed at the SCF level, which disappears when MP2 electron correlation is introduced. On the contrary, calculations on the dihydrated complex $HCl{\cdot}(H_2O)_2$ show that this second minimum is present, both at the SCF and the MP2 level, and even with continuum media of moderate polarity. The interpretation given by the authors emphasizes the contemporary presence of two important effects, namely the polarizing influence of the medium and the role of a water molecule which acts as a catalyst but it is not involved in a cyclic TS structure. This catalytic role is enhanced by the full QM description of the $(H_2O)_2$ dimer which allows geometry relaxation and electron charge rearrangement, effects that magnify the acid–base properties of the water molecules.

We have already commented on the convenience of using a full QM description of the interaction between HX and an electron–donor/proton–acceptor A to describe the catalytic effect of A on the heterolytic cleavage of the H–X bond. Rivail's study stresses the importance of using a similar level of description to represent the increased ability of electron–acceptor/proton–donor of the water molecule when it is bound to form a dimer $H_2O{\cdot}H_2O$. In this meaning it can be said that both water molecules play an active role, and that both must be included in the QM solute.

It is interesting to compare this study with a parallel research on the ionization of HCl in water performed by Ando and Hynes (1994). These authors use a completely different approach. The model they use emphasizes the role played by the crossing of free energy surfaces corresponding to different VB diabatic states (in this case the states corresponding to the structures $ClH\cdots OH_2\cdots OH_2$ (1) and $Cl^-\cdots HOH_2^+\cdots OH_2$ (2)).

In the definition of the diabatic free energy surfaces use is made of the solvent dynamic coordinates we have discussed about in Sections 4.2, 5.2, and 9.1. The nature of the chemical system under investigation allows the introduction of some simplifications. In particular, the solvent coordinates can be reduced to just one, which measures the collective fluctuation of the polar environment, and the proton transfer can be studied at a fixed geometry of the other solute's atoms. The dimensionality of the problem is thus reduced to one, i.e. the solvent coordinate, and the diabatic surfaces intersection is reduced to the crossing of two free energy curves, referred to structure 1 and 2, both having concave shape. The minima of the two curves correspond to the value of the solvent coordinate for which the equilibrium with structure 1 and 2, respectively is reached. Proton transfer occurs at the crossing point of the curves, i.e. at the solvent configuration for which the free energy of the two structures is the same. At this specific value of the solvent coordinate the one–dimensional proton transfer potential along the Cl–H distance is almost symmetric, and exhibits two minima. The same curve becomes more and more asymmetric when the solvent coordinate assumes values nearer to the ones of the minima of the two diabatic free energy curves. The energetic parameters for the passage from structure 1 to structure 2 are given by: a) the height of the crossing point of the two curves with respect to the minimum of curve 1; b) the barrier for the proton transfer. The computational steps used by Ando and Hynes are the following:
1) determination of the geometry of structures 1 and 2 *via ab initio* SCF calculations in a water cluster which corresponds to the first solvation shell;
2) Monte Carlo calculations for structure 1 and 2, in order to obtain the two free energy curves. MM interaction potentials of the type shown in eq.(119) are used here. The solute parameters (atomic charges and van der Waals coefficients) are drawn from SCF calculation on the trimer 1 and 2, as well as on other small solvation clusters;
3) determination of the proton transfer potential *via* SCF calculations on the trimer $ClH\cdots OH_2\cdots OH_2$, in which the position of H is varied, under the influence of a fixed solvent configuration which satisfy the condition on the solvent coordinate indicated above. The calculation is then repeated with other solvent configurations (all corresponding to the same solvent coordinate) to measure the variance.

The study we are resuming here continues, considering other aspects of the ionic dissociation, i.e. dynamics of the proton transfer, tunneling effects, and also further steps in the so–called "Grotthuss mechanism" which assumes that the solvent water molecules themselves can be involved in the process as proton acceptors. Regarding the latter point the question is whether the whole proton transfer process, of which we have considered here only the "first" transfer (the second one is a parallel passage of H from structure 2 to $Cl^- \cdots HOH \cdots HOH_2^+$) is "step–wise" or "concerted". The conclusion the authors obtain is that the concerted pathway is unfavorable compared to the stepwise mechanism, at least for the case of HCl ionization in water.

By comparing Hynes's and Rivail's models some common points and some differences are evident. Both studies use an $HCl{\cdot}H_2O{\cdot}H_2O$ unit as "solute" which is treated as a supermolecule in a QM description. Both of them introduce a one–dimensional adiabatic representation of the proton transfer potential but, while Rivail assumes solvent equilibration along the whole proton transfer path, Hynes uses a fixed, but appropriate, solvent configuration. The definition of appropriate configuration is reached using some concepts that are not present in Rivail's treatment, i.e. the crossing of two VB structures ("reactant" and "product") in the space of solvent dynamical coordinates.

The information on the reaction that can be drawn from the two studies we are comparing partially concerns the same points, on which there is a reasonable agreement and partially contains features which are not considered in both approaches. Hence there is a considerable complementarity between the two approaches. Other comments are left to the reader.

10.3. FINAL COMMENTS.

The comparison of results obtained with different methods could be extended to other classes of reactions. There is a considerable wealth of results for several classes of reactions with simple mechanisms, like S_N2, S_N1, ET reactions, which are the favourite examples selected by theoreticians to test new models. There is a rapid increase in the number of reactions for which a comparison among different methods is possible, and there is also an increase in the complexity of the studied mechanisms. We quote, as examples, Menshutkin and Friedel–Craft's reactions, Claisen's rearrangement, Diels–Alder's and other pericyclic reactions.

Several authors have given an account of their own work, supplementing it with some comparisons. We shall give here a very partial list: Warshel and Parson, (1991), Hynes *et al.* (1994), Bertrán *et al.* (1994), Severance and Jorgensen (1994), Jorgensen *et al.* (1994), Gao and Xia (1994), Gao (1995). There is not, at the best of our knowledge, a general review on this subject, and reasons of space forbid us to do it here. We shall confine ourselves to some final remarks, drawn from the few comparisons reported above as well as from the other non reviewed reactions.

In studying chemical reactions attention must be paid to get a balanced description of both solvation effects and intrinsic solute's properties. The safer way is to use high levels of description for both aspects; several solvation procedures among those we have examined seem to be accurate enough to give a chemically acceptable description, the same holds for standard good level quantum calculations (SCF or MCSCF plus correlation contributions, with extended basis sets). Methods using a lower accuracy for both aspects may be used when the goal is to model some specific aspects of the reaction; for example simplified models, as those based on a simplified VB description of the reacting system, supplemented by a guess of solvation effect, may play an important role to get a rationale of more complex reactions.

Accurate solvation procedures can be found in the family of continuum methods as well as in that of discrete methods. Now, the number of cases in which the application of methods belonging to the two families has given very similar results (with a good agreement with experimental data) is large. Continuum methods must take into account all the components of G, and they must use a realistic description of the cavity. *Ad hoc* parametrizations of cavities of simpler shapes such as to reproduce, for example, the desired value of an energy difference, often lead to considerable deformations of the reaction potential, and thus of the solute properties, which the interpretation of the phenomenon depends on. On the other side, discrete methods depend on the quality of the intermolecular potential as well as of the simulation procedure; both are critical parameters. The simulation should also include the solvent electronic polarization, or some estimates of its effect.

This said, it is difficult to suggest a precise strategy to be followed in the computational study of such complex events as reaction mechanisms often are. Generally speaking, a sequence of steps at increasing levels of sophistication seems to us the best choice. We would like to remark that the available experience, although large, is limited to relatively simple reaction mechanisms; nevertheless several "surprises" have been found, i.e. changes in the mechanism passing from gas–phase to solution, and from a solvent to another. The possibility of finding other "surprises" suggests caution in relying on a single strategy, approximation or method. Therefore, we cannot suggest, for example, to rely only on the strategy of determining the reaction coordinate *in vacuo*, and then to apply solvation, or to rely on the use of a low level of approximation to determine the basic features of the mechanism, and then complement it with a few *ab initio* calculations using a good basis set.

All methods or approximations we have examined may play a role in the sequence of steps we are envisaging here. Of course, computational costs are an important parameter. In general, continuum methods are by far less costly than others; we shall quote here COSMO, AMSOL, semiempirical SCRF with elliptical cavities, semiempirical PCM methods (we have quoted only methods which rely on a QM description of the solute). These approximate methods are less reliable than others in the study of reaction mechanisms, and the potential user has to decide if it is convenient to spend computational time to get a detailed description which perhaps requires some important corrections.

In conclusion, we think that both the researcher's ingenuity and chemical intuition must play a decisive role in the definition of the computational strategy. In our opinion, this is a good conclusion because it shows that the study of reaction mechanisms with theoretical tools is not reduced to a dull repetition of a standard computational procedure, but presents the more challenging aspect of an activity open to an innovative use of the instruments provided by theoretical chemistry.

References

Adamo, C. and Lelj, F. (1984) *Chem. Phys. Lett.* **223**, 164.
Ågren, H., Mikkelsen, K.V. (1991) *J. Mol. Struct.(Theochem)* **234**, 425.
Aguilar, M.A. and Olivares del Valle, F.J. (1989) *Chem. Phys.* **129**, 439.
Aguilar, M.A., Bianco, R., Miertuš, S. and Tomasi, J. (1993a), *Chem. Phys.* **174**, 397.
Aguilar, M.A., Olivares del Valle, F.J. and Tomasi, J. (1993b), *J. Chem. Phys.* **98**, 7375.
Alagona, G., Cimiraglia, R., Scrocco, E. and Tomasi, J. (1972) *Theor. Chim. Acta* **25**, 103.
Alagona, G., Pullman, A., Scrocco, E. and Tomasi, J. (1973) *Int. J. Peptide Protein Res.* **5**, 251.
Alagona, G., Scrocco, E. and Tomasi, J. (1978) *Theor. Chim. Acta* **47**, 133.
Alagona, G., Bonaccorsi, R., Ghio, C. and Tomasi, J. (1986) *J. Mol. Struct. (Theochem)* **135**, 39.
Albery, W.J. (1980) *Ann. Rev. Phys. Chem.* **31**, 227.
Allen, M. and Tildesley, D.J. (1987) *Computer Simulation of Liquids*, Clarendon, Oxford.
Amovilli, C. (1994) *Chem. Phys. Lett.* **229**, 244.
Amovilli, C. and Mc Weeny, R. (1991) *J. Mol. Struct. (Theochem)* **227**, 1.
Ángyán, J.G. (1995) *Chem. Phys. Lett.* **241**, 51.
Antonczack, S., Ruiz-López, M.F. and Rivail, J.L. (1994) *J. Am. Chem. Soc.* **116**, 3912.
Åqvist, J. and Warshel, A. (1993) *Chem. Rev.* **93**, 2523.
Bachs, M., Luque, F.J. and Orozco, M. (1994) *J. Comp. Chem.* **15**, 446.
Bader, R.F., Carrol, M.T., Cheeseman, J.R. and Chany, C. (1987) *J. Am. Chem. Soc.* **109**, 7968.
Banerjee, P.K. and Butterfield, R. (1981) *Boundary Element Methods in Engineering Science*, McGraw-Hill, London.
Barker, J.A. and Henderson, D. (1976) *Rev. Mod. Phys.* **48**, 587.
Barker, J.A. and Watts, R.O. (1969) *Chem. Phys. Lett.* **3**, 144.
Bash, P.A., Field, M.J., Davenport, R.C., Petsko, G.A., Ringe, D. and Karplus, M. (1991) *Biochemistry* **30**, 5826.
Bash, P.A., Field, M.J. and Karplus, M. (1993) *J. Am. Chem. Soc.* **109**, 8092.
Bashford, D. and Karplus, M. (1990) *Biochemistry* **29**, 10219.
Bashford, D. and Gewert, K.J. (1992) *J. Mol. Biol.* **224**, 473.
Basilevsky, M.V. and Chudinov, G.E. (1992) *Chem. Phys.* **165**, 213.
Basilevsky, M.V., Chudinov, G.E. and Napolov, D.V. (1993) *J. Phys. Chem.* **97**, 3270.
Basilevsky, M.V., Chudinov, G.E. and Newton, M.D. (1995), to be published.
Ben-Naim, A. (1974) *Water and Aqueous Solutions*, Plenum Press, New York.
Ben-Naim, A. (1987) *Solvation Thermodynamics*, Plenum Press, New York.
Bernardi, F. and Robb, M.A. (1987) in Lowley, K.P. (ed.), *Ab Initio Methods in Quantum Chemistry. Advances in Chemical Physics*, Wiley, J., New York, Vol. 67, pp. 155.
Bertrán, J., Lluch, J.M., González-Lafont, A., Dillet, V. and Pérez, V. (1994) in Cramer, C.J. and Truhlar, D.G. (eds.), *Structure and Reactivity in Aqueous Solution*, ACS Symp. Ser. 568, Washington, pp. 168.
Beskos, D.E. (1987) *Boundary Element Methods in Mechanics*, North-Holland, Amsterdam.
Beveridge, D.L. and Jorgensen, W.L. (eds.) (1986) *Computer Simulation of Chemical and Biochemical Systems*, Ann. N. Y. Acad. Sci., New York.
Biagi, A., Bonaccorsi, R. and Cammi, R. (1989) *Technical Report*, Istituto di Chimica Quantistica ed Energetica Molecolare, Pisa.
Bianco R., Juanós i Timoneda, J. and Hynes J.T. (1994) *J. Phys. Chem.* **98**, 12104.
Bianco R., Miertuš, S., Persico, M. and Tomasi J. (1992) *Chem. Phys.* **168**, 281.
Bonaccorsi, R., Ghio, C. and Tomasi, J. (1984a), *Int. J. Quantum Chem.* **26**, 637.
Bonaccorsi, R., Palla, P. and Tomasi, J. (1984b), *J. Am. Chem. Soc.* **106**, 1945.
Bonaccorsi, R., Floris, M.F. and Tomasi, J. (1990) *J. Mol. Liq.* **47**, 25.

Bonaccorsi, R., Petrongolo, C., Scrocco, E. and Tomasi, J. (1969) *Atti I Congr. Intern. Chim. Quant. Espr. Lat.*, Univ. of Modena, Italy, pp. 15.
Bonaccorsi, R., Petrongolo, C., Scrocco, E. and Tomasi, J. (1971) *Theor. Chim. Acta* **20**, 237.
Bonaccorsi, R., Scrocco, E. and Tomasi, J. (1986) *Int. J. Quantum Chem.* **29**, 717.
Born, M. (1920) *Z. Phys.* **1**, 45.
Breneman, C.M. and Wiberg, K.B. (1990) *J. Comp. Chem.* **11**, 361.
Buff, F.P. (1955) *J. Chem. Phys.* **23**, 419.
Burton, N.A., Green, D.V.S., Hillier, I.H., Taylor, P.J., Vincent, M.A. and Woodcock, S. (1993) *J. Chem. Soc., Perkin Trans. 2*, 331.
Cammi, R. and Tomasi, J. (1994) *J. Chem. Phys.* **101**, 3888; **100**, 7495.
Cammi, R. and Tomasi, J. (1995a) *Int. J. Quant. Chem. Symp.* **29**, 465.
Cammi, R. and Tomasi, J. (1995b) *J. Comput. Chem.* **16**, 1449.
Cammi, R., Cossi, M., Mennucci, B., Pomelli, C. and Tomasi, J. (1995a) *Int. J. Quantum Chem.*, submitted.
Cammi, R., Cossi, M. and Tomasi, J. (1995b) *J. Chem. Phys.*, in press.
Cammi, R., Cossi, M., Mennucci, B. and Tomasi, J. (1996) *J. Chem. Phys.*, submitted.
Car, R. and Parrinello, M. (1985) *Phys. Rev. Lett.* **55**, 2471.
Chandler, D. and Andersen, H.C. (1972) *J. Chem. Phys.* **57**, 1930.
Chen, J.L. Noodleman, L., Case, D.A. and Bashford, D. (1994) *J. Phys. Chem.* **98**, 11059.
Chiles, R.A. and Rossky, P.J. (1984) *J. Am. Chem. Soc.* **106**, 6867.
Chipot, C., Gorb, L.G. and Rivail, J.L. (1992) *Chem. Phys. Lett.* **191**, 287.
Chipot, C., Gorb, L.G. and Rivail, J.L. (1994) *J. Phys. Chem.* **98**, 1601.
Choi, P.S., Jhon, M.S. and Eyring, H. (1970) *J. Chem. Phys.* **53**, 2608.
Chudinov, G.E. and Napolov, D.Y. (1993) *Chem. Phys. Lett.* **201**, 250.
Chudinov, G.E., Napolov, D.Y. and Basilevsky, M.V. (1992) *Chem. Phys.* **160**, 41.
Ciccotti, G., Ferrario, M., Hynes, J.T. and Kapral, R. (1990) *J. Chem. Phys.* **93**, 6261.
Clementi, E. (1976), *Lecture Notes in Chemistry*, Springer, Berlin, Vol. 2.
Clementi, E. (1980), *Lecture Notes in Chemistry*, Springer, Berlin, Vol. 19.
Clementi, E., Kołos, W., Lie, G.C. and Raughin, G. (1980) *Int. J. Quantum Chem.* **17**, 377.
Coitiño, E.L., Tomasi, J. and Ventura, O.N. (1994) *J. Chem. Soc. Farady Trans.* **90**, 1745.
Coitiño, E.L., Cammi, R. and Tomasi, J. (1995a) *J. Comput. Chem.* **16**, 20.
Coitiño, E.L., Tomasi, J. and Ventura, O.N. (1995b) *J. Phys. Chem.*, submitted.
Coitiño, E.L., Tomasi, J. and Ventura, O.N. (1995c) *J. Am. Chem. Soc.*, submitted.
Colvin, M.E. and Melius, C.F. (1993) *Sandia Nat. Labor. Techn. Rep. SAND93-8239*; available upon request to the authors.
Colvin, M.E., Evleth, E. and Akacem, Y. (1995) *J. Am. Chem. Soc.* **117**, 4357.
Connolly, M.L. (1983) *J. Appl. Cryst.* **16**, 548.
Connolly, M.L. (1985) it J. Am. Chem. Soc. **107**, 1118; *J. Appl. Cryst.* **18**, 499.
Contreras,R.R. and Aizman, A.J. (1993) *J. Mol. Struct. (Theochem)* **282**, 143.
Contreras,R.R. (1994) *J. Mol. Struct. (Theochem)* **310**, 295.
Cossi, M., Mennucci, B. and Tomasi, J. (1994) *Chem. Phys. Lett.* **228**, 165.
Cossi, M., Cammi, R. and Tomasi, J. (1995) *Int. J. Quant. Chem. Symp.* **29**, 695.
Cossi, M., Mennucci, B. and Cammi, R. (1996a) *J. Comp. Chem.* **17**, 57.
Cossi, M., Barone, V., Cammi, R. and Tomasi, J. (1996b) *Chem. Phys. Lett.*, submitted.
Cramer, C.J. and Truhlar, D.G. (1991) *J. Am. Chem. Soc.* **113**, 8305; 8552; 9901(E).
Cramer, C.J. and Truhlar, D.G. (1992) *J. Comp. Chem.* **13**, 1089.
Cramer, C.J. and Truhlar, D.G. (1992) *J. Comput-Aided Mol. Des.* **6**, 629.
Cramer, C.J., Lynch, G.C., Hawking, G.D. and Truhlar, D.G. (1993) *QCPE Bull.: AMSOL, version 4.0* **13**, 78.
Cramer, C.J. and Truhlar, D.G. (1994) in Politzer, P. and Murray, J.S. (eds.), *Quantitative Treatment of Solute/Solvent Interactions*, Elsevier, Amsterdam.
Cramer, C.J. and Truhlar, D.G. (eds.) (1994), *Structure and Reactivity in Aqueous So-*

lution, ACS Symp. Ser. 568, Washington.
Cramer, C.J. and Truhlar, D.G. (1995a) in Boyd, D.B. and Lipkowitz, K.B. (eds.), *Reviews in Computational Chemistry*, VCH, New York, Vol. 6.
Cramer, C.J. and Truhlar, D.G. (1995b) in Tapia, O. and Bertrán J. (eds.), *Solvent Effects and Chemical Reactivity*, Kluver, Dordrecht.
Cremaschi, P., Gamba, A. and Simonetta, M. (1972) *Theor. Chim. Acta* **25**, 237.
Davis, M.E. and McCammon, J.A. (1989) *J. Comp. Chem.* **10**, 387.
Davis, M.E. and McCammon, J.A. (1990) *Chem. Rev.* **90**, 509.
de Vries, A.H., van Duijnen, P.Th., Juffer, A.H., Rullmann, J.A.C., Dijkman, J.P., Merenga, H. and Thole, B.T. (1995) *J. Comp. Chem.* **16**, 37.
Dewar, M.J., Zoebisch, E.G., Healy, E.F. and Stewart, J.J.P. (1985) *J. Am. Chem. Soc.* **107**, 3902.
Dillet, V., Rinaldi, D., Ángyán, J.G. and Rivail, J.L. (1993) *Chem. Phys. Lett.* **2025**, 18.
Dillet, V., Rinaldi, D. and Rivail, J.L. (1994) *J. Phys. Chem.* **98**, 5034.
Dillet, V., Rinaldi, D., Bertrán, J. and Rivail, J.L. (1995), *J. Phys. Chem.*, to be published.
ogonadze, R.R. (1971) in Hush, N.S. (ed.), *Reactions of Molecules at Electrodes*, Wiley, London, pp. 135.
Dreyfus, M. and Pullman, A. (1970) *Theor. Chim. Acta* **19**, 20.
Drummond, M.L (1988) *J. Chem. Phys.* **88**, 5014; 5021.
Dunning Jr., T.H. (1990) *Calculation and Characterization of Molecular Potential Energy Surfaces, Advances in Molecular Electronic Structure Theory*, JAI Press, Greenwhich.
Eckert, J., Kubas, G.J., Hall, J.P., Hay, P.J. and Boyle, C.M. (1990) *J. Am. Chem. Soc.* **112**, 2324.
Eigen, M. (1967) *Discuss. Faraday Soc.* **39**, 7.
Eksterowicz, J.E. and Hank, K.N. (1993) *Chem. Rev.* **93**, 2439.
Evleth, E., Akacem, Y. and Colvin, M.E. (1994) *Chem. Phys. Lett.* **227**, 412.
Ferenczy, G.G., Rivail, J.L. Surján, P.R. and Náray–Szabó, G. (1992) *J. Comp. Chem.* **13**, 830.
Field H.J., Bash, P.A. and Karplus, M. (1990) *J. Comp. Chem.* **11**, 700.
Floris, F.M. and Tomasi, J. (1989) *J. Comp. Chem.* **10**, 616.
Floris, F.M., Tomasi, J. and Pascual-Ahuir, J.L. (1991) *J. Comp. Chem.* **12**, 784.
Floris, F.M., Tani, A. and Tomasi, J. (1993) *Chem. Phys. Lett.* **169**, 11.
Floris, F.M., Persico, M., Tani, A. and Tomasi, J. (1995) *Chem. Phys.* **195**, 207.
Floris, F.M., Selmi, M., Tani, A. and Tomasi, J. (1996), manuscript in preparation.
Fois, E.S., Sprik, M. and Parrinello, M. (1994) *Chem. Phys. Lett.* **223**, 411.
Ford, G.P. and Wang, B. (1992) *J. Am. Chem. Soc.* **114**, 10563.
Fortunelli, A. (1995a) *J. Mol. Struct.(Theochem)*, in press.
Fortunelli, A. (1995b) *J. Phys. Chem.* **99**, 9056.
Fortunelli, A. and Tomasi, J. (1994) *Chem. Phys. Lett.* **34**, 231.
Fox, T. and Rösch, N. (1992) *J. Mol. Struct. (Theochem)* **276**, 279.
Fox, T., Rösch, N. and Zauhar R.J. (1993) *J. Comp. Chem.* **14**, 253.
Francl, M.M., Hout, R.F. and Hehre, W.J. (1984) *J. Am. Chem. Soc.* **106**, 563.
Frecer, V., Májeková, M. and Miertuš, S. (1989) *J. Mol. Struct. (Theochem)* **189**, 443.
Frecer, V., Miertuš, S. and Májeková, M. (1991) *J. Mol. Struct. (Theochem)* **227**, 157.
Friesner, R.A. (1991) *Ann. Rev. Phys. Chem.* **42**, 341.
Frisch, M.J., Trucks, G.W., Head–Gordon, M., Gill, F.M.W., Wong, M.W., Foresman, J.B., Johnson, B.G., Schlegel, H.B., Robb, M.A., Replage, E.S., Gomperts, R., Andres, J.L., Raghavachari, K., Binkley, J.S., Gonzalez, C., Martin, R.L., Fox, D.J., Defrees, D.J., Baker, J., Stewart, J.J.P. and Pople, J.A. (1992) *GAUSSIAN92, version A*; Gaussian Inc., Pittsburgh, PA.
Frisch, M.J., Trucks, G.W., Schlegel, H.B., Gill, F.M.W., Johnson, B.G., Robb, M.A., Cheeseman, J.R., Keith, T., Petersson, G.A., Montgomery, J.A., Raghavachari, K., Al–Laham, M.A., Zarkzewski, V.G., Ortiz, J.V., Foresman, J.B., Peng, C.Y., Ay-

ala, P.Y., Chen, W., Wong, M.W., Andres, J.L., Binkley, J.S., Defrees, D.J., Baker, J., Stewart, J.J.P., Head-Gordon, M., Gonzalez, C. and Pople, J.A. (1994) *GAUSSIAN94, Revision B.3*; Gaussian Inc., Pittsburgh, PA.

Fritsch, V. and Westhof, E. (1991) *J. Am. Chem. Soc.* **113**, 8271.

Fukui, K. (1970) *J. Phys. Chem.* **74**, 4161.

Furuki, K., Sakurai, M., Inoue, Y., Chûjô, R. and Harota, K. (1994) *J. Comp. Chem.* **115**, 90.

Gao J. (1992) *J. Phys. Chem.* **96**, 6432.

Gao J. (1993) *J. Am. Chem. Soc.* **115**, 2930.

Gao, J. (1995) in Boyd, D.B. and Lipkowitz, K.B. (eds.), *Reviews in Computational Chemistry*, VCH, New York, Vol. 7.

Gao, J. and Xia, X. (1994) in Cramer, C.J. and Truhlar, D.G. (eds.), *Structure and Reactivity in Aqueous Solution*, ACS Symp. Ser. 568, Washington, pp. 212.

Gavish, B. (1980) *Phys. Rev. Lett.* **44**, 1160.

Gehlen, J.N., Chandler, D., Kim, H.J. and Hynes, J.T. (1992) *J. Phys. Chem.* **96**, 1748.

German, E.D., Dogonadze, R.R., Kuznetsov, A.M., Levich, V.G. and Y.I. Kharkats (1971) *J. Res. Catlysis Hokkaido Univ.* **19**, 99; 115.

Gertner, B.J., Wilson, K.R. and Hynes, J.T. (1989) *J. Chem. Phys.* **90**, 3537.

Giesen, D.J., Storer, J.W., Cramer, C.J. and Truhlar, D.G. (1995) *J. Am. Chem. Soc.* **117**, 1057.

Gilson, M.K., Sharp, K.A. and Honig, B.H. (1988) *J. Comp. Chem.* **9**, 327.

Gould, I.R., Green, D.V.S., Young, P. and Hillier, I.H. (1992) *J. Org. Chem.* **57**, 4434.

Grant, J.A., Williams, R.L. and Scheraga, H.A. (1990) *Biopolymers* **30**, 929.

Green, D.V.S. (1991) Ph.D. Thesis, University of Manchester.

Gresh, N., Claverie, P. and Pullman, A. (1984) *Theor. Chim. Acta* **66**, 1.

Gresh, N., Claverie, P. and Pullman, A. (1986) *Int. J. Quant. Chem.* **29**, 101.

Grote, R.F. and Hynes, J.T. (1980) *Chem. Phys.* **73**, 2715.

Guest, H.F. and Kendrick. J. (1986): GAMESS-UK, Darensbury Laboratories.

Hall, R.J., Davidson, M.M., Burton, N.A. and Hillier, I.H. (1995) *J. Phys. Chem.*, in press.

Halicioglu, T. and Sinanoglu, O. (1969) *Ann. N. Y. Acad. Sci.* **158**, 308.

Haynes, G.R., Voth, G.A. and Pollak, E. (1994) *J. Chem. Phys.* **101**, 7811.

Heidrich, D., Kliesch, W. and Quapp, W. (1991) *Properties of Chemically Interesting Potential Energy Surface*, Lect. Notes in Chem, Springer, Berlin.

Helgaker, T. and Jorgensen, P. (1988) *Adv. Quantum Chemistry* **19**, 183.

Hirata, F. and Rossky, P.J. (1981) *Chem. Phys. Lett.* **84**, 329.

Honig, B.H., Sharp, K.A. and Yang, A.S. (1993) *J. Phys. Chem.* **97**, 1101.

Hoshi, H., Sakurai, M., Inoue, Y. and Chûjô, R. (1987) *J. Chem. Phys.* **87**, 1107.

Hoshi, H., Sakurai, M., Inoue, Y. and Chûjô, R. (1988) *J. Mole. Struct.(Theochem)* **180**, 267.

Huron, M.J and Claverie, P. (1972) *J. Phys. Chem.* **76**, 2123.

Huron, M.J and Claverie, P. (1974) *J. Phys. Chem.* **78**, 1853; 1862.

Hylton McCreery, J., Christoffersen, R. and Hall, G.G. (1976) *J. Am. Chem. Soc.* **98**, 7191.

Hynes, J.T. (1985) in Baer, M. (ed.), *The Theory of Chemical Reactions in Dynamics*, CRC Press, Boca Raton, Vol. IV, pp 171.

Hynes, J.T., Kim, H.J., Mathis, J.R., Bianco, R., Koji A. and Gertner B.J. (1994) in Jortner J. et al. (eds.), *Reaction Dynamics in Cluster and Condensed Phases*, Kluwer, Dordrecht, pp 286.

Hwang, J.C., King, G., Creighton, S. and Warshel, A. (1988) *J. Am. Chem. Soc.* **110**, 5297.

Jorgensen, W.L. (1989a) *J. Am. Chem. Soc.* **111**, 3770.

Jorgensen, W.L. (1989b) *Acc. Chem. Res.* **22**, 184.

Jorgensen, P. and Simons, J. (1986) *Geometrical Derivatives of Energy Surfaces and Molecular Properties*, Reidel, Dordrecht.

Jorgensen, W.L., Blake, J.F., Lin, D. and Severance, D.L. (1994) *J. Chem. Soc., Faraday Trans.* **90**, 1727.
Jorgensen, W.L., Buckner, J.K., Huston, S.E. and Rossky, P.J. (1987) *J. Am. Chem. Soc.* **109**, 1891.
Juffer, A.H., Botta, E.F.F., van Keulen, B.A.M., van der Ploeg, A. and Berendsen, H.J.C. (1991) *J. Comp. Phys.* **97**, 144.
Karelson, M.M., Tamm, T. and Zerner, M.C. (1993) *J. Phys. Chem.* **97**, 11907.
Kawata, M., Ten-no, S., Kato, S. and Hirata, F. (1995) *J. Am. Chem. Soc.* **117**, 1638.
Kawamura–Kuribayashi, H., Koga, N. and Morokuma, K. (1992) *J. Am. Chem. Soc.* **114**, 8687.
Kim, H.J. and Hynes, J.T. (1992) *J. Chem. Phys.* **96**, 5088.
Kim, H.J., Bianco, R., Gertner, B.J. and Hynes, J.T. (1993) *J. Phys. Chem.* **97**, 1724.
King, G. and Warshel, A. (1989) *J. Chem. Phys.* **91**, 3647.
Kirkwood, J.G. (1934) *J. Chem. Phys.* **2**, 351.
Kirkwood, J.G. and Buff, F.P. (1950) *J. Chem. Phys.* **18**, 991.
Kitaura, K. and Morokuma, K. (1976) *Int. J. Quant. Chem.* **10**, 325.
Klamt, A. and Schüürmann, G. (1993) *J. Chem. Soc., Perkin Trans. 2*, 799.
Klopman, G. (1967) *Chem. Phys. Lett.* **1**, 200.
Kölle, C. and Jug, K. (1995), to be published.
Kollman, P.A. (1993) *Chem. Rev.* **93**, 2395.
Kollman, P.A., Allen. L.C. (1970) *Theor. Chim. Acta* **18**, 399.
Kołos, W. (1979) *Theor. Chim. Acta* **51**, 219.
Kong, Y.S. and Warshel, A. (1995) *J. Am. Chem. Soc.* **117**, 6234.
Köster, A.M., Kölle, C. and Jug, K. (1993) *J. Chem. Phys.* **99**, 1224.
Kraka, E. and Dunning Jr., T.H. (1990) in Dunning Jr., T.H. (ed.) *Calculation and Characterization of Molecular Potential Energy Surfaces, Advances in Molecular Electronic Structure Theory*, JAI Press, Greenwhich, pp. 129.
Kramers, H.A. (1940) *Physica* **7**, 284.
Krishnan, R., Singh, S. and Robinson, G.W. (1992) *J. Chem. Phys.* **97**, 5516.
Kuharski, R.A. and Chandler, D. (1987) *J. Phys. Chem.* **90**, 6015.
Kushick, J. and Berne, B.J. (1977) in Berne, B.J. (ed.), *Statistical Mechanics. Part A*, Plenum Press, New York, pp. 41.
Kuznetsov, A.M. (1992) *J. Phys. Chem.* **96**, 3337.
Laasonen, K., Sprik, M., Parrinello, M. and Car, R. (1993) *J. Chem. Phys.* **99**, 9080.
Langlet, J., Claverie, P., Caillet, J. and Pullman, A. (1988) *J. Phys. Chem.* **92**, 1617.
Langlet, J., Gresh, N. and Gressner–Prettre, C. (1995) *Biopolymers*, in press.
Lavery, R., Sklenar, H., Zakrzewska, K. and Pullman, B. (1986) *J. Biomol. Struct. Dyn.* **3**, 989.
Lee, B. and Richards, F.M. (1971) *J. Mol. Biol.* **55**, 379.
Lee, S. and Hynes, J.T. (1988) *J. Chem. Phys.* **88**, 6853.
Lee, F.S. and Warshel, A. (1992) *J. Chem. Phys.* **97**, 3100.
Lee, F.S., Chu, Z.T. and Warshel, A. (1993) *J. Comp. Chem.* **14**, 161.
Levich, V.G. (1966) *Adv. Electrochem. Electrochem. Eng.* **4**, 249.
Levich, V.G. and Dogonadze, R.R. (1959) *Dokl. Akad. Nauk.* **124**, 123.
Li, J., Correa de Mello, P. and Jug, K. (1992) *J. Comp. Chem.* **13**, 85.
Lin, H., Müller-Plathe, F. and van Gunsteren, W.F. (1995) *J. Chem. Phys.* **102**, 1722.
Lindberg, K. and Cortes, E. (1984) *Physica A* **126**, 489.
Linder, B. (1967) *Adv. Chem. Phys.* **12**, 225.
Liotard, D.A., Hawking, G.D., Lynch, G.C., Cramer, C.J. and Truhlar, D.G. (1995) *J. Comp. Chem.* **16**, 422.
Lorensen, W.E. and Cline, H.E. (1987) *Comp. Graph.* **21**, 163.
Luque F.J. and Orozco M. (1995), to be published.
Luque F.J., Alemán, C. and Orozco M. (1995), to be published.
Luque F.J., Bachs, M. and Orozco M. (1994) *J. Comp. Chem.* **15**, 847.
Luque F.J., Negre, M.J. and Orozco M. (1993) *J. Phys. Chem.* **97**, 4386.

Luty, B.A., Davis, M.E. and McCammon, J.A. (1992) *J. Comp. Chem.* **13**, 768.
Luzhkov, V. and Warshel A. (1992) *J. Comp. Chem.* **13**, 199.
Marcus, R.A. (1956) *J. Chem. Phys.* **24**, 966, 979.
Marcus, R.A. (1960) *Discuss. Faraday Soc.* **29**, 21.
Marcus, R.A. (1963) *J. Phys. Chem.* **67**, 893.
Marcus, R.A. (1965) *J. Chem. Phys.* **43**, 679.
Maseras, F. and Morokuma, K. (1995) *J. Comp. Chem.*, in press.
Mathis, J.R. and Hynes, J.T. (1994) *J. Phys. Chem.* **98**, 5460.
Mathis, J.R., Bianco R. and Hynes, J.T. (1994) *J. Mol. Liq.* **61**, 81.
Mathis, J.R., Kim, H.J. and Hynes, J.T. (1993) *J. Am. Chem. Soc.* **115**, 8248.
McKee, M.L. and Page, M. (1993) in Lipkowitz, K.B. and Boyd, D.B. (eds.), *Reviews in Computational Chemistry*, VCH, New York, Vol. 4, pp. 35.
Mennucci, B., Cossi, M. and Tomasi, J. (1995) *J. Chem. Phys.* **102**, 6837.
Mennucci, B., Cossi, M. and Tomasi, J. (1996) *J. Phys. Chem.*, in press.
Mezey, P.G. (1987) *Potential Energy Hypersurfaces*, Elsevier, Amsterdam.
Miertuš, S. and Tomasi, J. (1982) *J. Chem. Phys.* **65**, 239.
Miertuš, S., Frecer, V., and Májeková, M. (1988) *J. Mol. Struct. (Theochem)* **179**, 353.
Miertuš, S., Scrocco, E. and Tomasi, J. (1981) *Chem. Phys.* **55**, 117.
Mikkelsen, K.V., Ågren, H., Jensen, H.J. and Helgaker, T. (1988) *J. Chem. Phys.* **89**, 3086.
Mikkelsen, K.V., Luo, Y., Ågren, H. and Jørgensen, P. (1994) *J. Chem. Phys.* **100**, 8240.
Mikkelsen, K.V., Jørgensen, P. and Jensen, H.J. (1994) *J. Chem. Phys.* **100**, 6597.
Miller, W.H., Handy, N.C. and Adams, J.E. (1980) *J. Chem. Phys.* **72**, 99.
Miller, W.H., Ruf, B.A. and Chang, Y.T. (1988) *J. Chem. Phys.* **89**, 6298.
Momany, F.A. (1978) *J. Phys. Chem.* **82**, 592.
Montagnani, R., Tomasi, J. and Demetropoulos, I. (1992) *Mol. Eng.* **2**, 123.
Morokuma, K. (1971) *J. Chem. Phys.* **55**, 1236.
Muller, R.F. and Warshel, A. (1995) *J. Phys. Chem.* **99**, 17516.
Negre, M.J., Orozco M., and Luque F.J. (1992) *Chem. Phys. Lett.* **196**, 27.
Noell, J.O. and Morokuma, K. (1976) *J. Phys. Chem.* **80**, 2675.
Nicholls, A. and Honig, B.H. (1991) *J. Comp. Chem.* **12**, 435.
Northrup, S.H. and Hynes, J.T. (1980) *J. Chem. Phys.* **73**, 2700.
Olivares del Valle, F.J. and Aguilar, M.A. (1993) *J. Mol. Struct. (Theochem)* **280**, 25.
Olivares del Valle, F.J. and Tomasi, J. (1991) *Chem. Phys.* **150**, 139.
Onsager, L. (1936) *J. Am. Chem. Soc.* **58**, 1486.
Orozco M., and Luque F.J., (1994) *Chem. Phys.* **182**, 237.
Orozco M. and Luque F.J., (1995) *J. Am. Chem. Soc.* **117**, 1378.
Orozco M., Bachs, M. and Luque F.J., (1995) *J. Comp. Chem.* **16**, 563.
Orozco M., Jorgensen, W.L. and Luque F.J., (1993) *J. Comp. Chem.* **14**, 1498.
Pacher, T., Cederbaum, L.S. and Köppel (1993) *Adv. Chem. Phys.* **84**, 293.
Pappalardo, R.R. and Sánchez Marcos, E. (1995) *J. Phys. Chem.*, in press.
Pappalardo, R.R., Martinez, J.M. and Sánchez Marcos, E. (1993) *J. Phys. Chem.* **97**, 4500.
Parchment, O.G., Green, D.V.S., Taylor, P.J. and Hillier, I.H. (1993) *J. Am. Chem. Soc.* **115**, 2352.
Parchment, O.G., Hillier, I.H., Green, D.V.S., Burton, N.A., Morley, J.O. and Scaefer, H.F. III (1992) *J. Chem. Soc., Perkin Trans. 2*, 1681.
Pascual-Ahuir, J.L. and Silla, E. (1990) *J. Comp. Chem.* **11**, 1047.
Pascual-Ahuir, J.L., Silla, E., Tomasi, J. and Bonaccorsi, R. (1987) *J. Comp. Chem.* **8**, 778.
Pascual-Ahuir, J.L., Silla, E. and Tuñón, I. (1994) *J. Comp. Chem.* **15**, 1127.
Pekar, S.I. (1951) *Introduction to Electron Theory of Crystals*, Technical Lit. Publ., Moscow.
Persico, M. and Tomasi, J. (1984) *Croat. Chim. Acta* **57**, 1395.

Pierotti, R.A. (1963) *J. Phys. Chem.* **67**, 1840.
Pierotti, R.A. (1965) *J. Phys. Chem.* **69**, 281.
Pierotti, R.A. (1976) *Chem. Rev.* **76**, 717.
Politzer, P. and Murray, J.S. (eds.) (1994), *Quantitative Treatment of Solute/Solvent Interactions*, Elsevier, Amsterdam.
Pomelli, C. and Tomasi, J. (1995), manuscript in preparation.
Popkie, H., Kistenmacher, K. and Clementi, E. (1973) *J. Chem. Phys.* **59**, 1325.
Postma, J.P.M., Berendsen, H.J.C. and Haak, J.R. (1982) *Faraday Symp. Chem. Soc.* **17**, 55.
Pulay, P. (1987) *Adv. Chem. Phys.* **69**, 241.
Purisima, E.O. and Nilar, S.H. (1995) *J. Comp. Chem.* **16**, 681.
Rahman, A. and Stillinger, F.H. (1971) *J. Chem. Phys.* **55**, 3336.
Rashin, A.A. (1990) *J. Phys. Chem.* **94**, 1725.
Rashin, A.A. and Namboodiri, K. (1988) *J. Phys. Chem.* **91**, 6063.
Rashin, A.A., Bukatin, M.A., Andrelm, J.A. and Hagler, A.T. (1994) *Biophysical Chemistry* **51**, 375.
Rauhut, G., Clark, T. and Steinke, K. (1993) *J. Am. Chem. Soc.* **115**, 9174.
Reichart, C. (1990) *Solvents and Solvent Effects in Organic Chemistry*, VCH, New York.
Reiss, H., Frish, H.L. and Lebowitz, J.L. (1959) *J. Chem. Phys.* **31**, 369.
Reiss, H., Frish, H.L., Helfand, E. and Lebowitz, J.L. (1960) *J. Chem. Phys.* **32**, 119.
Richards, F.M. (1977) *Ann. Rev. Biophys. Bioeng.* **6**, 151.
Rinaldi, D. and Rivail, J.L. (1973) *Theor. Chim. Acta* **32**, 57.
Rinaldi, D., and Pappalardo, R.R. (1992) *QCPE Bull.* **12**, 69.
Rinaldi, D., Costa Cabral, B.J. and Rivail, J.L. (1986) *Chem. Phys. Lett.* **125**, 495.
Rinaldi, D., Rivail, J.L. and Rguini, N. (1992) *J. Comp. Chem.* **13**, 675.
Rinaldi, D., Ruiz–López, M.F. and Rivail, J.L. (1983) *J. Chem. Phys.* **78**, 834.
Ringnalda, M., Langlois, J.M., Greeley, B., Murphy, R., Russo, T., Cortis, C., Muller, R., Marten, B., Donnelly, R.Jr., Mainz, D., Wright, J., Polland, W.T., Cao, Y., Won, Y, Miller, G., Goddard, W.A. III and Friesner, R.A. (1994), *PS–GVB*, Schrodinger Inc.
Rivail, J.L. and Rinaldi, D. (1976) *Chem. Phys.* **18**, 233.
Rivail, J.L., Antonczack, S., Chipot, C., Ruiz–López, M.F. and Gorb, L.G. (1994) in Cramer, C.J. and Truhlar, D.G. (eds.), *Structure and Reactivity in Aqueous Solution*, ACS Symp. Ser. 568, Washington, pp. 155.
Rivail, J.L., Terryn, B., Rinaldi, D. and Ruiz–López, M.F. (1985) *J. Mol. Struct. (Theochem)* **120**, 387.
Rossky, P.J. (1985) *Ann. Rev. Phys. Chem.* **36**, 321.
Ruiz–López, M.F. and Rinaldi, D. (1983) *J. Mol. Struct. (Theochem)* **93**, 277.
Ruiz–López, M.F., Bohr, F., Martins Costa, M.T.C. and Rinaldi, D. (1994) *Chem. Phys. Lett.* **22**, 105.
Ruiz–López, M.F., Rinaldi, D. and Bertrán, J. (1995), in press.
Ruf, B.A. and Miller, W.H. (1988) *J. Chem. Soc., Faraday Trans.* **84**, 1523.
Salahub, O.R. (1987), in Lawley, K.P. (ed.), *Ab Initio Methods in Quantum Chemistry*, Wiley, J., New York.
Sanhueza, J. E., Tapia, O., Laidlaw, W. G. and Trsic, M. (1979) *J. Chem. Phys.* **70**, 3096.
Schlegel, H.B. (1987) in Lowley, K.P. (ed.), *Ab Initio Methods in Quantum Chemistry. Advances in Chemical Physics*, Wiley, J., New York, Vol. 67, pp. 245.
Schlick, T. (1992) in Lipkowitz, K.B. and Boyd, D.B. (eds.), *Reviews in Computational Chemistry*, VCH, New York, pp. 1.
Severance, D.L. and Jorgensen, W.L. (1994) in Cramer, C.J. and Truhlar, D.G. (eds.), *Structure and Reactivity in Aqueous Solution*, ACS Symp. Ser. 568, Washington, pp. 243.
Shaik, S.S. and Hiberty, P.C. (1991) in Maksič, Z. (ed.), *Theoretical Models of Chemical Bonding*, Springer, Berlin, Vol. 4, pp. 269.

Sharp, K.A. (1991) *J. Comp. Chem.* **12**, 454.
Sharp, K.A., Jean–Charles, A. and Honig, B.H. (1992) *J. Phys. Chem.* **96**, 3822.
Sidis, V. (1992) *Adv. Chem. Phys.* **82**, 73.
Silla, E., Tuñón, I. and Pascual-Ahuir, J.L. (1991) *J. Comp. Chem.* **12**, 1077.
Simkin, B.Y. and Sheiker I. (1995), *Quantum Chemical and Statistical Theory of Solutions*, Ellis-Horwood, London.
Sinanoglu, O. (1967) *Chem. Phys. Lett.* **1**, 283.
Sinanoglu, O. (1974) *Theor. Chim. Acta* **33**, 279.
Sinanoglu, O. (1981) *J. Chem. Phys.* **75**, 463.
Singh, U.C. and Kollmann, P.A. (1986) *J. Comp. Chem.* **7**, 718.
Smith, F.T. (1969) *Phys. Rev.* **179**, 111.
Sokalski, W.A., Shibata, M., Ornstein, R.L. and Rein, R. (1992) *J. Comp. Chem.* **13**, 883.
Sörensen, P.E. and Jencks, W.P. (1987) *J. Am. Chem. Soc.* **109**, 7.
Stanton, R.V., Hartsough, D.S. and Merz Jr., K.M. (1995) *J. Comp. Chem.* **16**, 113.
Stefanovich, E.V. and Truong, T.N. (1995) *J. Comp. Chem.*, submitted.
Stewart, J.W. (1989) *J. Comp. Chem.* **10**, 221.
Stewart, J.P. (1990) *QCPE Bull.* **10**, 86.
Still, W.C., Tempczyk, A., Hawley, R.C. and Hendrickson, T. (1991) *J. Am. Chem. Soc.* **112**, 6127.
Storer, J.W., Giesen, D.J., Cramer, C.J. and Truhlar, D.G. (1995) *J. Comp.-Aided Molec. Des.* **9**, 87.
Straub, J.E., Berne, B.J. and Roux, B. (1990) *J. Chem. Phys.* **93**, 6804.
Szafran, M., Karelson, M.M., Katritzky, A.R., Koput, J. and Zerner, M.C. (1993) *J. Comp. Chem.* **14**, 371.
Tannor, D.J., Marten, B., Murphy, R., Friesner, R.A., Sitkoff, D., Nicholls, A., Ringnalda, M., Goddard III, W.A. and Honig, B. (1994) *J. Am. Chem. Soc.* **116**, 11875.
Tapia, J. and Goschinski, O. (1975) *Mol. Phys.* **29**, 1653.
Tapia, O. (1982) in Ratajczac, H. and Orville–Thomas, W.J. (eds.), *Molecular Interactions*, Wiley, J., New York, Vol. 3.
Ten-no, S., Hirata, F. and Kato, S. (1993) *Chem. Phys. Lett.* **214**, 391.
Ten-no, S., Hirata, F. and Kato, S. (1994) *J. Chem. Phys.* **100**, 7443.
Te Velde, G. and Baerends, E.J. (1992) *J. Comp. Phys.* **99**, 84.
Théry, V., Rinaldi, D., Rivail, J.L., Maigret, B. and Ferenczy, G.G. (1994) *J. Comp. Chem.* **15**, 269.
Tomasi, J. (1982) in Ratajczac, H. and Orville–Thomas, W.J. (eds.), *Molecular Interactions*, Wiley, J., New York, Vol. 3.
Tomasi, J. (1994) in Cramer, C.J. and Truhlar, D.G. (eds.), *Structure and Reactivity in Aqueous Solution*, ACS Symp. Ser. 568, Washington, pp. 10.
Tomasi, J. and Persico, M. (1994) *Chem. Rev.* **94**, 2027.
Tomasi, J., Alagona, G., Bonaccorsi, R., Ghio, C. and Cammi, R. (1991) in Maksič, Z. (ed.), *Theoretical Models of Chemical Bonding*, Springer, Berlin, Vol. 3, pp. 546.
Truhlar D.G. and Garrett, B.C. (1984) *Ann. Rev. Phys.* **36**, 159.
Truhlar, D.G., Isaacson, A.D. and Garrett, B.C. (1985) in Baer, M. (ed.), *The Theory of Chemical Reactions in Dynamics*, CRC Press, Boca Raton, Vol IV.
Truhlar D.G., Schenter, G.K. and Garrett, B.C. (1993) *J. Chem. Phys.* **98**, 5756.
Truhlar D.G., Steckler, R. and Gordon, M.S. (1987) *Chem. Rev.* **85**, 217.
Truong, T.N. and Stefanovich, E.V. (1995a) *J. Phys. Chem.*, in press.
Truong, T.N. and Stefanovich, E.V. (1995b) *Chem. Phys. Lett.* **240**, 253.
Truong, T.N. and Stefanovich, E.V. (1995c) *Chem. Phys. Lett.* **244**, 65.
Valleau, J.P. and Torrie, G.M. (1977) in Berne, B.J. (ed.), *Statistical Mechanics. Part A*, Plenum Press, New York, pp. 169.
Valleau, J.P. and Whittington, S.G. (1977) in Berne, B.J. (ed.), *Statistical Mechanics. Part A*, Plenum Press, New York, pp. 137.
van Duijnen, P.Th. and de Vries, A.H. (1995) *Int. J. Quant. Chem. Symp.* **29**, 523.

van Duijnen, P.Th., Juffer, A.H. and Dijkman, J.P. (1992) *J. Mol. Struct. (Theochem)* **260**, 195.
van Gunsteren, W.F., Weiner, P.K. and Wilkinson, A.J. (1993) *Computer Simulation of Biomolecular Systems*, ESCOL, Leiden, Vol. 2.
van der Zwan, G. and Hynes, J.T. (1984) *Chem. Phys.* **90**, 21.
Varnek, A.A., Glebov, A.S., Wipff, G. and Feil, D. (1995) *J. Comp. Chem.* **16**, 1.
Vigné–Maeder, F. and Claverie, P. (1988) *J. Chem. Phys.* **88**, 4934.
Voth, G.A. (1992) *J.Chem. Phys.* **97**, 5908.
Wang, B. and Ford, G.P. (1992) *J. Chem. Phys.* **97**, 4162.
Warshel, A. (1978) *Chem. Phys. Lett.* **55**, 454.
Warshel, A. (1979) *J. Phys. Chem.* **83**, 1640.
Warshel, A. (1982) *J. Phys. Chem.* **86**, 2218.
Warshel, A. (1991) *Computer Modeling of Chemical Reactions in Enzymes and Solutions*, Wiley–Interscience, New York.
Warshel, A. and Åqvist, J. (1991) *Ann. Rev. Biophys. Chem.* **20**, 267.
Warshel, A. and Levitt, M. (1976) *J. Mol. Biol.* **103**, 227.
Warshel, A. and King, G. (1985) *Chem. Phys. Lett.* **121**, 124.
Warshel, A. and Parson, W.W. (1991) *Ann. Rev. Chem. Phys.* **42**, 279.
Warshel, A. and Russel, S.T. (1984) *Quart. Rev. Bioph.* **17**, 3.
Warshel, A. and Weiss, R.M. (1980) *J. Am. Chem. Soc.* **102**, 6218.
Warwicker, J. and Watson, H.C. (1982) *J. Mol. Biol.* **157**, 671.
Waszkowycz, B., Hillier, I.H., Gensmantel, N. and Payling, D.W. (1991) *J. Chem. Soc., Perkin Trans. 2*, 225.
Weaver, M.J. (1992) *Chem. Rev.* **92**, 463.
Weiner, S.J., Kollman, P.A., Nguyen, D.T. and Case, D.A. (1986) *J. Comp. Chem.* **7**, 230.
Wong, M.W., Frisch, M.J. and Wiberg, K.B. (1991) *J. Am. Chem. Soc.* **113**, 4776.
Wong, M.W., Wiberg, K.B. and Frisch, M.J. (1992) J. Am. Chem. Soc., **114**, 523; 1645.
Woodcock, S., Green, D.V.S., Vincent, M.A., Hillier, I.H., Guest, M.F. and Sherwood, P. (1992) *J. Chem. Soc., Perkin Trans. 2*, 2151.
Xu, Y.W., Wang, C.X. and Shi, Y.Y. (1992) *J. Comp. Chem.* **13**, 1109.
Yamaguchi, Y., Osamura, Y., Goddard, J.D. and Schaefer III. H.E. (1994) *A New Dimension to Quantum Chemistry: Analytical Derivative Methods in Ab Initio Molecular Electronic Structure Theory*, Oxford Univ. Press, New York.
Yomosa, S. (1974) *J. Phys. Soc. (Japan)* **36**, 1655.
Yomosa, S. (1978) *J. Phys. Soc. (Japan)* **44**, 602.
You, T.J. and Harvey, S.C. (1993) *J. Comp. Chem.* **14**, 484.
Yu, H. and Karplus, M. (1990) *J. Am. Chem. Soc.* **112**, 5706.
Zauhar, R.J. (1991) *J. Comp. Chem.* **12**, 575.
Zauhar, R.J. and Morgan, R.S. (1985) *J. Mol. Biol.* **186**, 815.
Zauhar, R.J. and Morgan, R.S. (1988) *J. Comp. Chem.* **9**, 171.
Zauhar, R.J. and Morgan, R.S. (1990) *J. Comp. Chem.* **11**, 603.
Zusman, I. (1980) *Chem. Phys.* **49**, 295.

FREE ENERGY PERTURBATION CALCULATIONS WITHIN QUANTUM MECHANICAL METHODOLOGIES

ROBERT V. STANTON
STEVEN L. DIXON
KENNETH M. MERZ JR.*
Department of Chemistry
The Pennsylvania State University
152 Davey Laboratory
University Park, PA 16802

ABSTRACT Newly developed techniques for use in quantum free energy perturbation calculations are presented. These techniques represent the first means of performing arbitrary perturbations in quantum mechanical systems. The methods that have been developed include single and dual-toplogy approaches as well as strategies that imploy molecular mechanical intermediates. The theoretical and practical considerations for carrying out the associated molecular dynamics simulations are also discussed. Simulations are presented for systems that are purely quantum mechanical, and for systems that involve a combination of quantum mechanical and molecular mechanical atoms. Preliminary results demonstrate these procedures to constitute a powerful tool in free energy calculations, with the potential to significantly increase the accuracy of simulations on both large-scale and small-scale systems.

G. Náray-Szabó and A. Warshel (eds.), Computational Approaches to Biochemical Reactivity, 103–123.

1. Introduction

Free energy perturbation (FEP) (Lybrand, Ghosh et al., 1985; Mezei and Beveridge, 1986; Jorgensen, 1989) and potential of mean force (PMF) (van Eerden, Briels et al., 1989; Elber, 1990) calculations within molecular mechanical (MM) force fields have proven to be very powerful computational methodologies. These techniques, which have been developed and refined over the last two decades, allow access to important thermodynamic quantities, such as the relative binding free energies of ions within an ionophore(Marrone and Merz, 1992), or inhibitors to an enzyme (Merz and Kollman, 1989). While other techniques, such as linear response theory (Åqvist, Carmen et al., 1994), have been developed to calculate these same quantities they may require additional parameterization. FEP and PMF calculations, in contrast, are done using statistical averages of the energy differences between the various intermediate states considered in the calculation and thus may be applied to any system for which a potential energy surface is known. Many elegant applications of these theoretical techniques within molecular dynamics (MD) calculations currently exist within the literature (Bash, Fields et al., 1987; Grootenhuis and Kollman, 1989; Merz and Kollman, 1989; van Eerden, Briels et al., 1989; Åqvist, 1990; Cieplak and Kollman, 1990; Dang and Kollman, 1990; Elber, 1990; Kollman, 1993; Hartsough and Merz, 1995).

Recently, quantum mechanical (QM) MD (Car and Parrinello, 1985; Field, 1991; Hartke and Carter, 1992; Zhao, Cramer et al., 1993), as well as coupled potential MD (Singh and Kollman, 1986; Field, Bash et al., 1990; Gao, 1992; Stanton, Hartsough et al., 1995) calculations have become computationally practical. Typically, in such calculations a system is modeled either fully or in part by a QM Hamiltonian, within the Born-Oppenheimer approximation, and is propagated classically through time using Newtonian mechanics. Coupled potentials (CP), in addition to the QM portion of the system, also include atoms which are modeled using a classical force field. The division of a large system into QM and MM portions allows for a part of the system to be modeled using the more accurate QM potential function while still including environmental effects from the MM region of the system. Typically, the QM calculation encompasses a region of particular interest such as the solute within a solvent system or the active site within an enzyme. Ideally, the entire system would be modeled using quantum mechanics, however due to the scaling of the computational expense of QM calculations with the size of the system (between N^3 and N^7, where N is the number of basis functions) fully QM representations of large systems remain very expensive to carry out.

With the increased use of quantum mechanics for MD simulations, a natural extension is their use within FEP and PMF calculations. Unfortunately, though, the extrapolation of the FEP and PMF methods used with classical force field calculations to QM potentials is not straightforward. In a classical FEP calculation the relative free energy difference between two states is calculated by slowly perturbing the parameters appropriate for one state into those of the other. For example, if a ketone oxygen were being perturbed into an hydroxyl oxygen, the bond, angle, torsional, and van der Waals (vdW) parameters as well as the atomic charges, would be coupled to a perturbation parameter λ such that $\lambda = 1$ corresponds to the ketone and $\lambda = 0$ the alcohol. Two approaches are available for the calculation of the intermediate states in classical FEP simulations. These are referred to as the single and dual topology methods. In the dual topology method the forces and energy of the fractional lambda state are generated by weighted averages of these quantities obtained from independent calculations at $\lambda = 1$ and $\lambda = 0$. The single topology method, alternatively, uses weighted averages of the parameter values to calculate forces and energy. A study contrasting the single and dual topology methods showed the single topology method to converge more rapidly than the dual in many cases (Pearlman, 1994).

If single topology methods are to be used within QM based simulations techniques must be developed which allow for the calculation of states which correspond to fractional values of the perturbation parameter. Four possible system perturbations can be defined within QM-FEP calculations: 1) changing an atom type; 2) changing the number of atoms; 3) changing the number of electrons; and 4) changing the number of atomic orbital basis functions. Perturbations between systems can, and usually do, incorporate many of these changes simultaneously. Recently, within our lab we have developed methods allowing for each of these four types of perturbations within the single topology formalism (Stanton, Dixon et al., 1995; Stanton, Little et al., 1995).

In the remainder of this paper we will outline both single and dual topology QM-FEP methods. This review will mostly focus on single-topology approaches since our efforts have mostly involved this approach. Nonetheless, we will briefly describe the dual-topology approach as well as an alternative method proposed by Warshel and Wesolowski. (Wesolowski and Warshel, 1994). In the latter scheme the free energy is determined by changing the QM starting and end points into their MM equivalent. Then by calculating the corresponding MM FEP they are able to complete a free energy cycle and obtain the relative free energy difference between the two QM systems. In contrast to this method the single and dual topology schemes presented here involve a direct QM conversion and requires no MM intermediate.

2. Dual Topology QM-FEP Methods

Of the three QM-FEP methods which will be outlined here the QM implementation of the dual topology method is the most similar to the techniques which are used with MM systems. However, as opposed to a dual-topology classical force field calculation in which the interactions are pairwise additive such that only a limited subset of terms needs to be recalculated, a QM method requires two full SCF calculations. This method is readily understood by considering basic FEP theory. In the equation below we give the basic master equations for the FEP method.

$$G_B - G_A = \Delta G = \sum_{\lambda=0}^{1} -RT \ln \left\langle e^{-\Delta H(\lambda)/RT} \right\rangle_\lambda \qquad (1)$$

$$\Delta H(\lambda) = H_{\lambda \pm \Delta\lambda} - H_\lambda \qquad (2)$$

$$H_\lambda = \lambda H_B - (1-\lambda) H_A \qquad (3)$$

In these equations H_A (G_A) and H_B (G_B) are the Hamiltonian (Gibbs free energy) for state A and state B of the system, R is the gas-constant, T is the absolute temperature and λ is the parameter that is used to couple states A and B together by taking on values ranging from 0 to 1. In the dual topology approach the Hamiltonians (*i.e.*, the total energy of the system interacting with its environment) for states A and B are explicitly evaluated at λ=0 and λ=1. To then determine the value at, for example, λ=0.5 equation 3 is applied. Then $\Delta H(\lambda)$ is evaluated, such that $\Delta\lambda$ is a small increment, whose size depends on the method being used (*i.e.*, window *versus* slow growth FEP methods (Kollman and Merz, 1990; Mitchell and McCammon, 1991; Kollman, 1993) in the FEP calculation. This is a straightforward procedure which does not involve calculation of the Hamiltonian of intermediate states as is done in the single topology methods.

In their application of the QM-FEP dual topology methodology Bash *et. al.* (Bash, Fields et al., 1987) applied it to the S_N2 reaction of the chloride ion with chloromethane. One state within their calculation was chloride and chloromethane separated by 6Å while the second state used was the optimized symmetric gas-phase transition state. Their calculations were done with a coupled potentials using AM1 (Dewar, Zoebisch et al., 1985) or MNDO (Dewar and Thiel, 1977; Dewar and Thiel, 1977) for the QM part and a modified version of CHARMM (Brooks, Bruccoleri et al., 1983) for the classical portion. The simulations were done within an 11Å sphere of water with a stochastic boundary condition. The results of their calculations gave a barrier height of 27.0-27.7 kcal/mole using the AM1 Hamiltonian (depending on the van der Waals parameters used on the QM atoms) and 30.6 kcal/mole if the MNDO Hamiltonian was used. The values correlate well with the experimental value of ~26 kcal/mole. The agreement with experiment is a very positive sign for the utility of this method whose other benefit is its conceptual and operational simplicity. It is important, though, to keep in mind some of the methodological drawbacks of the dual topology approach. These include the large memory demands necessary for keeping all of the variables needed for the two separate QM calculations and the slower convergence rate of dual *versus* single topology methods (Pearlman, 1994).

3. Single Topology QM-FEP Methods

In the next four subsections we will discuss the strategies we have employed to use the single topology method to carry out QM-FEP studies that involve changing the identity of atoms, the number of electrons, atoms and orbitals within a system. All of the derivations and tests shown here are done within the semiempirical PM3 Hamiltonian (Stewart, 1989; Stewart, 1989), although they could easily be extended for use with other QM potential functions.

3.1 VARIATION OF THE NUMBER OF ELECTRONS

Here we formulate a fractional electron method (FEM) which allows for non-integer numbers of electrons in a QM system. The approach relies on the use of a pseudo-closed-shell expression for the electronic energy, where fractional occupation numbers for the MO's are assumed at the outset. If M is the number of atomic orbitals (AO), n the total number of electrons in the system and n_k the number of electrons in the kth molecular orbital (MO) then it is possible to define the electronic energy of a system using the pseudo-closed shell formula:

$$E_{elec} = \sum_{\mu=1}^{M}\sum_{v=1}^{M}\left(H_{\mu v} + G_{\mu v}\right)P_{\mu v} \tag{4}$$

where,

$$H_{\mu v} = \int \chi_{\mu}^{\dagger}(1)\left[-\frac{1}{2}\nabla_1^2 - \sum_{A} V_A(r_1)\right]\chi_v(1)\delta\tau_1 \tag{5}$$

$$G_{\mu v} = \sum_{\lambda=1}^{M}\sum_{\sigma=1}^{M}\left[(\mu v|\lambda\sigma) - \frac{1}{2}(\mu\sigma|\lambda v)\right]P_{\lambda\sigma} \tag{6}$$

$$P_{\mu v} = \sum_{k=1}^{M} n_k \mathbf{c}_{\mu k}^{\dagger}\mathbf{c}_{vk} \tag{7}$$

$$(\mu v|\lambda\sigma) = \int\int \chi_{\mu}^{\dagger}(1)\chi_v(1)\frac{1}{r_{12}}\chi_{\lambda}^{\dagger}(2)\chi_{\sigma}(2)d\tau_1 d\tau_2 \tag{8}$$

$H_{\mu v}$, $G_{\mu v}$ and $P_{\mu v}$ are the one-electron matrix, two-electron matrix and the density mtrix, respectively. Note that in this derivation the number of electrons (n_k) in a MO may take on a non-integer value ($0 \le n_k \le 2$). While not representative of a physical system non-integer electrons have been used previously in, for example, the so-called half-electron method (Dewar, Hashmall et al., 1968).

Energy gradients and thus atomic forces can be computed analytically only if the electronic energy is variationally optimized with respect to the MO's. Further, the energy expression (4) implicitly assumes that the MO's are orthonormal. Thus we require,

$$\frac{\delta \mathrm{E}}{\delta \mathbf{c}_k} = 0 \tag{9}$$

$$\mathbf{c}_i^{\dagger}\mathbf{c}_j = \delta_{ij} \tag{10}$$

$$\mathbf{c}_i^{\dagger}S\mathbf{c}_j = \delta_{ij} \tag{11}$$

where $\mathbf{S}$ is the overlap matrix and C_i and C_j are the LCAO expansion coefficients for the ith and jth molecular orbital. Equation 11 is the standard *ab initio* Hartree-Fock formulation, while equation 10 is employed with semiempirical treatments. To keep the discussion general we will focus on the Hartree-Fock procedure and not semiempirical methods. Extension to semiempirical approaches is straightforward.

The quantity which must be minimized is the Lagrangian L,

$$L = E_{elec} - 2\sum_{i}^{occ}\sum_{j}^{occ}\varepsilon_{ij}\left(\mathbf{c}_i^{\dagger}S\mathbf{c}_j - \delta_{ij}\right) \tag{12}$$

where ε_{ij} are Lagrange multipliers. The minimization is most conveniently carried out with respect to $\mathbf{c}_k{}^{\dagger}$ while treating $\mathbf{c}_k$ as formally independent. Thus we have,

$$0 = \frac{\delta L}{\delta \mathbf{c}_k^{\dagger}} = \frac{\delta}{\delta \mathbf{c}_k^{\dagger}}\left\{\sum_{\mu=1}^{M}\sum_{\nu=1}^{M}\left[H_{\mu\nu} + G_{\mu\nu}\right]P_{\mu\nu} - 2\sum_{i}^{occ}\sum_{j}^{occ}\varepsilon_{ij}\left(\mathbf{c}_k^{\dagger}S\mathbf{c}_j - \delta_{ij}\right)\right\} \tag{13}$$

Formal differentiation, followed by a series of algebraic manipulations yields,

$$0 = \sum_{\mu=1}^{M}\sum_{\nu=1}^{M}\left(H_{\mu\nu} + G_{\mu\nu}\right)\frac{\delta P_{\mu\nu}}{\delta \mathbf{c}_k^{\dagger}} + \sum_{\lambda=1}^{M}\sum_{\sigma=1}^{M}\frac{\delta P_{\lambda\sigma}}{\delta \mathbf{c}_k^{\dagger}}\left[G_{\sigma\lambda}\right]^{\dagger} - 2\sum_{j}^{occ}\varepsilon_{kj}S\mathbf{c}_j \tag{14}$$

Using the Hermitian nature of G the subscripts of G can be interchanged and the expression can be rewritten as,

$$0 = \sum_{\mu=1}^{M}\sum_{\nu=1}^{M}\left(H_{\mu\nu} + G_{\mu\nu}\right)\frac{\delta P_{\mu\nu}}{\delta \mathbf{c}_k^{\dagger}} + \sum_{\lambda=1}^{M}\sum_{\sigma=1}^{M}\frac{\delta P_{\lambda\sigma}}{\delta \mathbf{c}_k^{\dagger}}G_{\lambda\sigma} - 2\sum_{j}^{occ}\varepsilon_{kj}S\mathbf{c}_j \tag{15}$$

Rearranging terms gives,

$$0 = \sum_{\mu=1}^{M}\sum_{\nu=1}^{M}\left(H_{\mu\nu} + 2G_{\mu\nu}\right)\frac{\delta P_{\mu\nu}}{\delta \mathbf{c}_k^{\dagger}} - 2\sum_{j}^{occ}\varepsilon_{kj}S\mathbf{c}_j \tag{16}$$

$$0 = \sum_{\mu=1}^{M}\sum_{\nu=1}^{M}\left(H_{\mu\nu} + 2G_{\mu\nu}\right)n_k c_{\nu k}\mathbf{e}_{\mu} - 2\sum_{j}^{occ}\varepsilon_{kj}S\mathbf{c}_j \tag{17}$$

$$0 = n_k(\mathbf{H} + 2\mathbf{G})\mathbf{c}_{\mathbf{k}} - 2\sum_{j}^{occ}\varepsilon_{kj}\mathbf{S}\mathbf{c}_j \tag{18}$$

And finally,

$$n_k \mathbf{F}\mathbf{c}_\mathbf{k} = 2\sum_j^{occ} \varepsilon_{kj} S \mathbf{c}_j \tag{19}$$

where the Fock matrix **F** is defined as (**H**+2**G**). For a closed-shell system, the familiar equation

$$\mathbf{F}\mathbf{c}_\mathbf{k} = \sum_j^{occ} \varepsilon_{kj} \mathbf{S}\mathbf{c}_j \tag{20}$$

would result because n_k=2 for all occupied MO's. In the case of the fractional electron method, however, at least one occupation number will be less than 2 and equation 20 is retained.

At this point it is customary to define a unitary transformation **U** of the occupied MO's that would diagonalize the matrix of Lagrange multipliers ε_{kj}, so that the problem to solve has the general form,

$$\mathbf{F}\mathbf{c}_\mathbf{k} = \varepsilon_k \mathbf{S}\mathbf{c}_\mathbf{k} \tag{21}$$

This is easily justified in the case of a closed-shell system because the wavefunction and all the molecular properties calculated from it are unaffected by a unitary transformation of the occupied MO's. Unfortunately, there may be no actual wavefunction that corresponds to the fractional electron system, and the energy will be affected by any unitary transformation that mixes fully occupied MO's with partially occupied MO's. So the unitary transformation argument cannot be used to convert equation 20 to equation 21. However, as long as equation 20 is satisfied for some set of Lagrange multipliers that preserve the orthonormality of the MO's, the corresponding vectors c_k will be nontrivial solutions that coincide with an extremum of E_{elec}. This is all that is required in the fractional electron method, thus a diagonal matrix of Lagrange multipliers may be assumed, and the MO's are solutions of:

$$n_k \mathbf{F}\mathbf{c}_\mathbf{k} = 2\varepsilon_k \mathbf{S}\mathbf{c}_\mathbf{k} \tag{22}$$

Note that equation 20 can be related to equation 19 by defining an F':

$$\mathbf{F}'_k \mathbf{c}_k = \varepsilon'_k \mathbf{S}\mathbf{c}_k \tag{23}$$

$$\mathbf{F}'_k = \frac{n_k}{2}\mathbf{F} \tag{24}$$

The eigenvalues of equation 23 are related to those of equation 22 by,

$$\varepsilon_k' = \frac{n_k}{2}\varepsilon_k \tag{25}$$

Thus by making n_k a function of λ we can alter the number of electrons in a system. The states corresponding to fractional λ values will be not be real systems, however, since free energy is a state function all that is required is the initial and final states be real. This is satisfied in this approach.

3.2 VARIATION IN THE NUMBER OF ATOMIC ORBITALS

Changing the number of atomic orbitals can be done by scaling the Fock matrix elements corresponding to these orbitals with the perturbation parameter λ. This requires that the perturbations be done in the direction in which the number of atomic orbitals decreases. As λ approaches zero, the energy levels of the disappearing atomic orbitals becomes very high so that they no longer contribute to the occupied molecular orbitals, and thus do not affect the energy of the system. This simple method allows the deletion of atomic orbitals without the generation of a new independent Fock matrix as would be required for a dual topology QM approach.

3.3 CHANGING THE IDENTITY OF AN ATOM

If the number of valence electrons and the locations of the atomic orbitals on two molecular species are identical, the energies for these two systems will exhibit the same mathematical dependence on the atomic parameters, independent of the identity of the element. A brief derivation of the PM3 approximation (Stewart, 1989; Stewart, 1989; Stewart, 1991), which is largely based on the AM1 (Dewar, Zoebisch et al., 1985) and MNDO (Dewar and Thiel, 1977) methods, is given here to illustrate the dependency of the parameters during a QM-FEP simulation. The total heat of formation of a molecule, within PM3, is given as a combination of electronic and nuclear repulsion energies E_{elect} and E_{nuc} along with the experimental heat of formation of the atoms, and the calculated electronic energy for the gaseous atom E_{el}^A.

$$\Delta H_f = E_{elect} + E_{nuc} - \sum_A E_{el}^A + \sum_A \Delta H_f^A \tag{26}$$

E_{el}^A is taken to be a function of (1) the ground state atomic orbital (AO) population of ν, (given by the density matrix element $P_{\nu\nu}$), (2) the one-electron energies for AO ν of the ion resulting from removal of the valence electrons, $U_{\mu\nu}$, and (3) the two-electron one-center integrals, $\langle\nu\nu|\mu\mu\rangle$ and $\langle\nu\mu|\nu\mu\rangle$.

$$E_{el}^A = f\left(P_{\nu\nu}, U_{\nu\nu}, (\nu\nu|\mu\mu), (\nu\mu|\nu\mu)\right) \tag{27}$$

The two electron one center integrals are given as <ss|ss>=G_{ss}, <ss|pp>=G_{sp}, <pp|pp>=G_{pp}, <pp|p'p'>=G_{p2} and <sp|sp>=H_{sp}, while U_{vv} is taken to be U_{ss} or U_{pp}. For G_{p2} the p and p' indices represent different Cartesian p orbitals on the same center. The nuclear energy is expressed as the summation over the individual nuclear-nuclear interactions for atoms i and j.

$$E_{nuc} = \sum_{i<j} E_n(i,j) \tag{28}$$

Within PM3 $E_n(i,j)$ is given as:

$$E_n(i,j) = Z_i Z_j \left(s_i s_i | s_j s_j\right) \bullet \left(1 + e^{(-\alpha_i R_{ij})} + e^{(-\alpha_j R_{ij})}\right) + \frac{Z_i Z_j}{R_{ij}} \bullet \left(\sum_k a_{ki} \exp\left[-b_{ki}(R_{ij} - c_{ki})^2\right] + \sum_k a_{kj} \exp\left[-b_{kj}(R_{ij} - c_{kj})^2\right]\right) \tag{29}$$

The simple exponential forms in equation 29 were used in the original MNDO formulation (Dewar and Thiel, 1977; Dewar and Thiel, 1977), while the Gaussian terms were added within AM1 (Dewar, Zoebisch et al., 1985) and PM3 (Stewart, 1989; Stewart, 1989; Stewart, 1991) to facilitate the representation of hydrogen bonded systems by reducing the long-range repulsion of the core-core term used in MNDO. The adjustable semiempirical parameters associated with the exponential and Gaussian terms in equation 29 are α_i, a_{ki}, b_{ki} and c_{ki}, while Z_i is the number of valence electrons on atom i and <$s_i s_i | s_j s_j$> is a two center two electron repulsion integral.

Recasting equation 4 we can express the total energy of a system as

$$E_{elect} = \frac{1}{2} \sum_{\mu=1} \sum_{\nu=1} P_{\mu\nu}\left(H_{\mu\nu} + F_{\mu\nu}\right) \tag{30}$$

Where $P_{\mu\nu}$ is the density matrix, $H_{\mu\nu}$ is the one-electron matrix and $F_{\mu\nu}$ is the Fock matrix. The diagonal Fock matrix elements are given by,

$$\mathbf{F}_{\mu\mu} = U_{\mu\mu} + \sum_{B \neq A} V_{\mu\mu,B} + \sum_{\nu}^{A} P_{\nu\nu}\left[(\mu\mu|\nu\nu) - \frac{1}{2}(\mu\nu|\mu\nu)\right] + \sum_{B \neq A} \sum_{\lambda}^{B} \sum_{\sigma}^{B} P_{\lambda\sigma}(\mu\mu|\lambda\sigma) \tag{31}$$

where μ is centered on atom A. If μ and ν are both on atom A the off diagonal elements are,

$$\mathbf{F}_{\mu\nu} = \sum_{B\neq A} V_{\mu\nu,B} + \frac{1}{2}P_{\mu\nu}\left[3(\mu\nu|\mu\nu) - (\mu\mu|\nu\nu)\right] + \sum_{B\neq A}\sum_{\lambda}^{B}\sum_{\sigma}^{B} P_{\lambda\sigma}(\mu\nu|\lambda\sigma) \quad (32)$$

While if μ is on A and ν is on B the off diagonal elements take the form,

$$\mathbf{F}_{\mu\nu} = \frac{1}{2}\left(\beta_\mu + \beta_\nu\right)S_{\mu\nu} - \frac{1}{2}\sum_{\lambda}^{A}\sum_{\sigma}^{B}(\mu\lambda|\nu\sigma) \quad (33)$$

In equations 31 and 32 β_ν and β_μ are semiempirical parameters, $<\mu\nu|\lambda\sigma>$ represents the two electron two center repulsion integrals and $S_{\mu\nu}$ is the overlap of Slater type orbitals of the form,

$$\phi_\mu = N r^{n-1} e^{-r\xi} Y_l^m(\theta,\phi) \quad (34)$$

Here ξ is an AO exponent which can be labeled as ξ_s or ξ_p, depending on whether it corresponds to a s-type or p-type AO.

In PM3 18 parameters are defined for each heavy atom (*i.e.* C, O, N, *etc.*) while hydrogen has only 11 because the p AO parameters are absent. The parameters for the heavy atoms are U_{ss}, U_{pp}, β_s, β_p, ξ_s, ξ_p, α_i, G_{ss}, G_{sp}, G_{pp}, G_{p2}, H_{sp}, a_{1i}, b_{1i}, c_{1i}, a_{2i}, b_{2i}, and c_{2i}. If the parameters are linearly scaled from the values appropriate for one element to those of another an interconversion of element types can be achieved (*i.e.* $H(\lambda) = \lambda H_A + (1-\lambda)H_B$, where H_A and H_B are Hamiltonians with parameter sets appropriate for system A and B, respectively).

One additional complication does arise if overlap integrals are required for an atom that is undergoing a change in principal quantum number (e.g., Zn^{+2} -> Cd^{+2}). Since the Slater orbitals depend on the principal quantum number the overlap integral must reflect the change in this parameter. This may be done rather abruptly using integer steps in discrete windows of the simulation, or it can be done gradually by calculating the overlap integral for each of the principal quantum numbers and then assuming a linear interpolation between the two values. An alternative method which would alleviate this problem altogether would be to employ a numerical quadrature scheme to estimate overlap integrals between pseudo-Slater-type orbitals with non-integer quantum numbers. The linear interpolation is used here, however, because it is straightforward to apply within the existing framework of MOPAC.

3.4 CHANGING THE NUMBER OF ATOMS

Changing the number of atoms in a system is a special case of changing atomic identities. Atoms being deleted are perturbed into dummy particles for which all parameter values are zero. Typically, this technique would be combined with one or more of the methods described above as the number of orbitals and electrons is quite likely to also change when an atom is deleted.

Except for numerical instabilities, the non-physical nature of states arising from such a treatment is not a concern because the free energy which is being calculated in the simulation is a state function and, hence, it is independent of the path between the initial and final states. As long as the initial and final states remain are real the path between them will have no effect on the overall result.

3.5 EXAMPLES OF SINGLE TOPOLOGY QM-FEP SIMULATIONS

The simulations in this study were done using the semiempirical MO method PM3(Stewart, 1989; Stewart, 1989; Stewart, 1991) within the context of the QM/MM coupled potential recently implemented within our lab.(Hartsough and Merz, 1995) A modified version of MOPAC 5.0 (Besler, Merz et al., 1990; Merz and Besler, 1990), was used for the QM calculations and the driver MD program was a customized version of AMBER (Pearlman, Case et al., 1991). The PM3 Hamiltonian was selected because of its superior correlation with experimental heats of formation for the compounds studied herein (Stewart, 1989). From a methodological perspective, AM1 (Dewar, Zoebisch et al., 1985), or MNDO (Dewar and Thiel, 1977), or any other LCAO based method could have been used.

The gas phase fractional electron method simulations were done with 0.5 ps of initial equilibration and 9.0 ps for the perturbation using a 0.1 fs time step. A very short timestep was employed to ensure an accurate integration of the equations of motions during the perturbation from one QM state to another. The simulations were done using the slow growth (Kollman, 1993) methodology in which the perturbation parameter lambda (λ) controlled the number of electrons present in the system. Perturbations between element types and to dummy atoms were controlled through the direct scaling of the semiempirical parameters (Stanton, Little et al., 1995). Each system was tightly coupled to a temperature bath which was maintained at 300K using the Berendsen method (Berendsen, Potsma et al., 1984). In addition, a holonomic constraint technique was employed to help maintain the geometry of the compounds as atoms were deleted (van Gunsteren and Berendsen, 1977). Bonds which changed lengths between the starting state (λ=0) and the final state (λ=1) were coupled to λ and constrained to change smoothly between the appropriate gas-phase values during the simulation.

In the calculation of the relative solvation free energies the classical solute molecules were placed within an initial solvent box of dimension 21 x 21 x 21Å with approximately 275 TIP3P (Jorgensen, Chandrasekhar et al., 1983) water molecules. The simulations used constant pressure (1 atm.) and temperature (300K) conditions (Berendsen, Potsma et al., 1984). The time step used for the two systems studied here were different due to the presence of high frequency motions within the ammonium ion and methane which are not present in chloride or fluoride. A time step of 1.5 fs seconds was used in the perturbation of chloride to fluoride while a time step of 0.5 fs was used for the perturbation of ammonium ion to methane. In both cases SHAKE was used to constrain the bond lengths of the solvent molecules at their equilibrium values (van Gunsteren and Berendsen, 1977). The resulting systems where equilibrated for 30 ps and free energy data was collected for an additional 60 ps using double wide sampling. To examine the hysteresis of the FEP calculations a second simulation was conducted using the end point of the initial 30 ps of equilibration to begin a second 30 ps of equilibration and then an additional 60 ps of double wide sampling. During the sampling phase for the chloride to fluoride ion interconversion eleven simulations were used to obtain ten free energy windows to complete the perturbation from λ=0 to λ=1. For methane to the ammonium ion twenty-one windows were used to affect the interconversion.

Another aspect of the calculations done in solvent is the treatment of long range interactions, which is qualitatively handled through the use of a Born correction (Born, 1920; Straatsma and Berendsen, 1988). It is important to note that while the Born corrections are equivalent for two identically charged atoms (which are taken to be point charges), this will not be the case when the charge of the system is changed along with the atomic identity. Where this occurs the Born correction must be added to our calculated free energies. In the case of a point charge using a 9.0Å cutoff the Born correction is 18.2 kcal/mole. The value of the Born correction for a non-point ion is more difficult to evaluate. For example, when considering the ammonium ion the actual sphere in which solvent is "seen" by the solute is ~10Å in radius because of the 1Å N-H bond. For this size sphere a Born correction of 16.4 kcal/mole is calculated. However, the correct value to use is not obvious as pointed out in a similar study by Boudon and Wipff(Boudon and Wipff, 1991) where they used a point charge-derived Born correction for the ammonium ion.

The thermodynamic cycle used for the calculation of the relative solvation free energies is given in Figure 1. Since the Gibb's free energy is a state function this cycle gives,

$$\Delta\Delta G_{gas}^{x\rightarrow y} + \Delta G_{solv}^{y} - \Delta\Delta G_{solv}^{x\rightarrow y} - \Delta G_{solv}^{x} = 0 \qquad (35)$$

Upon rearrangement $\Delta\Delta G_{solv}^{x\rightarrow y}$ can be calculated as:

$$\Delta\Delta G_{solv}^{x\rightarrow y} = \Delta G_{solv}^{y} - \Delta G_{solv}^{x} + \Delta\Delta G_{gas}^{x\rightarrow y} \qquad (36)$$

As mentioned above when the charge of the system changes a Born correction must be added to $\Delta\Delta G_{solv}^{x\rightarrow y}$ to account for long range interactions. Consideration of this gives:

$$\Delta\Delta G_{solv}^{x\rightarrow y} = \Delta G_{solv}^{y} - \Delta G_{solv}^{x} + \Delta\Delta G_{gas}^{x\rightarrow y} - \Delta\Delta G_{born} \qquad (37)$$

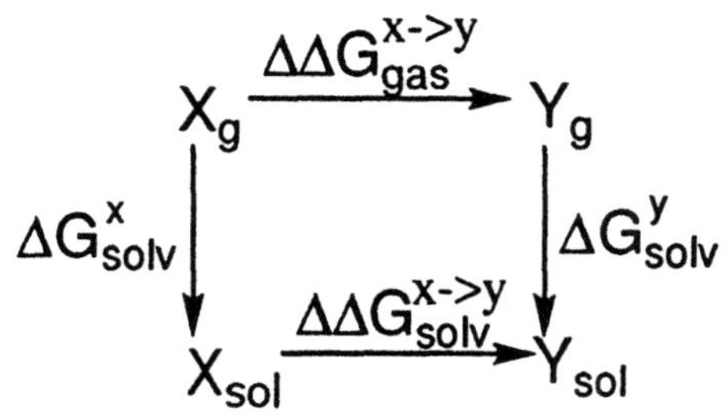

Figure 1: Thermodynamic Cycle for Single and Dual Topology Methods.

The results from the FEP simulations where the atom types are altered are summarized in Table 1. The first involved the perturbation of a chloride ion into a fluoride ion when solvated by classical water. The $\Delta\Delta G_{solv}^{Cl^- \to F^-}$ was calculated to be -4.8±0.6 kcal/mole through the perturbation of the QM atoms within our methodology. Equation 35 can be used to compare this calculated number with an estimated experimental value. The free energy of solvation for chloride has been reported experimentally to be -75 (Pearson, 1986), -75.8 (Friedman and Krishnan, 1973) and -82.9 kcal/mole (Marcus, 1985) while the value for fluoride has been reported to be -103.8 (Friedman and Krishnan, 1973), -105(Pearson, 1986) and -112.8 kcal/mole (Marcus, 1985). The value of $\Delta\Delta G_{gas}^{Cl^- \to F^-}$ is the energetic difference in the gas phase between the two atomic species or simply the difference between their heats of formation (*i.e.*, $\Delta H_f^{Cl^-}$ = -51.2 kcal/mole, $\Delta H_f^{F^-}$ = -31.2 kcal/mole and $\Delta\Delta G_{gas}^{Cl^- \to F^-}$ = 20.0 kcal/mol (within PM3). This gives a $\Delta\Delta G_{solv}^{Cl^- \to F^-}$ (from equation 35) of between -8.0 and -10.0 kcal/mole (average value of -9.3 kcal/mol) depending on the set of experimental solvation free energies chosen. In this case we have only evaluated $\Delta\Delta G_{solv}^{Cl^- \to F^-}$ using experimental solvation free energies obtained from the same literature source (Friedman and Krishnan, 1973; Marcus, 1985; Pearson, 1986). If the solvation free energy values obtained from the various sources (using the same standard state) were mixed a larger range of values for $\Delta\Delta G_{solv}^{Cl^- \to F^-}$ would result. The experimental value for $\Delta\Delta G_{solv}^{Cl^- \to F^-}$ is in reasonable accord with our calculated value of -4.8±0.6 kcal/mole. The difference between the estimated average experimental value and the calculated $\Delta\Delta G_{solv}^{Cl^- \to F^-}$ can be improved by deriving Lennard-Jones parameters that are suitable for coupled potential simulations of chloride and fluoride (Stanton, Hartsough et al., 1993).

quantity	Cl^- to F^-	NH_4^+ to CH_4
$\Delta\Delta G_{solv}^{x->y}(calc)$	-4.8±0.6	-94.6±1.5
$\Delta\Delta G_{solv}^{x->y}(\exp)$ b	-8.0- -10.0	-102.8- -111
$\Delta\Delta G_{gas}^{x->y}$	20.0	-166.4

a) All energies are given in kcal/mole.
b) Estimated from equations 36 and 37.

Table 1: Free Energies[a]

The second system studied involved the perturbation of the ammonium ion into methane in aqueous solution and the results for these simulations are summarized in Table 1. The number of valence electrons in this perturbation is maintained even though the overall charge changes. The calculated $\Delta\Delta G_{solv}^{NH_4^+\rightarrow CH_4}$ was -94.6±1.5 kcal/mole. For comparison an estimate of the experimental value for $\Delta\Delta G_{solv}^{NH_4^+\rightarrow CH_4}$ can be determined using equation 37. In this equation $\Delta\Delta G_{gas}^{NH_4^+\rightarrow CH_4}$ can again be estimated from the difference of the heats of formation for the two compounds in the gas phase ($\Delta H_f^{NH_4^+}$=153.4 kcal/mole and $\Delta H_f^{CH_4}$=-13.0 kcal/mole) as evaluated by PM3 (Stewart, 1989). This results in a $\Delta\Delta G_{gas}^{NH_4^+\rightarrow CH_4}$ value of -166.4 kcal/mole, which was additionally verified through a gas phase free energy simulation as a check of our methodology. The experimental $\Delta G_{solv}^{CH_4}$ is 1.93 (Ben-Naim and Marcus, 1984) kcal/mole while $\Delta G_{solv}^{NH_4^+}$ has been reported to be -69.8 (Marcus, 1985), -74.8 (Ford and Scribner, 1983), and -78 kcal/mole (Aue, Webb et al., 1976). Using equation 37 and the values for $\Delta G_{solv}^{CH_4}$, $\Delta G_{solv}^{NH_4^+}$, $\Delta\Delta G_{gas}^{NH_4^+\rightarrow CH_4}$ and a Born correction of 16.4 kcal/mole we estimate that $\Delta\Delta G_{solv}^{NH_4^+\rightarrow CH_4}$ is between -102.8 to -111 kcal/mole. Our calculated value (-94.6±1.5 kcal/mol) is several kcal/mole more positive than the estimated experimental value. However, modification of the Lennard-Jones parameters for CH_4 and NH_4^+ used in the coupled potential calculations should bring the calculated results into better agreement with the estimated experimental value.

To test the accuracy of the fractional electron method, as well as our ability to account for changes in the number of orbitals in a system, three perturbations were carried out. These were the interconversion of ethane to methane, methanol to methane and methane to fluoromethane. The results of these simulations are given in Table 2. How to compare our calculated results to that determined using other approaches is complicated. We could compare to free energies calculated using the appropriate partition functions(Hehre, Radom et al., 1986); however, since we are restraining the molecule in the FEP simulation, the vibrational, rotational and translational contributions are all affected. Given this uncertainty we have presented free energy differences using the partition function approach (Hehre, Radom et al., 1986) with all terms considered and with only the vibrational contribution considered. In Table 2 we also present the calculated gas-phase heat of formation difference between the two molecules.

	$\Delta G_{gas}^{x\rightarrow y}$ (FEP)	$\Delta G_{gas}^{x\rightarrow y}$ (partition)[b]	$\Delta G_{gas}^{x\rightarrow y}$ (partition)[c]	$\Delta\Delta H_{f}^{x\rightarrow y}$ (gas)
C_2H_6->CH_4	5.3±01	8.2	5.8	5.1
CH_3OH->CH_4	35.0±0.1	42.6	39.4	39.0
CH_4->CH_3F	-44.5±0.1	-43.4	-40.8	-40.8

a) All energies are given in kcal/mol.
b) Calculated from the vibrational frequencies and the appropriate partition functions.
c) Calculated from the vibration frequencies and the appropriate partition functions but rotational and translational contributions have been removed.

Table 2: Free Energies[a]

The free energy difference calculated from the gradual perturbation of ethane to methane in the gas phase was 5.3 kcal/mol. This is in reasonable agreement with the heat of formation difference as well as the partition function approach where the rotational and translational terms have been neglected. The agreement with the full partition function approach is not as good. In the case of the conversion of methane to fluoromethane, the FEP and partition function method were in closer agreement with the calculated free energy changes of -44.5 and -43.4 kcal/mol, respectively. However, the FEP value was ~4 kcal/mol higher than the value given by the heat of formation and partial partition function differences. The conversion of methane to methanol was the last of the test perturbations examined. In this system our FEP value was 35 kcal/mol, while the other calculated values were 4-7 kcal/mol higher. Overall, the agreement between the various types of calculated values was reasonable and it is clear that the FEP approach is giving results that are comparable to energy differences computed in other ways. Thus, if an accurate QM model is employed, the fractional electron approach should be a reliable method for estimating free energy differences between two molecules. Disagreement between the FEP and partition function energies can be associated with several factors (time-scales, use of MD constraints, etc.) but further study is needed to pinpoint problem areas.

4. Molecular Mechanical Intermediate

The QM-FEP method developed by Warshel and Wesolowski (Wesolowski and Warshel, 1994) differs from both the standard single and dual topology methods in that it uses a classical intermediate. Their methodology has distinct advantages and disadvantages over the fully QM methods discussed above. The most prominent of the benefits being its reduced computational expense. The use of classical intermediates means that a greatly diminished number of QM energy evaluations need to be done and these only at the beginning and end points of the calculation to obtain the relative free energies of the QM and classical representations of a system. Another advantage is that no states corresponding to intermediate QM structures need to be calculated. In this way all of the relatively unstable fractionally occupied structures of the single-topology method and dual-existence QM structures of the dual topology method can be avoided. One of the disadvantages of using classical intermediates is in the multiple intermediate states which appear within the calculation. If the relative free energy is not calculated accurately between any of these states then the total free energy between the two quantum mechanical structures of interest will be in error. The single and dual topology methods avoid this problem by directly interconverting between the initial and final states of the quantum mechanical system to determine the free energy.

A diagram of the free energy cycle used within the classical intermediate method is shown in Figure 2. The rational behind the technique can be understood much in the same way as that of the single-topology method. Since the free energy is a state function and path independent the path taken to move from the initial to final states being considered is unimportant as long as it is continuous. The calculated difference in free energies between two quantum mechanical systems should therefore be completely independent of the accuracy of the classical potential used as long as the free energy differences between the classical states and between the QM and classical states are calculated accurately. The difference in free energies between classical states was calculated using standard FEP techniques while the QM-classical relative free energies where calculated using the expression,

$$\exp\left(-\Delta G_{QM-CL}/kT\right) = \exp\left\langle \left(U_{QM} - U_{Cl}\right)/kT \right\rangle \qquad (38)$$

In this expression k is the Boltzman constant and T the temperature of the system in degrees Kelvin. The (U_{QM}-U_{CL}) term gives the differences between the energies of the quantum and classical systems at a particular configuration.

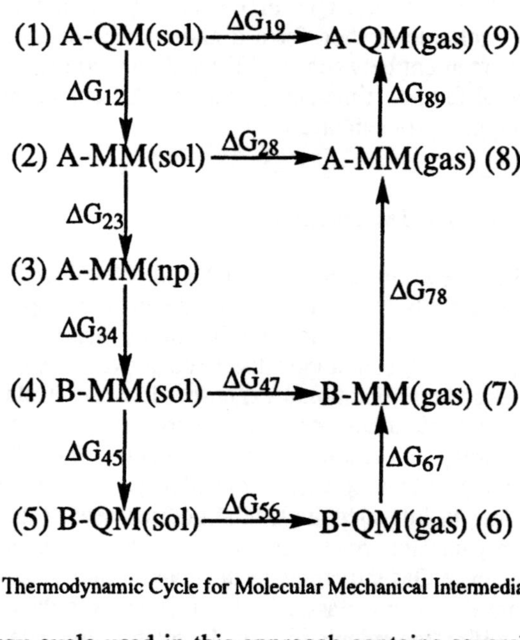

Figure 2: Thermodynamic Cycle for Molecular Mechanical Intermediate Method.

The free energy cycle used in this approach contains several intermediates (see Figure 2). State 1 is the solvated QM representation of atom A, state 2 is a solvated classical atom A, state 3 a solvated non polar representation of atom A (*i.e.*, the nuclear charges have been zeroed to allow for solvent reorganization), state 4 is a solvated classical atom B, state 5 is a solvated QM atom B, states 6 and 7 are gas phase QM and classical models of atom A, while analogously states 8 and 9 are classical and QM gas phase models of atom B, respectively. The free energy can be written as,

$$\Delta\Delta G_{sol}^{QM} = \Delta\Delta G_{sol}^{MM} + \Delta G_{12} + \Delta G_{45} \tag{39}$$

In eqn. 39 $\Delta\Delta G_{sol}^{MM}$ refers to the intermolecular contribution from the classical part of the system (*i.e.*, A-MM(sol) -> B-MM(sol)) to the solvation free energy. From consideration of the thermodynamic cycle in Figure 2 and assuming that ΔG_{67} and ΔG_{89} are zero, we can recast this equation as:

$$\Delta\Delta G_{sol}^{QM} = \Delta G_{23} + \Delta G_{34} - \Delta G_{78} + \Delta G_{12} + \Delta G_{45} \tag{40}$$

Finally, realizing that the intramolecular energy (*i.e.*, ΔG_{78}) is also contained in the solution phase simulations (*i.e.*, $\Delta G_{23} + \Delta G_{34}$) allows us to recast eqn. 40 as:

$$\Delta\Delta G_{sol}^{QM} = \Delta G_{23,sol} + \Delta G_{34,sol} + \Delta G_{12} + \Delta G_{45} \tag{41}$$

where $\Delta G_{23,sol}$ and $\Delta G_{34,sol}$ represent the intermolecular solvation interactions only. Using eqn. 41 the relative solvation free energies between QM atoms in solution can be readily determined.

As an example of their classical intermediate methodology Warshel and Wesolowski (Wesolowski and Warshel, 1994) examined the interconversion of water and methane in aqueous solution. Their simulations were carried using the frozen density functional theory (FDFT) approach (Wesolowski and Warshel, 1993) developed within their laboratory to model the QM system while the trajectories and the MM intermediates where simulated using the ENZYMIX program and force field associated with this program. The solute atom was imbedded in a 9Å sphere of explicit water molecules which was further surrounded by a grid of Langevin dipoles (Wesolowski and Warshel, 1994). The free energy of solvation for the quantum mechanical perturbation of water to methane was calculated to be -9.6kcal/mol using their methodology. This is in good correlation with the experimentally predicted value of -8.3kcal/mol. The authors also report their results as a function of the α parameter of the nonadditive Thomas-Fermi-Weizsaecker (TFW) kinetic energy functionals (Wesolowski and Warshel, 1994). Their results were found to be highly dependent on the α parameter and ranged from a $\Delta\Delta G$ (H_20->CH_4) of -12.4 with α equal to 1 and -4.1 with α equal to zero. Additional tests of the methodology would seem necessary to determine the optimal value of α.

5. Conclusions

We have discussed three methods by which quantum mechanical-free energy perturbation simulations can be carried out on molecules in the gas and condensed phases. Each of these methods has its own strengths and weaknesses which will only be fully understood with further tests. Only in the past few years has the calculation of extensive QM-MD trajectories and QM-FEP simulations become computationally feasible. It is for this reason that the field remains primarily uninvestigated and in need of significant development and testing. The next decade should bring rapid progress in QM-MD studies on large systems, and through the development of QM-FEP methodologies we will also be able to calculate free energies from condensed phase QM and QM/MM simulations.

6. Acknowledgments

We would like to thank the NIH (GM-29072) for their generous support of this research. The Pittsburgh Supercomputer Center and The Cornell Theory Center through a MetaCenter grant are acknowledged for generous allocations of computer time.

7. Literature Cited

Åqvist, J. (1990). "Ion-Water Interaction Potentials Derived from Free Energy Perturbation Simulations." J. Phys. Chem. **94**: 8021.

Åqvist, J., M. Carmen and J. Samuelsson (1994). "A New Mthod for Predicting Binding Affinity in Computer-aided Drug Design." Prot. Eng. **7**: 385-391.

Aue, D. H., H. M. Webb and M. T. Bowers (1976). "A Thermodynamic Analysis of Solvation Effects on the Basiciites of Alkyamines. An Electrostatic Analysis of Subsituent Effects." J. Am. Chem. Soc. **98**: 318-329.

Bash, P., M. Fields and M. Karplus (1987). "Free Energy Perturbation Method for Chemical Reactions in the Condensed Phase: A Dynamical Approach Based on the Combined Quantum and Molecular Mechanical Potential." J. Am. Chem. Soc. **109**(26): 8092-8094.

Ben-Naim, A. and Y. Marcus (1984). "Solvation Thermodynamics of Nonionic Solutes." J. Chem. Phys. **81**: 2016-2027.

Berendsen, H. J. C., J. P. M. Potsma, W. F. van Gunsteren, A. D. DiNola and J. R. Haak (1984). "Molecular Dynamics with Coupling to an External Bath." J. Chem. Phys. **81**: 3684-3690.

Besler, B. H., K. M. J. Merz and P. A. Kollman (1990). "Atomic Charges Derived from Semiempirical Methods." J. Comput. Chem. **11**(4): 431-439.

Born, M. (1920). Z. Phys. **1**: 45.

Boudon, S. and G. Wipff (1991). "Free Energy Calculations Involving NH_4^+ in water." J. Comput. Chem. **12**: 42.

Brooks, B. R., R. E. Bruccoleri, B. D. Olafson, D. J. States, S. Swaminathan and M. Karplus (1983). "CHARMM: A Program for Macromolecular Energy, Minimization, and Dynamics Calculations." J. Comput. Chem. **4**(2): 187-217.

Car, R. and M. Parrinello (1985). "Unified Approach for Molecular Dynamics and Density-Functional Theory." Phys. Rev. Lett. **55**: 2471.

Cieplak, P. and P. Kollman (1990). "Monte Carlo Simulation of Aqueous Solutions of Li+ and Na+ Using Many-Body Potentials. Coordination Numbers, Ion Solvation Enthalpies, and the Relative Free Energy of Solvation." J. Chem. Phys. **92**(11): 6761.

Dang, L. X. and P. A. Kollman (1990). "Free Energy of Association of the 18-Crown-6:K+ Complex in Water: A Molecular Dynamics Simulation." J. Am . Chem. Soc. **112**(15): 5716-5720.

Dewar, M. J. S., J. A. Hashmall and C. G. Venier (1968). "Ground States of Molecules. IX. Hydrocarbon Radicals and Radical Ions." J. Am. Chem. Soc. **90**: 1953-1957.

Dewar, M. J. S. and W. Thiel (1977). "Ground States of Molecules. 38. The MNDO Method, Approximations and Parameters." J. Am. Chem. Soc. **99**(15): 4899-4907.

Dewar, M. J. S. and W. Thiel (1977). "Ground States of Molecules. 39. MNDO Results for Molecules Containing Hydrogen, Carbon, Nitrogen, and Oxygen." J. Am. Chem. Soc. **99**(15): 4907-4917.

Dewar, M. J. S., E. G. Zoebisch, E. F. Healy and J. J. P. Stewart (1985). "AM1: A New General Purpose Quantum Mechanical Molecular Model." J. Am. Chem. Soc. **107**(13): 3902-3909.

Elber, R. (1990). "Calculation of the Potential of Mean Force Using Molecular Dynamics with Linear Constraints: An Application to a Conformational Transition in a Solvated Peptide." J. Chem. Phys. **93**(6): 4312-4321.

Field, M. J. (1991). "Constrained Optimization of Ab Initio and Semiemperical Hartree-Fock Wave Functions Using Direct Minimization or Simulated Annealing." J. Phys. Chem. **95**: 5104-5108.

Field, M. J., P. A. Bash and M. Karplus (1990). "A Combined Quantum Mechanical and Molecular Mechanical Potential for Molecular Dynamics Simulations." J. Comput. Chem. **11**: 700-733.

Ford, G. P. and J. D. Scribner (1983). "A Simple Method for Predicting Hydration Energies of Organic Cations Derived from Protonation or Alkylation of Neutral Oxygen and Nitrogen Bases." J. Org. Chem. **48**: 2226-2233.

Friedman, H. L. and C. V. Krishnan (1973). Thermodynamics of Ion Hydration. Water a Comprehensive Treatise. New York, Plenum Press. 1-118.

Gao, J. (1992). "Comparison of the Hybrid AM1/TIP3P and the OPLS Functions Through Monte Carlo Simulations of Acetic Acid in Water." J. Phys. Chem. **96**: 6432-6439.

Grootenhuis, P. D. J. and P. A. Kollman (1989). "Crown Ether-Neutral Molecule Interactions Studied by Molecular Mechanics, Normal Mode Analysis, and Free Energy Perturbation Calculations. Near Quantitative Agreement between Theory and Experimental Binding Free Energies." J. Am. Chem. Soc. **111**: 4046-4051.

Hartke, B. and E. A. Carter (1992). "Ab Initio Molecular Dynamics with Correlated Molecular Wave Functions: Generalized Valence Bond Molecular Dynamics and Simulated Annealilng." J. Chem. Phys. **97**(9): 6569-6578.

Hartsough, D. and K. M. J. Merz (1995). "Potential of Mean Force Calculation on the S_N1 Fragmentation of tert-Butyl Chloride." J. Phys. Chem. **99**: 384-390.

Hehre, W. J., L. Radom, P. R. van Schleyer and J. A. Pople (1986). Ab Initio Molecular Orbital Theory. New York, John Wiley.

Jorgensen, W. L. (1989). "Free Energy Calculations: A Breakthrough for Modeling Organic Chemistry in Solution." Acc. Chem. Res. **22**: 184-189.

Jorgensen, W. L., J. Chandrasekhar, J. Madura, R. W. Impey and M. L. Klein (1983). "Comparison of Simple Potential Functions for the Simulation of Liquid Water." J. Chem. Phys. **79**: 926.

Kollman, P. A. (1993). "Free Energy Calculations: Applications to Chemical and Biochemical Phenomena." Chem. Rev. **93**(7): 2395-2417.

Kollman, P. A. and K. M. J. Merz (1990). "Computer Modeling of the Interactions of Complex Molecules." Acc. Chem. Res. **23**: 246-252.

Lybrand, T. P., I. Ghosh and J. A. McCammon (1985). "Hydration of Chloride and Bromide Anions: Determination of Relative Free Energy by Computer Simulation." J. Am. Chem. Soc. **107**: 7793-7794.

Marcus, Y. (1985). Ion Solvation. New York, John Wiley.

Marrone, T. J. and K. M. Merz Jr. (1992). "Molecular Recogntion of Potassium Ion by the Naturally Occurring Ionophore Nonactin." J. Am. Chem. Soc. **114**: 7542.

Merz, K. M., Jr. and B. H. Besler (1990). "MOPAC 5.0 ESP." QCPE Bull. **10**: 15.

Merz, K. M., Jr. and P. A. Kollman (1989). "Free Energy Perturbation Simulation of the Inhibition of Thermolysin: Prediction of the Free Energy of Binding of a New Inhibitor." J. Am. Chem. Soc. **111**(15): 5649-5658.

Mezei, M. and D. L. Beveridge (1986). "Free Energy Simulations." Ann. NY. Acad. Sci. **482**: 1-23.

Mitchell, M. J. and J. A. McCammon (1991). "Free Energy Difference Calculations by Thermodynamic Integration: Difficulties in Obtaining a Precise Value." J. Comput. Chem. **12**(2): 271-275.

Pearlman, D. A. (1994). "Comparison of Alternative Approaches to Free Energy Calculations." J. Phys. Chem. **98**(5): 1487-1493.

Pearlman, D. A., D. A. Case, J. C. Caldwell, G. L. Seibel, U. C. Singh, P. Weiner and P. A. Kollman (1991). AMBER 4.0. University of California, San Francisco.

Pearson, R. G. (1986). "Ionization Potentials and Electron Affinities in Aqueous Solution." J. Am. Chem. Soc. **108**: 6109-6114.

Singh, U. C. and P. A. Kollman (1986). "A Combined *Ab initio* Quantum Mechanical and Molecular Mechanical Method for Carrying out Simulations on Complex Molecular Systems: Applications to the $CH_3Cl + Cl^-$ Exchange Reaction and Gas Phase Protonation of Polyethers." J. Comput. Chem. **7**(6): 718-730.

Stanton, R. V., S. L. Dixon and K. M. Merz Jr. (1995). "A General Formulation for a Quantum Free Energy Perturbation Study." J. Phys. Chem. **99**(27): 10701-10704.

Stanton, R. V., D. S. Hartsough and J. K. M. Merz (1995). "An Examination of a Density Functional/Molecular Mechanical Coupled Potential." J. Comput. Chem. **16**(1): 113-128.

Stanton, R. V., D. S. Hartsough and K. M. Merz (1993). "Calculation of Solvation Free Energies Using a Density Functional / Molecular Dynamics Coupled Potential." J. Phys. Chem. **97**(46): 11868-11870.

Stanton, R. V., L. R. Little and K. M. Merz (1995). "Quantum Free Energy Perturbation Study within a PM3/MM Coupled Potential." J. Phys. Chem. **99**(2): 483-486.

Stewart, J. J. P. (1989). "Optimization of Parameters for Semiempirical Methods I. Method." J. Comput. Chem. **10**(2): 209-220.

Stewart, J. J. P. (1989). "Optimization of Parameters for Semiempirical Methods II. Applications." J. Comput. Chem. **10**(2): 221-264.

Stewart, J. J. P. (1991). "Optimization of Parameters for Semiempirical Methods. III Extension of PM3 to Be, Mg, Zn, Ga, Ge, As, Se, Cd, In, Sn, Sb, Te, Hg, Tl, Pb and Bi." J. Comput. Chem. **12**(3): 320-341.

Straatsma, T. P. and H. J. C. Berendsen (1988). "Free Energy of Ionic Hydration: Analysis of a Thermodynamic Technique to Evaluate Free Energy Differences by Molecular Dynamics Simulations." J. Chem. Phys. **89**(9): 5876.

van Eerden, J., W. J. Briels, S. Harkema and D. Feil (1989). "Potential of Mean Force by Thermodynamic Integration: Molecular-Dynamics Simulation of Decomplexation." Chem. Phys. Lett. **164**(4): 370-376.

van Gunsteren, W. F. and H. J. C. Berendsen (1977). "Algorithm for Macromolecular Dynamics and Constraint Dynamics." Mol. Phys. **34**: 1311.

Wesolowski, T. and A. Warshel (1994). "Ab Initio Free Energy Perturbation Calculations of Solvation free Energy Using the Frozen Density Functional Approach." J. Phys. Chem **98**: 5183-5187.

Wesolowski, T. A. and A. Warshel (1993). "Frozen Density Functional Approach for *Ab Initio* Calculations of Solvated Molecules." J. Phys. Chem. **97**: 8050-8053.

Zhao, X. G., C. S. Cramer, B. Weiner and M. Frenklach (1993). "Dynamics with the AM1 Potential: Reactions on Diamond Surfaces." J. Phys. Chem. **97**: 1639-1648.

HYBRID POTENTIALS FOR MOLECULAR SYSTEMS IN THE CONDENSED PHASE

MARTIN J. FIELD
Laboratoire de Dynamique Moléculaire
Institut de Biologie Structurale — Jean-Pierre Ebel
41 Avenue des Martyrs
38027 Grenoble Cedex 1
France

1. Introduction

This chapter gives an account of hybrid quantum mechanical and molecular mechanical potentials for investigating the properties of condensed phase molecular systems. These potentials can be applied to the study of a large variety of molecular systems, whether they are organic or inorganic, crystalline or liquid and they are particularly useful for investigating phenomena such as reactions or light-activated processes where a knowledge of the electronic structure of a portion of the system is essential.

This chapter concentrates on the form of the potentials and how they can be used to calculate the energy of the system and the forces on the particles in a precise and an efficient manner. The chapter does not explicitly discuss molecular dynamics, Monte Carlo or other types of simulation method because the implementation of these methods is, in principle, independent of the way in which the energy and forces are calculated [1, 2, 3]. Also it does not describe in detail any applications of these potentials as a comprehensive and up-to-date review has recently been published by Gao [4]. Another collection of papers covering all aspects of work with hybrid potentials has also recently been published and may be of interest [5].

The plan of the chapter is as follows. Section 2 outlines why the use of hybrid potentials is important if large systems are to be studied and section 3 desribes the basic form of the effective Hamiltonian for a hybrid potential. The next four sections, sections 4 to 7, each cover a separate term in the hybrid Hamiltonian and section 8 describes the implementation of the hybrid potentials for use in actual calculations. Section 9 concludes.

G. Náray-Szabó and A. Warshel (eds.), Computational Approaches to Biochemical Reactivity, 125–151.

2. Why are Hybrid Potentials Necessary?

One of the prerequisites when performing simulations of a molecular system is to have available a way of calculating the energy of the system and the forces on its constituent atoms. In principle, this can be done by finding solutions to the time-independent Schrödinger equation within the Born-Oppenheimer approximation [6]. The latter says that, because the masses of the electrons and nuclei are so different, the motions of the two types of particle can be treated separately. Application of the approximation then implies that the time-independent Schrödinger equation may be solved for the wavefunction describing the distribution of electrons with fixed values for the nuclear positions (or, in other words, at a fixed nuclear geometry). The time-independent Schrödinger equation is [7]:

$$\hat{H}_e(\{\mathbf{r}_i\},\{\mathbf{R}_\alpha\})\;\Psi_I(\{\mathbf{r}_i\},\{\mathbf{R}_\alpha\}) = E_I(\{\mathbf{R}_\alpha\})\;\Psi_I(\{\mathbf{r}_i\},\{\mathbf{R}_\alpha\}) \qquad (1)$$

where $\{\mathbf{r}_i\}$ and $\{\mathbf{R}_\alpha\}$ are the coordinates of the electrons and the nuclei respectively and Ψ_I and E_I are the wavefunction and the energy of the Ith electronic state of the system. $\hat{H}_e$ is the non-relativistic, electrostatic Hamiltonian for the electrons, which, in atomic units, is:

$$\hat{H}_e = -\frac{1}{2}\sum_i \nabla_i^2 - \sum_{i,\alpha} \frac{Z_\alpha}{|\mathbf{R}_\alpha - \mathbf{r}_i|} + \sum_{i>j} \frac{1}{|\mathbf{r}_i - \mathbf{r}_j|} + \sum_{\alpha>\beta} \frac{Z_\alpha Z_\beta}{|\mathbf{R}_\alpha - \mathbf{R}_\beta|} \qquad (2)$$

where the subscripts i and j and α and β refer to electrons and nuclei respectively.

Every solution of equation 1 defines a different electronic state of the molecular system. Associated with each state is its energy, E_I, which, because it is a function of the nuclear coordinates, defines a many-dimensional surface called the potential energy surface (PES) for that state. Physically, it represents the effective potential energy of interaction between the nuclei. Equation 1 may be rewritten to define the state energy as an expectation value of the state wavefunction over the electronic Hamiltonian:

$$E_I(\{\mathbf{R}_\alpha\}) = \frac{\left\langle \Psi_I \middle| \hat{H}_e \middle| \Psi_I \right\rangle}{\langle \Psi_I \mid \Psi_I \rangle} \qquad (3)$$

The expectation value integrals are performed with respect to the electronic variables only and not those of the nuclei.

Accurate potential energy surfaces are essential if quantitative studies of molecular systems are to be performed. In principle, it is possible to solve equation 1 for the energy of a system to arbitrary accuracy. In practice, however, it proves difficult, if not impossible, to treat a complete system in this way quite simply because the methods that exist to solve the Schrödinger equation for large systems are too slow. Thus, while it may be possible to solve the Schrödinger equation once or a small number of times for a large system, simulation studies generally require the energy and forces for thousands of different configurations of the nuclei (typically 10^6 or more).

To circumvent this problem researchers use empirical or molecular mechanical potentials. These are potentials which have been chosen to have a simple analytic form and which have been parameterized, using experimental or quantum mechanical data, to reproduce as closely as possible the "true" potential energy surface for the system. In practice, it is difficult to find simple analytic forms and parameterizations that are valid over a large region of the surface and empirical surfaces generally have a relatively restricted range of applicability — they can only be used for the systems for which they were parameterized and they can only be used to provide information about a relatively small number of points on the surface (such as around stable, equilibrium configurations or along a specific reaction path, for example).

To summarize, methods that directly solve the Schrödinger equation (and which give rise to quantum mechanical potential surfaces) are too slow to use directly for simulations of large systems although they are of wide applicability and they can be used to study processes such as reactions. Empirical potentials are fast enough to use in simulation studies of large systems but they have a limited applicability. Thus, there are two possible options if processes, such as reactions in large systems, are to be studied using quantum mechanical potentials:

- An empirical potential is developed and parameterized for the particular process of interest and used in the simulation. Quantum mechanical calculations are used to provide data for the parameterization procedure.
- A *hybrid potential* is used in which a small number of atoms are treated with a quantum mechanical method and the remainder with an empirical potential.

Both approaches have their disadvantages and advantages. Simulations with empirical potentials are undoubtedly quicker than those with hybrid potentials. However, with empirical potentials there is always the suspicion that the construction of the potential has biased the simulation to a particular result and that some of the important behaviours of the system have been ignored.

In the remainder of this chapter it is the second alternative, that of hybrid potentials, which is discussed in detail. The union of the two types of potential, molecular mechanical (MM) and quantum mechanical (QM), is crucial to the success of hybrid potentials and will be discussed in de-

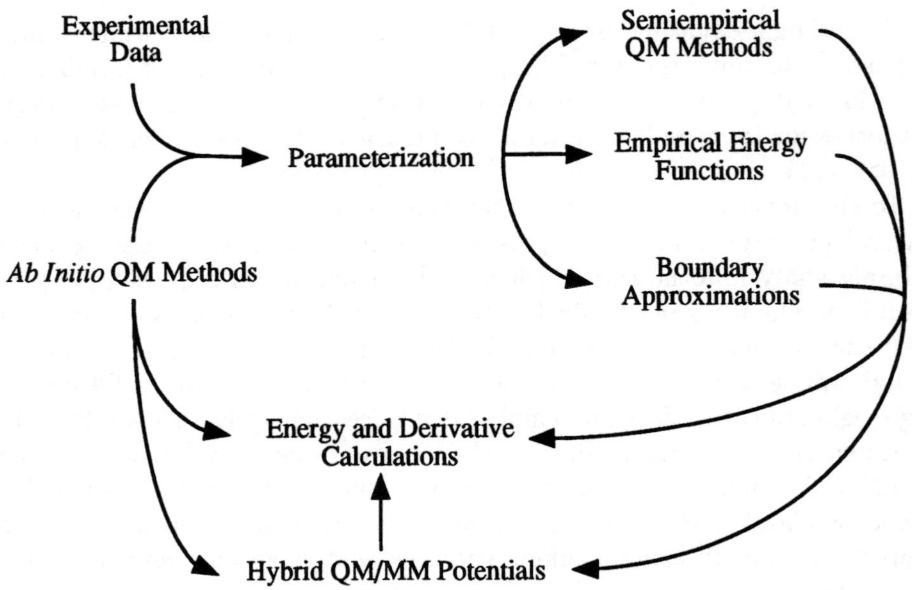

Figure 1. A diagram illustrating the relationships between the various potential methods discussed in the text.

tail in the following sections. Figure 1 illustrates in a schematic way the interrelation of all the potentials discussed in the text.

3. The Effective Hamiltonian for the Hybrid Potential

Hybrid potentials of the type discussed in this chapter have been in use for twenty five years or so. Some of the earliest examples were those designed to study conjugated organic molecules in which the π-electrons of the system were treated with a semiempirical QM method and the σ-bonding framework was described with a MM force field [8, 9]. The methods were used to study the structure and spectra of the molecules [10, 11] and photoisomerization processes [12, 13]. The first "true" combined potential in which both σ and π electrons were considered in the quantum mechanical region was also developed by Warshel, in collaboration with Levitt, which they used to study the mechanism of the enzyme, lysozyme [14]. They combined a semiempirical QM method to describe a portion of the enzyme and the substrate, and a standard MM force field to describe the rest of the atoms.

There has been a great deal of subsequent work in the area and a number of groups have actively developed and applied hybrid potentials. For example, more recently, Warshel has extensively developed his empirical valence bond/MM approach for studying reactions and electron transfer [15]. Another hybrid potential is due to Singh and Kollman [16] who combined an *ab initio* molecular orbital potential with the AMBER force field [17] and studied the $CH_3Cl + Cl^-$ S_N2 exchange reaction in solution, the protonation of polyethers in the gas phase and formamide hydrolysis [18]. Field *et al.* [19] developed a potential based on the semiempirical AM1 and MNDO methods of Dewar *et al.* [20, 21] and the CHARMM force field [22]. It has been used to perform free energy calculations of the same S_N2 reaction in solution [23], to study, using energy minimisation, the reaction mechanism of the enzyme, triose phosphate isomerase [24] and to investigate the solvation of ions and small organic molecules [25, 26, 27]. More recently, Warshel and co-workers have published details of QM/MM potentials which use both semiempirical AM1 [28] and *ab initio* potentials [29]. Other workers who have developed combined QM(molecular orbital)/MM potentials include Hillier *et al.* [30], Ferenczy, Rivail and co-workers [31, 32], Thompson [33, 34] and Maseras and Morokuma [35]. Various groups have also been working on combined QM(density functional theory)/MM methods, including Merz *et al.* [36] and Salahub *et al.* [37].

When using a hybrid potential to calculate the energy of a system it is necessary to partition the atoms in the system into various regions, each of which is described using a separate potential. For the class of hybrid potential that is discussed in this chapter there are three regions — a quantum mechanical region, a molecular mechanical region and a boundary region. A schematic illustrating the partitioning is given in figure 2. Whether a system can be partitioned in this fashion depends upon the nature of the problem. For many reactions, the portion of the system needing a quantum description is relatively small and localised but for other processes, such as the tunneling of electrons through proteins, many atoms may be involved and a hybrid potential model may not be appropriate.

A rigorous derivation of the form of the hybrid potentials treated in this chapter requires the construction an effective Hamiltonian, $\hat{H}_{\text{Eff}}$, for the system. This Hamiltonian can then be used as the Hamiltonian for the solution of the time-independent Schrödinger equation for the wavefunction of the electrons on the QM atoms, Ψ, and for the potential energy of the system, $E_{\text{QM/MM}}$. If $\{\mathbf{R}_m\}$ are the coordinates of the MM atoms, the Schrödinger equation (equation 1) becomes:

$$\hat{H}_{\text{Eff}}\ \Psi\left(\{\mathbf{r}_i\},\{\mathbf{R}_\alpha\},\{\mathbf{R}_m\}\right) = E\left(\{\mathbf{R}_\alpha\},\{\mathbf{R}_m\}\right)\ \Psi\left(\{\mathbf{r}_i\},\{\mathbf{R}_\alpha\},\{\mathbf{R}_m\}\right) \quad (4)$$

The effective Hamiltonian is the sum of four terms:

$$\hat{H}_{\text{Eff}} = \hat{H}_{\text{QM}} + \hat{H}_{\text{QM/MM}} + \hat{H}_{\text{MM}} + \hat{H}_{\text{Boundary}} \quad (5)$$

where:

- $\hat{H}_{\text{QM}}$ is the standard quantum mechanical Hamiltonian for the QM method being used.

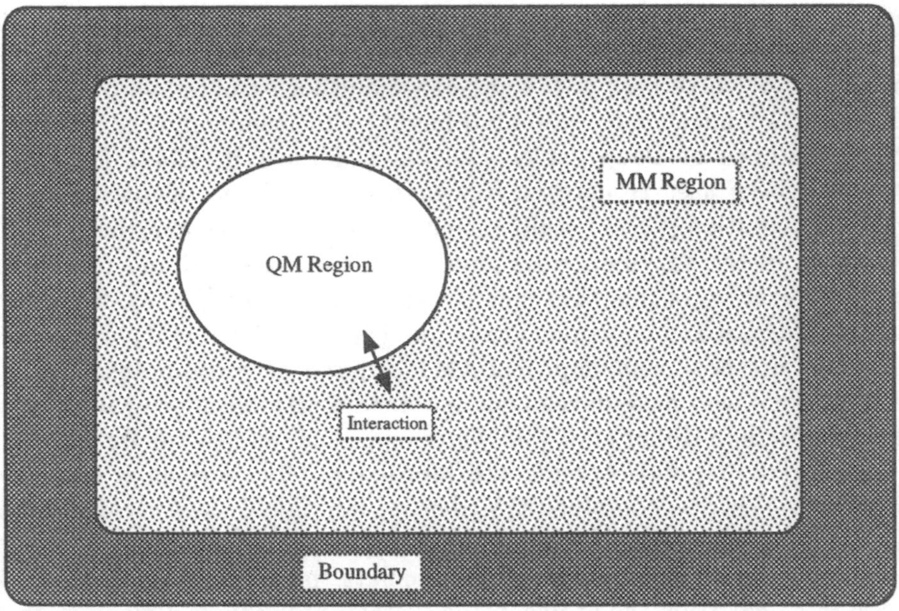

Figure 2. **Partitioning of a system into QM, MM and boundary regions.**

- $\hat{H}_{QM/MM}$ is the Hamiltonian for the interaction of the QM and MM regions.
- $\hat{H}_{MM}$ is the molecular mechanics energy. In most cases this term is not an operator and is equal to the energy on the molecular mechanics atoms, E_{MM}. It is calculated using a standard empirical potential.
- $\hat{H}_{Boundary}$ is the boundary Hamiltonian and it will consist of two parts — an operator for the interaction with the QM atoms, $\hat{H}_{Boundary(QM)}$, and an energy for the interaction with the MM atoms, $E_{Boundary(MM)}$.

Each of these terms will be discussed in detail in the following four sections. It is to be noted, however, that, in general, the purely QM and MM parts of the Hamiltonian are essentially the same as for calculations on purely QM or MM systems respectively. The most important differences arise in the form of the interaction Hamiltonian, $\hat{H}_{QM/MM}$, and, to a lesser extent, of the boundary terms.

4. The Quantum Mechanical Hamiltonian

There are many methods that are available to solve the time-independent Schrödinger equation for molecular systems within the Born-Oppenheimer approximation. The emphasis in this section is on those methods that have been used or would be suitable for use with hybrid potentials. Three classes are considered — molecular orbital (MO) methods, density functional theory (DFT) methods and empirical valence bond (EVB) techniques. Both *ab initio* and semi-empirical versions of MO methods exist and are widely used. By contrast, the DFT methods in use are primarily *ab initio* and EVB methods can be classified as empirical or semiempirical.

In general, *ab initio* QM methods are those, of whatever type, that seek to obtain solutions of the time-independent Schrödinger equation within a given theory without making any further approximations. In contrast, *semiempirical* QM methods are ones that use a quantum mechanical formalism to define the form of the potential and its fashion of calculation but which employ approximations to simplify and, hence, speed up numerical calculations. The methods must be parameterized with data from experiment or more sophisticated QM calculations to ensure that they give acceptable results [38, 39].

4.1. MOLECULAR ORBITAL METHODS

Molecular orbital (MO) methods seek a solution to the electronic wavefunction, Ψ, for a particular state of a system of the form:

$$\Psi = \sum_I A_I \Phi_I \tag{6}$$

The $\{\Phi_I\}$ are Slater determinants of one-electron orthonormal spin orbitals, $\{(\phi\sigma)_i\}$:

$$\Phi_I = \tilde{\mathcal{A}} \left|(\phi\sigma)_1(\phi\sigma)_2 \ldots (\phi\sigma)_M\right| \tag{7}$$

where M is the number of electrons and $\tilde{\mathcal{A}}$ is the antisymmetrization operator.

The spin-orbitals, themselves, are the product of a spin function, σ, and a space function, ϕ. The spin function can take either of two values, α or β, while the space function is expanded as a linear combination of basis functions, $\{\eta_\mu\}$:

$$\phi_i = \sum_\mu C_{\mu i} \eta_\mu \tag{8}$$

The expansion coefficients, $\{A_I\}$ and $\{C_{\mu i}\}$, in the sums in equations 6 and 8 are the variable parameters in the formalism. They are called the configuration interaction (CI) coefficients and the MO coefficients respectively. The optimum values of these parameters, at a particular nuclear geometry, are found by minimising the energy expression for the system with respect to the variables, subject to the constraints that the orbitals are orthonormal:

$$\langle \phi_i | \phi_j \rangle - \delta_{ij} = 0 \qquad \forall\ i,j \tag{9}$$

The energy expression itself is determined by substituting the expressions for the MO wavefunction (equations 6 to 8) into the expectation value expression of equation 3.

There are many variations on the procedure sketched above. The simplest MO wavefunctions are Hartree-Fock (HF) wavefunctions in which a single determinant appears in the expansion in equation 6 and the MO coefficients are the only variable parameters. More elaborate types are CI wavefunctions in which only the CI coefficients are allowed to vary. The values of the MO coefficients are fixed, having been taken, usually, from a previous HF calculation. The most general types are multiconfigurational wavefunctions, in which both sets of coefficients are treated variationally.

Non-variational wavefunctions, in which there is no optimisation of parameters, also exist. One approach is Møller-Plesset (MP) perturbation theory which, like CI methods, is employed most often to improve upon a previously determined HF wavefunction. Fuller details of all these approaches may be found in the monograph by Szabo and Ostlund [40].

Ab initio MO methods based on HF or small multiconfigurational wavefunctions have been the method of choice, up to the present, for studies of organic systems and other molecules with light nuclei. The properties of stable species on the PES are often reproduced very well by calculations with just HF wavefunctions. Studies of reactions usually require the more sophisticated and expensive techniques, such as CI or MP perturbation theory, that take into account the effects of the correlation between the electrons that is omitted from the HF approximation. The additional energy lowering computed with these methods with respect to that obtained with an HF calculation is called the *correlation energy.* A detailed and up-to-date discussion of the accuracy of state-of-the-art MO methods when applied to a variety of problems may be found in the book by Hehre *et al.* [41].

Calculations using *ab initio* HF MO theory of reasonable accuracy can be performed on systems up to, perhaps, 50 to 100 atoms. The limiting step is the calculation of the two-electron integrals which, formally, has a fourth order power dependence upon the size of the basis set (which is itself roughly proportional to the number of atoms in the system). An important algorithmic development is the pseudospectral method of Friesner *et al.* [42, 43, 44] which reduces the formal order dependence of a HF calculation from four to three, enabling larger problems to be tackled. The reduction is achieved by introducing a physical space, grid representation of the wavefunction as well as the normal spectral or basis-set one. With the grid, the evaluation of the two-electron integrals is not needed, being replaced by the calculation of one-electron integrals at each grid point. There is also the exciting possibility that the pseudospectral algorithms could be applied to CI and MP perturbation theory methods (which are yet more expensive and have order dependencies of five or more) resulting in yet more substantial savings in the computational time required for the calculations. The basis functions can take any analytical form but, for molecular calculations, they are most usually Gaussians, with centres placed on the nuclei.

MO-based semiempirical methods have proven extremely useful in studying organic systems. Most take as their basis HF theory but then approximate some or all of the electronic integrals, which are the most expensive part of an *ab initio* HF calculation, with simpler analytical forms. Many integrals, especially those involving basis functions on three or four different centres, are set equal to zero. A discussion of the many different types of approximation is given by Pople and Beveridge [45].

Some of the more widely used semiempirical MO methods in recent years have been the MNDO and AM1 approximations developed by Dewar and co-workers [20, 21]. They have been parameterized for calculations on systems containing first row elements and some heavier atoms and their reliability has been extensively documented, both in tests of the parameterization and in applications to real problems.

The methods have been very successful, but they do suffer drawbacks. The lack of parameters for many elements seriously limits the types of problems to which the methods can be applied and their accuracy for certain problems is not very good (for example, both MNDO and AM1 do not well describe water-water interactions). There are also questions about the theoretical foundations of the models. The parameterization is performed using experimental data at a temperature of 298K and implicitly includes vibrational and correlation information about the state of the system. Therefore, the parameterization is used, in part, to compensate for quantities that the HF method cannot, by itself, account for. But what happens if vibrational or correlation energy calculations are performed? With these caveats and if one can be certain of their accuracy in given circumstances, the methods are very useful as calculations can be performed with them much more quickly than *ab initio* QM calculations. Even so, they are probably still too computationally intensive to treat complete condensed phase systems in a routine manner.

Thiel and co-workers have partially overcome some of the faults of the MNDO and AM1 methods with their semiempirical potentials. The MNDOC potential has been parameterized to take into account correlation effects and gives good results for reactions [46, 47]. Another extends the MNDO method to include d-orbitals for some second row atoms which are essential if the bonding in compounds containing these elements is to be properly described [48]. Other improvements are also underway [49]. Another model which includes d-orbitals is the *semiempirical ab initio method*

1 (SAM1) developed by Dewar, Holder and co-workers [50]. It has been parameterized for use with some transition metals, including copper and iron.

Another approach that has been employed as a stop-gap alternative to developing a new method, is to alter the parameters of an existing one to obtain better agreement for a particular problem. Truhlar and co-workers have modified the AM1 method in this way to study reactions, the parameters being changed to obtain better agreement between the calculated and experimental values for some properties, including the reaction barriers [51].

4.2. DENSITY FUNCTIONAL THEORY

The foundations of modern density functional (DF) theory were laid in the middle 1960's when Hohenberg and Kohn [52] proved that the properties, including the energy, of a QM system in its ground state are wholly determined by the density of the particles. A knowledge of the many-particle wavefunction is unnecessary. Subsequently, Kohn and Sham [53] derived a set of one-particle equations, analogous to those obtained when solving for HF wavefunctions, which could be used to calculate the density and, hence the remaining properties, of a system [54].

If the electrons in a molecular wavefunction are assumed to occupy one-electron orbitals, $\{\phi_i\}$, then the total electron density, ρ, as a function of position, $\mathbf{r}$, is simply the sum, for all the electrons, of the square of the orbitals:

$$\rho(\mathbf{r}) = \sum_i |\phi_i(\mathbf{r})|^2 \tag{10}$$

The energy of the system is a functional of the electron density and is written:

$$\begin{aligned} E_{DF}[\rho(\mathbf{r})] &= -\frac{1}{2}\sum_i \int d\mathbf{r}\phi_i(\mathbf{r})\nabla^2\phi_i(\mathbf{r}) - \sum_\alpha Z_\alpha \int d\mathbf{r}\frac{\rho(\mathbf{r})}{|\mathbf{R}_\alpha - \mathbf{r}|} + \\ & \frac{1}{2}\int d\mathbf{r}\int d\mathbf{r}'\frac{\rho(\mathbf{r})\rho(\mathbf{r}')}{|\mathbf{r}-\mathbf{r}'|} + E_{xc}[\rho(\mathbf{r})] \end{aligned} \tag{11}$$

where E_{xc} is the *exchange-correlation energy.* The Kohn-Sham (KS) equations are derived by functionally varying the energy expression with respect to the form of the orbitals. The equations for the orbitals and their energies, $\{\epsilon_i\}$, that result, in atomic units, are:

$$\left(-\frac{1}{2}\nabla^2 - \sum_\alpha \frac{Z_\alpha}{|\mathbf{R}_\alpha - \mathbf{r}|} + \int \frac{\rho(\mathbf{r}')}{|\mathbf{r}'-\mathbf{r}|}d\mathbf{r}' + V_{xc}\right)\phi_i = \epsilon_i\phi_i \tag{12}$$

where the *exchange-correlation potential*, V_{xc}, is the functional derivative of the exchange-correlation energy, i.e. $V_{xc} = \delta E_{xc}[\rho]/\delta\rho(\mathbf{r})$.

The exchange-correlation potential is the source of both the strengths and the weaknesses of the DF approach. In HF theory, the analytical form of the term equivalent to V_{xc}, *the exchange potential*, arises directly during the derivation of the equations, but it depends upon the one-particle density matrix, making it expensive to calculate. In DF theory the analytical form of V_{xc} must be put into the calculations because it does not come from the derivation of the Kohn-Sham equations. Thus, it is possible to choose forms for V_{xc} that depend only upon the density and its derivatives and which are cheap to calculate (the so-called local and non-local density approximations). The V_{xc} factor can also be chosen to account for some of the correlation between the electrons, in contrast to HF methods for which additional calculations must be made. The drawback is that there does not appear to be any systematic way of improving the potential. Indeed, many such terms have been proposed.

Calculations using DF theory have been particularly popular for solids and species involving transition metals [54, 55]. Recently, though, they have started to be used to study organic molecules and have been found to be surprisingly accurate for the calculation of many properties, including reaction barriers, as long as the more sophisticated non-local exchange-correlation potentials are used. In fact, the results they give are often as accurate as HF calculations which have been followed by additional calculations to calculate a part of the correlation energy, and yet they can be performed at a fraction of the cost [56]. Recent algorithmic improvements hold out the possibility that these costs can be reduced still further especially for large systems [57, 58] and support the available evidence that, for calculations of the ground states of large systems, density functional theory will be the quantum mechanical method of choice. However, the calculation of excited state properties for such systems is currently not possible using density functional techniques. Equally, semi-empirical versions of DFT methods have not been widely used for tackling chemical problems.

4.3. EMPIRICAL VALENCE BOND METHODS

In MO theory the orbitals are usually very delocalised, containing contributions from many, if not all, of the atoms in the system. This contradicts the intuitive notion that the electronic structure for the full system should be only slightly perturbed from that of its constituent atoms. Valence bond methods seek to correct this by using orbitals that are highly localised and which often resemble atomic orbitals. Wavefunctions constructed from such orbitals have the conceptual advantage that they can often be identified directly with specific *resonant* electronic forms of a molecule.

VB functions have been championed by Warshel and his co-workers for use in studying reactions in enzymes and in solution. The method, which they term the empirical VB (EVB) method, supposes that the wavefunction for a particular problem, ψ, can be written as a linear combination of the wavefunctions of resonant forms, $\{\varphi_i\}$, which are postulated to be important in the process. For example, for a bond breaking reaction, $AB \to A^+ + B^-$, which produces ionic products, the contributing resonant forms could be the covalent, AB, and ionic, A^-B^+, states that dissociate to atoms and ions respectively. The total wavefunction is:

$$\psi = a_{AB}\varphi_{AB} + a_{A^+B^-}\varphi_{A^+B^-} \tag{13}$$

The potential energies of the ground and excited state surfaces, $E_\pm$, are obtained by solving the secular equation for the two-state problem and are:

$$E_\pm = \frac{1}{2}(E_1 + E_2) \pm \sqrt{\left(\frac{E_1 - E_2}{2}\right)^2 + H^2} \tag{14}$$

where E_1 and E_2 are the energies of the covalent and ionic resonant structures and H is the interaction matrix element over the electronic Hamiltonian between the two states. To use the method for calculation, analytical forms must be chosen for E_1, E_2 and H (as functions of the distances between the atoms, for example) and then parameterized to reproduce experimental or theoretical data for the system. A detailed review of the EVB method and his work with it has been published by Warshel and the reader is referred to that for further details [15].

The EVB scheme is, conceptually, one of the nicest QM methods as it provides an immediate picture of the chemical processes occurring during a reaction, for example. Unfortunately, it suffers from the same parameterization problem as empirical molecular mechanical methods. Warshel uses experimental gas and solution phase data but Chang and Miller have recently published a more systematic parameterization scheme for EVB models which is based upon the use of data from *ab initio* calculations [59]. A related method, the MM-VB scheme, has been extensively developed by Bernardi, Olivucci, Robb and co-workers and has been applied to a number of organic, primarily pericyclic, reactions [60, 61].

5. The Molecular Mechanical Energy Function

Empirical or molecular mechanics (MM) potentials or force fields are traditionally the ones used for MC and MD simulations of large condensed phase systems, such as liquids or proteins. A typical force field is a sum of internal and non-bonded energy terms. The internal terms describe the energies of the bonds, angles and torsions while the non-bonded part consists of electrostatic and Lennard-Jones contributions [62]:

$$E_{\mathrm{MM}} = E_{bond} + E_{angle} + E_{torsion} + E_{electrostatic} + E_{LJ} \qquad (15)$$

where :

$$E_{bond} = \frac{1}{2}\sum_i k_{b_i}\left(b_i - b_i^0\right)^2 \qquad (16)$$

$$E_{angle} = \frac{1}{2}\sum_i k_{\theta_i}\left(\theta_i - \theta_i^0\right)^2 \qquad (17)$$

$$E_{torsion} = \frac{1}{2}\sum_i k_{\phi_i}\left(1 + \cos\left(n_i\phi_i + \delta_i\right)\right) \qquad (18)$$

$$E_{electrostatic} = \sum_{m<n}\left\{\frac{Q_m Q_n}{\epsilon R_{mn}}\right\} \qquad (19)$$

$$E_{LJ} = \sum_{m<n}\left\{\frac{A_{mn}}{R_{mn}{}^{12}} - \frac{B_{mn}}{R_{mn}{}^{6}}\right\} \qquad (20)$$

The internal terms are summed over all bond, angle and torsions that contribute to the energy. For the bond and angle energies, b_i and θ_i are the bond length and angle of the ith bond or angle and b_i^0 and θ_i^0 are reference values. For the torsion terms ϕ_i is the torsion angle, n_i is the periodicity of the torsion and δ_i is its phase. All the internal terms have associated force constants, $k_{(b,\theta,\phi)_i}$. The non-bonded terms are summed over all pairs, $\{m, n\}$, of MM atoms that interact. R_{mn} is the distance between atoms m and n, Q_m is the partial charge on the mth atom, ϵ is the dielectric constant and A_{mn} and B_{mn} are Lennard-Jones parameters for the $\{m, n\}$ atom pair. Full definitions may be found in references [62, 17, 22].

Many variations of the basic force-field in equations 15 to 20 exist and extra terms are often added. One of the more important is the inclusion of a polarizability function which is needed if changes in the charge distribution on the atoms are to occur [63, 64]. In the simplest models each atom is assigned a dipole polarizability, γ_m. Dipoles, $\boldsymbol{\mu_m}$, are induced upon each atom that are proportional to the electric field, $\boldsymbol{\xi_m}$, at the atom due to the surroundings:

$$\boldsymbol{\mu_m} = \gamma_m \boldsymbol{\xi_m} \qquad (21)$$

The fields are themselves functions of the dipoles on the other atoms:

$$\boldsymbol{\xi_m} = \sum_{n \neq m}\left\{\frac{Q_n \mathbf{R}_{mn}}{R_{mn}^3} + \frac{\left(3\mathbf{R}_{mn}\mathbf{R}_{mn}^{\dagger} - R_{mn}^2\mathbf{I}\right)\boldsymbol{\mu_n}}{R_{mn}^5}\right\} \qquad (22)$$

where $\mathbf{R}_{mn} = \mathbf{R}_m - \mathbf{R}_n$, $\mathbf{I}$ is the 3×3 identity matrix and the superscript, $\dagger$, indicates the transpose operation. The dipoles are determined by solving equations 21 to 22 using a direct matrix inversion technique for small cases or an iterative approach for larger ones. Once the dipoles are known the polarisation energy, E_{pol}, is:

$$E_{pol} = -\frac{1}{2}\sum_m \gamma_m \boldsymbol{\xi}_m^{\dagger} \left(\sum_{n \neq m} \frac{Q_n \mathbf{R}_{mn}}{R_{mn}^3} \right) \tag{23}$$

Because of the expense in determining the dipoles, alternative models for the polarizability have been introduced. One, due to Sprik and Klein [65, 66], does not include dipole polarizabilities on the atoms. Instead it spreads the partial charge on an atom amongst a few sites that are very close to it. The position of the sites is fixed relative to the atom, but their charges are treated as dynamical variables that are allowed to change during an MD simulation, subject to the constraint that the total charge on the atom remains constant.

MM potentials have the advantage that the energy and forces for a large number of particles can be calculated rapidly. However, the parameterization of the potential is an onerous task, especially if the molecules to be studied contain atoms of several types of element in a number of different environments (as in proteins).

To study reactions, in which bonds are broken or formed, it is evident that the analytical form of the bond energy function in equation 16 is inappropriate. It can be modified by including terms, such as Morse functions, that permit bond dissociation, although no work appears to have been performed to introduce such terms into a MM potential function for reacting systems in a systematic way.

Several groups have parameterized force fields for special cases. Jorgensen and co-workers have investigated a large number of organic reactions in solution using empirical potentials [67]. In their approach they calculate the energy for the reaction along a specific reaction coordinate in the gas phase using an *ab initio* MO method and then parameterize a suitably chosen analytical function against the data they have generated. Further QM calculations are performed for the reacting species interacting with one or more solvent molecules to obtain information for the parameterization of extra functions that describe the interaction of the solute and solvent.

Stillinger *et al.* have developed a number of very original potentials to study reactive systems, including water dissociation [68], liquid sulphur [69] and fluorine [70]. The water potential is constructed as a sum of two-body terms for the interaction of bare protons, H^+, and doubly charged oxygen anions, O^{2-}, with each other and with themselves and a short-range many-body polarisation term that affects only the oxygens. For sulphur and fluorine, the atoms are taken to be the fundamental units for the energy calculation and the potential is built up as a sum of two- and three-body interactions.

Another approach is that of Houk and co-workers. They have determined transition structures for various organic reactions with *ab initio* MO calculations and used the data to obtain structural parameters (for example, bond lengths and bond angles) to model the transition states with the MM2 [62] and, more recently, the MM3 force fields [71]. If the assumption is made that the transition structure remains essentially unchanged, they can investigate in a semiquantitative fashion the stereochemistry of the reactions with their models [72, 73].

A problem related to obtaining MM potentials for reactions is to devise functions for the excited states of molecules. Blair *et al.* [74] and DeBolt *et al.* [75] have parameterized MM potential energy functions for the first excited states of formaldehyde and acetone respectively. The force fields in both cases were of the same form as for the ground states, except that the parameters, especially the partial atomic charges, were altered to reflect the changes that occur in the excited state.

6. The QM/MM Interaction Hamiltonian

The form of the interaction Hamiltonian will determine the accuracy of the full hybrid potential. It is necessary to ensure that the QM and MM regions interact in a seamless fashion which reproduces as well as possible the interactions that would be felt by the particles in the system if it was being described by the QM potential alone. For most purposes this will not be possible and so it is essential to check how the energies and forces are affected by the way in which the system is partitioned.

The interaction Hamiltonian for many hybrid potentials contains a term which accounts for the electrostatic interaction between the MM atoms and the QM electrons and nuclei and a Lennard-Jones term that mimics the effects of the exchange repulsion and dispersion interactions between the MM and QM atoms. For an *ab initio* QM method, it takes the form:

$$\hat{H}_{\mathrm{QM/MM}} = -\sum_{im} \frac{Q_m}{R_{im}} + \sum_{\alpha m} \frac{Z_\alpha Q_m}{R_{\alpha m}} + \sum_{\alpha m} \left\{ \frac{A_{\alpha m}}{{R_{\alpha m}}^{12}} - \frac{B_{\alpha m}}{{R_{\alpha m}}^{6}} \right\} \tag{24}$$

The last two sums in the Hamiltonian are purely additive and can be calculated and added separately to the total energy for the system. However, to properly calculate the distortion of the electron density of the QM region due to the field of the MM atoms, it is necessary to include terms arising from the first term directly in the QM calculation. For an MO or DF method, this entails the calculation of additional one-electron electrostatic integrals, $I_{\mu\nu}$, that are added into the one-electron matrix before the HF or KS equations are solved. They have the form:

$$I_{\mu\nu} = \sum_m Q_m \int d\mathbf{r} \, \frac{\eta_\mu(\mathbf{r})\eta_\nu(\mathbf{r})}{|\mathbf{R}_m - \mathbf{r}|} \tag{25}$$

In the case of *ab initio* methods the form of the electrostatic interaction integrals is fixed as the charges on the molecular mechanics atoms and the form of the basis functions are fully determined by the MM and QM models respectively. For semiempirical methods the interaction integrals between the QM and MM atoms must be parameterized to reproduce interactions obtained from quantum mechanical or experimental data (see reference [19] for details).

In many force fields, truncation schemes are often used to reduce the number of non-bonded electrostatic and Lennard-Jones interactions that need to be calculated. Such schemes are readily incorporated into the interaction Hamiltonian either by omitting all interactions that have a distance greater than some cutoff or by multiplying the appropriate interactions by a tapering function that reduces the interactions to zero beyond a certain distance. It is to be noted that in some hybrid force fields (see, for example, [35]) the electrostatic interaction terms are not included and the QM/MM interaction is due solely to the Lennard-Jones terms (and link-atoms if they are present). This could be a reasonable approximation in non-polar systems (such as the transition metal complexes for which some of these force fields were developed) but it will not be sufficiently accurate in the general case.

There are three variations to the form of the basis interaction Hamiltonian which are worth considering in more detail.

The first variant concerns what happens when the MM energy involves a term which must be calculated using some sort of iterative procedure. This is the case, for example, when polarizabilities are employed or when there is an external reaction field. The calculation of the QM and MM energies is now more complicated because the two procedures are coupled. In the case of the addition of polarizability, there are extra terms, $\hat{H}^{\rm pol}_{\rm QM/MM}$, in the interaction Hamiltonian of the form:

$$\hat{H}^{\rm pol}_{\rm QM/MM} = -\frac{1}{2} \sum_m \boldsymbol{\mu}_m \cdot \left\{ \sum_i \frac{\mathbf{r}_{im}}{r^3_{im}} - \sum_\alpha \frac{Z_\alpha \mathbf{R}_{\alpha m}}{R^3_{\alpha m}} \right\} \tag{26}$$

The induced dipoles on the MM atoms and the optimum wavefunction for the QM atoms are interdependent because the field at each MM atom depends upon the electron distribution of the QM atoms and vice-versa. In most studies the determination of the dipoles and wavefunction has been carried out by alternative iterations of the QM and MM optimization procedures (a self-consistent field iteration or other minimization step for the wavefunction and a single iteration for the determination of the dipoles). Such schemes converge rapidly. A number of workers have used polarizable MM force fields in their hybrid potentials. Warshel and Levitt [14] and Singh and Kollman [16] both had polarisable terms in their combined potentials although they used approximations to calculate the induced dipoles on the MM atoms for the complete system. An exact scheme and further approximations have been discussed by Luzhkov and Warshel [28] and Thompson has used polarizable force fields [34].

A second modification that can be made concerns the use of pseudopotentials on the MM atoms. These provide a more accurate way of accounting for the electrostatic and dispersion interactions between the QM and MM atoms. They enter into the calculation in the same way as the electrostatic interactions — there will be a term for interaction with the nuclei of the QM atoms which is additive and a term which is incorporated into the one-electron matrix. They have been employed to describe the core electrons of atoms in QM calculations and also by Sprik and Klein [76], Estrin *et al.* [77] and Vaidehi *et al.* [29] and other groups to describe interactions between QM and MM atoms. Both pseudopotentials and polarizabilities can be included simultaneously as pseudopotentials, in general, do not describe charge distortion effects on the MM atoms.

The third variation which needs discussing is what happens when atoms in the same molecule are split between the QM and MM regions (see Figure 3). As mentioned above, if the solvation of a small molecule or the reaction between two molecules in a solvent is being studied then the partitioning of the system into QM and MM regions is straightforward — the solutes can consist of QM atoms and the solvent of MM atoms. If, however, large molecules are being considered (as, for example, for a reaction in an enzyme in which it is necessary to treat the catalytic sidechains of some residues quantum mechanically) then partitioning of the same molecule into different regions is unavoidable. There are two problems. First the wavefunction on the QM atoms must be terminated properly because if these bonds are left unsatisfied the electronic state and the properties of the QM part of the system are drastically altered and the QM calculation becomes much more difficult. Second the interactions, particularly electrostatic, between the MM atoms and the QM atoms across the junction must be successfully reproduced. The solutions to neither of these problems is entirely straightforward which makes this area currently the most vexing one in the domain of hybrid potentials.

Two types of method have been proposed to cope with this situation.

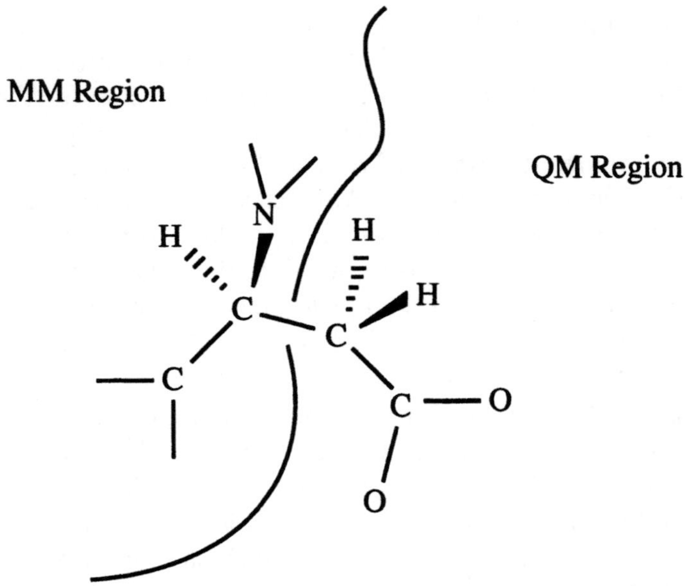

Figure 3. **Partitioning of atoms in the same molecule between the QM and MM regions.**

The simplest, due to Singh and Kollman [16] and used by Field *et al.* [19] and by Maseras and Morokuma [35], is to introduce extra QM *junction dummy* or *link* atoms that satisfy the broken bonds produced when the molecule is partitioned into QM and MM regions. The link atoms have generally been treated as hydrogens with one electron in a single 1s orbital, although halogens and other group types have also been used. The link atoms are placed along the QM/MM bond at a distance from the QM atom that is appropriate for the type of the QM atom. The link atoms are usually treated so that they are invisible to the MM atoms although they participate fully in the QM calculation. Their positions can be constrained or they can be allowed to move freely during a simulation. The addition of the link atom satisfies the bonds that are broken for the QM atoms but the electrostatic interactions between the MM atoms and the QM atoms must be specified too. Maseras *et al.* ignore these. Field *et al.* used an *ad hoc* approximation and tried various scaling schemes for the interaction of the MM partial charges and the QM atoms near the broken bond and then chose the one that most accurately reproduced various properties such as rotational barriers and ionization energies. A better solution would be to put a pseudopotential on the MM atoms nearest the broken bond although no work in this direction has yet been done.

A second and more elegant approach is to use localized hybrid orbitals for the QM atoms at the junction. This method was used by Warshel and Levitt in their original work on lysozyme [14] and has been further developed by Théry *et al.* [32]. Both groups use semiempirical QM models and have not extended their schemes to *ab initio* potentials. In the implementation of Théry *et al.* the QM atom at the border of the QM/MM region is given four orthogonal hybrid orbitals. One of these is frozen and not allowed to change and points along the bond to the MM atom. The others, which are constrained to be orthogonal to the excluded orbital, are allowed to participate in the QM calculation as normal. The form of the frozen hybrid orbital is predetermined for a given type of bond and is assumed to be transferable from one system to another. The MM atom of the broken bond interacts with the electrons and nuclei of the QM region in the same way as the other MM atoms. However, it was found that the results for the energy were very sensitive to the value that was used for the charge on the MM atom.

It is clear, at the moment, that the "junction problem" of QM/MM hybrid potentials is, as yet, unresolved. In all schemes the partitioning of a molecule should be done in a chemically sensible way — only single bonds should be broken and preferably those between non-polar atoms such as carbons. Additionally, if possible, the junction should be as far away as possible from the region of interest (e.g. where the reaction is occurring). And, above all, several partitionings should be tried so that the sensitivity of the results to the partitioning can be tested.

7. Boundary Approximations

A large number of boundary approximations exist for use in molecular simulations. All of these can be adapted for use with hybrid potentials.

The most common boundary representation is *periodic boundary conditions* which assumes that the system consists of a periodic array (or a "crystal") of identical systems [1]. Another common method, developed for the simulation of biomacromolecules, is the *stochastic boundary* approach, in which the influences of the atoms outside the boundary are replaced by a simple boundary force [78, 79, 80]. Warshel uses a *Langevin dipoles* model in which the solvent is explicitly replaced by a grid of polarizable dipoles. The energy is calculated in a similar way to the polarization energy in a molecular mechanics force field (see above) [15].

A less expensive technique often used for protein calculations is the *accessible surface area* method that supposes that the contribution of each atom to the total solvation free energy of the system is proportional to the solvent accessible surface area of the atom [81, 82].

Another model is the *reaction field* model [83]. Here, the solute is placed inside a cavity surrounded by a medium with a dielectric constant that is appropriate for the desired solvent and the effects on the energy of the system due to its reaction field are calculated. Reaction field models have been implemented with MM [84] and QM potentials [85]. Solvation models also exist which combine accessible surface area and reaction field terms. Kang *et al.* have outlined a model for MM potentials [86] and Cramer and Truhlar have developed one for use with the AM1 semiempirical QM method [87]. A more detailed review of reaction field methods can be found in the chapter by van Duijnen in this book.

The form of the boundary Hamiltonian for the QM atoms depends upon the boundary approximation being used. For periodic boundary conditions it will be similar in form to equation 24 where the sums now run over the atoms, both QM and MM, in the periodic images. The interactions between the QM atoms in the central box and the MM atoms of the periodic images can be handled in exactly the same way as for interactions with the MM atoms in the central box. The simplest way to handle the interactions between the QM atoms in the central box and those in the periodic images is to neglect them by use of an appropriate cutoff method. This is the method that has been used in the majority of applications. An alternative scheme is to calculate the effective charges on the QM atoms at each iteration of the QM calculation and then to treat the QM atoms in the outer boxes like MM atoms with an effective point charge. The method in which the full interactions are calculated has not, to the author's knowledge, been used in simulations. For stochastic boundary conditions the boundary energy for the QM atoms is exactly the same as for the MM atoms and can be computed in the same way.

8. Implementations of Combined Hamiltonians

Once the Hamiltonian is defined it is straightforward to calculate the energy of the system. For an MO-based method the total energy of the system (equation 3) is simply the expectation value of the effective Hamiltonian over the wavefunction. It is:

$$E_{\mathrm{QM/MM}} = \frac{\left\langle \Psi \left| \hat{H}_{\mathrm{QM}} + \hat{H}_{\mathrm{QM/MM}} + \hat{H}_{\mathrm{Boundary(QM)}} \right| \Psi \right\rangle}{\langle \Psi \mid \Psi \rangle} + E_{\mathrm{MM}} + E_{\mathrm{Boundary(MM)}} \qquad (27)$$

where the parts of the energy that are independent of QM particles have been separated out from the expectation value. For a density functional method the energy expression is formally the same as that of equation 12 except that the external potential, originally due to the nuclei, now contains terms due to the MM atoms. The extra terms, E_{MM} and $E_{\mathrm{Boundary(MM)}}$, must also be added in. The forces and higher derivatives on the QM and MM atoms are easily determined by differentiating the total energy expression (equation 27, for example) with respect to the coordinates of the QM nuclei and the MM atoms respectively. The energy and forces, once available, can be used in exactly the same way for simulations as those obtained for systems consisting solely of QM or MM atoms.

To clarify the scheme outlined above, it is useful to consider in detail the steps that are necessary to calculate the energy and forces using a hybrid QM/MM potential during a simulation:

1. Define the atoms in the system, their elemental type and their coordinates.
2. Divide the atoms in the system into QM and MM regions.
3. Define the QM method to be used to describe the QM atoms. This automatically defines the electronic Hamiltonian. For an *ab initio* method it is necessary to specify other information such as a basis set or an exchange-correlation functional while for a semiempirical model the parameters used for evaluation of the integrals are needed. The charge and the multiplicity on the QM atoms must also be given — these determine the number of electrons in the QM calculation (which must be integer!).
4. Define the empirical force field for the MM atoms. The atom types of each MM atom and the force field parameters must be supplied.
5. Define the boundary approximation.
6. Define the interaction Hamiltonian. For a Hamiltonian of the form of equation 24 and for *ab initio* QM methods the only parameters that need to be defined are the Lennard-Jones parameters for the QM atoms as the charges and Lennard-Jones parameters for the MM atoms are determined once the MM force field is specified. The QM atom parameters can be obtained from the MM force field or from a parameterization procedure. For semiempirical methods it is necessary to specify parameters for the evaluation of the one-electron electrostatic interaction integrals.
7. Choose a starting wavefunction (e.g. a set of initial MO vectors) for the QM atoms using a standard technique.
8. Perform the simulation. Each time an energy is required the QM/MM potential energy calculation will consist of the following steps:
   (a) Evaluate the electronic integrals that are needed for a QM calculation. In addition to the normal integrals there will be the one-electron electrostatic interaction integrals, $I_{\mu\nu}$, between the QM and MM atoms which are added into the one-electron matrix.
   (b) Perform the QM calculation using a self-consistent field procedure or another type of method. An optimum wavefunction or density will result which can then be used to calculate the energy of the QM atoms and the QM/MM energy due to electrostatic interac-

tions. Note that the method of performing the QM calculation is unaltered and all types of calculation can be performed, including calculations on excited states.

(c) Calculate the nuclear interaction energy between the QM atoms.

(d) Calculate the Lennard-Jones energy between the QM and MM atoms and the QM(boundary) energy if this has not already been determined.

(e) Calculate the energy between the MM atoms and between the MM atoms and the boundary.

(f) Add all the energy terms together.

(g) Calculate the derivatives.

The above scheme is a general one and is easy to modify to take into account different QM or MM potentials. Once the energy and the forces for the system are available they can be used in a molecular dynamics simulation, for example, in exactly the same way as those obtained using a purely MM approach.

9. Summary

This chapter has reviewed the form and various implementations of hybrid quantum mechanical and molecular mechanical potentials. There have been many studies using such potentials, including work on solvation energies, spectral shifts, reactions in solution and in enzymes. The reader is referred to the review by Gao for an extensive discussion [4].

Most studies to date have used semiempirical models as the QM portion of the potential. However, *ab initio* QM combined potentials which overcome some of the limitations of the semiempirical methods are beginning to be employed as the power of computers increases and as experience with combined potentials grows.

It is clear that hybrid potentials are a powerful tool in molecular modeling and they will continue to be developed and applied in an increasing number of ways in the future.

Acknowledgements

The author would like to thank the Institut de Biologie Structurale — Jean-Pierre Ebel for support.

References

1. Allen, M.P. and Tildesley, D.J. (1987) *Computer Simulations of Liquids*, Oxford University Press, London.
2. McCammon, J.A. and Harvey, S. (1987) *Dynamics of Proteins and Nucleic Acids*, Cambridge University Press, Cambridge.
3. Brooks III, C.L., Karplus, M. and Pettitt (1988) Proteins: A Theoretical Perspective of Dynamics, Structure and Thermodynamics, *Advs. Chem. Phys.* **71**, 1–259.
4. Gao, J. (in press) Methods and Applications of Combined Quantum Mechanical and Molecular Mechanical Potentials, in K.B. Lipkowitz and D.B. Boyd (eds.), *Reviews in Computational Chemistry*, VCH, New York.
5. See the special issue of the *International Journal of Quantum Chemistry* devoted to the proceedings of the CECAM workshop on: Hybrid Quantum and Classical Mechanical Methods for the Simulation of Biopolymers in Solution. Held in Lyon, France, 9-11 May, 1995. M.J. Field and J. Gao (eds.) To be published, 1996.
6. Sutcliffe, B.T. (1995) The Idea of a Potential Energy Surface, *J. Mol. Struct. (Theochem)* **341**, 217–235.
7. Davydov, A.S. (1991) *Quantum Mechanics*, 2nd edit., Pergamon Press, Oxford.
8. Warshel, A. and Karplus, M. (1972) Calculation of Ground and Excited State Potential Surfaces of Conjugated Molecules. I. Formulation and Parameterization, *J. Am. Chem. Soc.* **94**,, 5612–5625.
9. Allinger, N.L. and Sprague, J.T. (1973) Calculations of the Structures of Hydrocarbons Containing Delocalized Electronic Systems by the Molecular Mechanics Method, *J. Am. Chem. Soc.* **95**, 3893–3907.
10. Warshel, A. and Huler, E. (1974) Theoretical Evaluation of Potential Surfaces, Equilibrium Geometries and Vibronic Transition Intensities of Excimers: The Pyrene Crystal Excimer, *Chem. Phys.* **6**, 463–468.
11. Dauber, P., Brith, M., Huler, E. and Warshel, A. (1975) The Vibronic Structure of Crystalline Ethylene, *Chem. Phys.* **7**, 108–115.
12. Warshel, A. and Karplus, M. (1975) Semiclassical Trajectory Approach to Photoisomerization, *Chem. Phys. Letts.* **32**, 11–17.
13. Warshel, A. (1976) Bicycle-Pedal Model for the First Step in the Vision Process, *Nature* **260**, 679–683.
14. Warshel, A. and Levitt, M. (1976) Theoretical Studies of Enzymic Reactions: Dielectric, Electrostatic and Steric Stabilization of the Carbonium Ion in the Reaction of Lysozyme, *J. Mol. Biol.* **103**, 227–249.
15. Warshel, A. (1992) *Computer Modeling of Chemical Reactions in Enzymes and Solutions*, J. Wiley and Sons, New York.
16. Singh, U.C. and Kollman, P.A. (1986) A Combined Ab Initio Quantum Mechanical and Molecular Mechanical Method for Carrying out Simulations on Complex Molecular Systems: Applications to the $CH_3 + Cl^-$ Exchange Reaction and Gas Phase Protonation of Polyethers, *J. Comput. Chem.* **7**, 718–730.
17. Weiner, S.J., Kollman, P.A., Case, D.A., Singh, U.C., Ghio, C., Alagona, G., Profeta, S. and Weiner, P. (1984) A New Force Field for Molecular Mechanical Simulation of Nucleic Acids and Proteins, *J. Am. Chem. Soc.* **106**, 765–784.
18. Weiner, S.J., Singh, U.C. and Kollman, P.A. (1985) Simulation of Formamide Hydrolysis by Hydroxide Ion in the Gas Phase and in Aqueous Solution, *J. Am. Chem. Soc.* **107**, 2219–2229.
19. Field, M.J., Bash, P.A. and Karplus, M. (1990) A Combined Quantum Mechanical and Molecular Mechanical Potential for Molecular Dynamics Simulations, *J. Comput. Chem.* **11**, 700–733.
20. Dewar, M.J.S. and Thiel, W. (1977) Ground States of Molecules. 38. The MNDO Method. Approximations and Parameters, *J. Am. Chem. Soc.* **99**, 4899–4907.
21. Dewar, M.J.S., Zoebisch, E.G., Healy, E.F. and Stewart, J.J.P. (1985) AM1: A New General Purpose Quantum Mechanical Molecular Model, *J. Am. Chem. Soc.*

107, 3902–3909.

22. Brooks, B.R., Bruccoleri, R.E., Olafson, B.D., States, D.J., Swaminathan, S. and Karplus, M. (1983) CHARMM: A Program for Macromolecular Energy, Minimization and Dynamics Calculations, *J. Comput. Chem.* **4**, 187–217.
23. Bash, P.A., Field, M.J. and Karplus, M. (1987) Free Energy Perturbation Method for Chemical Reactions in the Condensed Phase: A Dynamical Approach Based on a Combined Quantum and Molecular Mechanics Potential, *J. Am. Chem. Soc.* **109**, 8092–8094.
24. Bash, P.A., Field, M.J., Davenport, R.C., Petsko, G.A., Ringe, D. and Karplus, M. (1991) Computer Simulation and Analysis of the Reaction Pathway of Triosephosphate Isomerase, *Biochemistry* **30**, 5826–5832.
25. Gao, J. (1992) Absolute Free Energy of Solvation from Monte Carlo Simulations using Combined Quantum and Molecular Mechanical Potentials, *J. Phys. Chem.* **96**, 537–540.
26. Gao, J. and Pavelites, J.J. (1992) Aqueous Basicity of the Carboxylate Lone Pairs and the C-O Barrier in Acetic Acid: A Combined Quantum and Statistical Mechanical Study, *J. Am. Chem. Soc.* **114**, 1912–1914.
27. Gao, J. and Xia, X. (1992) A Priori Evaluation of Aqueous Polarization Effects Through Monte Carlo QM/MM Simulations, *Science* **258**, 631–635.
28. Luzhkov, V. and Warshel, A. (1992) Microscopic Models for Quantum Mechanical Calculations of Chemical Processes in Solutions: LD/AMPAC and SCAAS/AMPAC Calculations of Solvation Energies, *J. Comput. Chem.* **13**, 199–213.
29. Vaidehi, N., Wesolowski, T.A. and Warshel, A. (1992) Quantum Mechanical Calculation of Solvent Free Energies. A Combined Ab Initio Pseudopotential Free Energy Perturbation Approach, *J. Chem. Phys.* **97**, 4264–4271.
30. Waszkowycz, B., Hillier, I.H., Gensmantel, N. and Payling, D.W. (1991) A Combined Quantum Mechanical/Molecular Mechanical Model of the Potential Energy Surface of Ester Hydrolysis by the Enzyme Phospholipase A_2, *J. Chem. Soc., Perkin Trans., 2*, 225–231.
31. Ferenczy, G.G., Rivail, J.-L., Surján, P.R. and Náray-Szabó, G. (1992) NDDO Fragment Self-Consistent Field Approximation for Large Electronic Systems, *J. Comput. Chem.* **13**, 830–837.
32. Théry, V., Rinaldi, D., Rivail, J.-L., Maigret, B. and Ferenczy, G.G. (1994) Quantum Mechanical Computations on Very Large Molecular Systems: The Local Self-Consistent Field Method, *J. Comput. Chem.* **15**, 269–282.
33. Thompson, M.A., Glendening, E.D. and Feller, D.F. (1994) The Nature of K^+/Crown-ether Interactions: A Hybrid Quantum Mechanical/Molecular Mechanical Study, *J. Phys. Chem.* **98** , 10465–10476.
34. Thompson, M.A. and Schenter, G.K. (1995) Excited States of the Bacteriochlorophyll b Dimer of Rhodopseudomonas viridis: A QM/MM Study of the Photosynthetic Reaction Center that Includes MM Polarization, *J. Phys. Chem.* **99**, 6374–6386.
35. Maseras, F. and Morokuma, K. (in press) A New "Integrated Ab Initio + Molecular Mechanics" Geometry Optimization Scheme of Equilibrium Structures and Transition States, *J. Comp. Chem.*.
36. Stanton, R.V., Hartsough, D.S. and Merz, Jr., K.M. (1993) Calculation of Solvation Free Energies Using a Density Functional/Molecular Dynamics Coupled Potential, *J. Phys. Chem.* **97**, 11868–11870.
37. Wei, D. and Salahub, D.R. (1994) A Combined Density Functional and Molecular Dynamics Simulation of a Quantum Water Molecule in Aqueous Solution, *Chem. Phys. Letts.* **224**, 291–296.
38. See the articles in (1977) *Semiempirical Methods of Electronic Structure Calculation, Modern Theoretical Chemistry*, vols. 7 and 8, G.A. Segal (ed.), Plenum Press, New York.

39. Freed, K. (1983) Is There a Bridge between Ab Initio and Semiempirical Theories of Valence?, *Acc. Chem. Res.* **16**, 137–144.
40. Szabo, A. and Ostlund, N. (1989) *Modern Quantum Chemistry: Introduction to Advanced Electronic Structure Theory*, McGraw-Hill, New York.
41. Hehre, W.J., Radom, L., Schleyer, P.v.R. and Pople, J.A. (1986) *Ab Initio Molecular Orbital Theory*, J. Wiley and Sons, New York.
42. Langlois, J.-M., Muller, R.P., Coley, T.R., Goddard III, W.A., Ringnalda, M.N., Won, Y. and Friesner, R.A. (1990) Pseudospectral Generalized Valence-Bond Calculations: Applications to Methylene, Ethylene and Silylene, *J. Chem. Phys.* **92**, 7488–7497.
43. Ringnalda, M.N., Belhadj, M. and Friesner, R.A. (1990) Pseudospectral Hartree-Fock Theory: Applications and Algorithmic Improvements, *J. Chem. Phys.* **93**, 3397–3407.
44. Won, Y., Lee. J.-G., Ringnalda, M.N. and Friesner, R.A. (1991) Pseudospectral Hartree-Fock Gradient Calculations, *J. Chem. Phys.* **94**, 8152–8157.
45. Pople, J.A. and Beveridge, D.L. (1970) *Approximate Molecular Orbital Theory*, McGraw-Hill, New York.
46. Schroder, S. and Thiel, W. (1985) Comparison of Semiempirical and Ab Initio Transition States, *J. Am. Chem. Soc.* **107**, 4422–4430.
47. Schroder, S. and Thiel, W. (1986) Correlation Effects on Semiempirical Transition States, *J. Am. Chem. Soc.* **108**, 7985-7989.
48. Thiel, W. and Voityuk, A.A. (1992) Extension of the MNDO Formalism to d-Orbitals: Integral Approximations and Preliminary Numerical Results, *Theor. Chim. Acta* **81**, 391–404.
49. Thiel, W. (1988) Semiempirical Methods: Current Status and Perspectives, *Tetrahedron* **44**, 7393–7408.
50. Holder, A.: Semichem Inc., 12715 West 66th Terrace, Shawnee, KS 66216, USA.
51. Gonzalez-Lafont, A., Truong, T.N. and Truhlar, D.G. (1991) Direct Dynamics Calculations with Neglect of Diatomic Differential Overlap Molecular Orbital Theory with Specific Reaction Parameters, *J. Phys. Chem.* **95**, 4618–4627.
52. Hohenberg, P. and Kohn, W. (1964) Inhomogeneous Electron Gas, *Phys. Rev.* **136**, B864–B871.
53. Kohn, W. and Sham, L.J. (1965) Self-Consistent Equations Including Exchange and Correlation Effects, *Phys. Rev.* **140**, A1133–A1138.
54. See the articles in (1990) *Advs. in Quantum Chemistry* **21**, S.B. Trickey (ed.), Academic Press, New York.
55. Salahub, D. (1987) Transition-Metal Atoms and Dimers, *Adv. Chem. Phys.* **69**, 447–520.
56. Andzelm, J. and Wimmer, E. (1992) Density Functional Gaussian-Type-Orbital Approach to Molecular Geometries, Vibrations and Reaction Energies, *J. Chem. Phys.* **96**, 1280–1303.
57. Yang, W. (1991) Direct Calculation of Electron Density in Density Functional Theory, *Phys. Rev. Letts.* **66**, 1438–1441.
58. Gallia, G. and Parrinello, M. (1992) Large Scale Electronic Structure Calculations, *Phys. Rev. Letts.* **69**, 3547–3550.
59. Chang, Y.-T. and Miller, W.H. (1990) An Empirical Valence Bond Model for Constructing Global Potential Energy Surfaces for Chemical Reactions of Polyatomic Molecular Systems, *J. Phys. Chem.* **94**, 5884–5888.
60. Bernardi, F., Olivucci, M., McDouall, J.J.W. and Robb, M.A. (1988) Parameterization of a Heitler-London Valence Bond Hamiltonian from Complete-Active-Space Self-Consistent-Field Computations: An Application to Chemical Reactivity, *J. Chem. Phys.* **89**, 6365–6375.
61. Bernardi, F., Olivucci, M. and Robb, M.A. (1992) Simulation of MC-SCF Results on Covalent Organic Multi-Bond Reactions: Molecular Mechanics with Valence Bond (MM-VB), *J. Am. Chem. Soc.* **114**, 1606–1616.

62. Burkert, U. and Allinger, N.L. (1982) *Molecular Mechanics*, Am. Chem. Soc., Washington, D.C..
63. Russel, S.T. and Warshel, A. (1985) Calculations of Electrostatic Energies in Proteins: The Energetics of Ionized Groups in Bovine Pancreatic Trypsin Inhibitor, *J. Mol. Biol.* **185**, 389–404.
64. Van Belle, D., Couplet, I., Prevost, M. and Wodak, S.J. (1987) Calculations of Electrostatic Properties of Proteins: Analysis of Contributions from Induced Protein Dipoles, *J. Mol. Biol.* **198**, 721–735.
65. Sprik, M. and Klein, M.L. (1988) A Polarizable Model for Water using Distributed Charge Sites, *J. Chem. Phys.* **89**, 7556–7560.
66. Sprik, M. (1991) Computer Simulation of the Dynamics of Induced Polarization Fluctuations in Water, *J. Phys. Chem.* **95**, 2283–2291.
67. Jorgensen, W.L., Chandrasekar, J., Buckner, J.K. and Madura, J.D. (1986) Computer Simulations of Organic Reactions in Solution, in D.L. Beveridge and W.L. Jorgensen (eds.), *Computer Simulation of Chemical and Biochemical Systems, Ann. N.Y. Acad. Sci.* **482**, 198–209.
68. Stillinger, F.H. and David, C.W. (1978) Polarizable Model for Water and its Ionic Dissociation Products, *J. Chem. Phys.* **69**, 1473–1484.
69. Stillinger, F.H. and Weber, T.A. (1987) Molecular Dynamics Study of Chemical Reactivity in Liquid Sulfur, *J. Phys. Chem.* **91**, 4899–4907.
70. Stillinger, F.H. and Weber, T.A. (1988) Molecular Dynamics Simulation for Chemically Reactive Substances: Fluorine, *J. Chem. Phys.* **88**, 5123–5133.
71. See, Allinger, N.L., Chen, K., Rahman, M. and Pathaseril, A. (1991) Molecular Mechanics (MM3) Calculations on Aldehydes and Ketones, *J. Am. Chem. Soc.* **113**, 4505–4517, and references therein.
72. Houk, K.N., Paddon-Row, M.N., Rondan, N.G., Wu, Y.-D., Brown, F.K., Spellmeyer, D.C., Metz, J.T., Li, Y. and Loncharich, R.J. (1986) Theory and Modeling of Stereoselective Organic Reactions, *Science* **231**, 1108–1117.
73. Raimondi, L., Brown, F.K., Gonzalez, J. and Houk, K.N. (1992) Empirical Force-Field Models for the Transition States of Intramolecular Diels-Alder Reactions Based upon Ab Initio Transition Structures, *J. Am. Chem. Soc.* **114**, 4796–4804.
74. Blair, J.T., Levy, R.M. and Krogh-Jespersen, K. (1990) Molecular Mechanics Parameters for Electronically Excited States: The (n,π^*) Singlet State of Formaldehyde, *Chem. Phys. Letts.* **166**, 429–436.
75. DeBolt, S.E. and Kollman, P.A. (1990) A Theoretical Examination of Solvatochromism and Solute-Solvent Structuring in Simple Alkyl Carbonyl Compounds. Simulations Using Statistical Mechanical Free Energy Perturbation Methods, *J. Am. Chem. Soc.* **112**, 7515–7524.
76. Sprik, M. and Klein, M.L. (1988) Optimization of a Distributed Gaussian Basis Set using Simulated Annealing: Application to the Adiabatic Dynamics of the Solvated Electron, *J. Chem. Phys.* **89**, 1592–1607.
77. Estrin, D.A., Tsoo, C. and Singer, S.J. (1991) Accurate Non-Local Electron-Argon Pseudopotential for Condensed Phase Simulation, *Chem. Phys. Letts.* **184**, 571–578.
78. Brooks III, C.L. and Karplus, M. (1983) Deformable Stochastic Boundaries in Molecular Dynamics, *J. Chem. Phys.* **79**, 6312–6325.
79. Brooks III, C.L., Brünger, A.T. and Karplus, M. (1985) Active Site Dynamics in Protein Molecules: A Stochastic Boundary Molecular Dynamics Approach, *Biopolymers* **24**, 843–865.
80. Beglov, D. and Roux, B. (1994) Finite Representation of an Infinite Bulk System: Solvent Boundary Potential for Computer Simulations, *J. Chem. Phys.* **100**, 9050–9063.
81. Eisenberg, D. and McLachlan, A.D. (1986) Solvation Energy in Protein Folding and Binding, *Nature* **319**, 199–203.
82. Wesson, L. and Eisenberg, D. (1992) Atomic Solvation Parameters Applied to

Molecular Dynamics of Proteins in Solution, *Protein Science* **1**, 227–235.
83. Böttcher, C.J.F. (1952) *Theory of Electric Polarisation*, Elsevier, Amsterdam.
84. Still, W.C., Tempczyk, A., Hawley, R.C and Hendrickson, T. (1990) Semianalytical Treatment of Solvation for Molecular Mechanics and Dynamics, *J. Am. Chem. Soc.* **112**, 6127–6129.
85. Tomasi, J. and Persico, M. (1994) Molecular Interactions in Solution: An Overview of Methods Based on Continuous Distributions of the Solvent, *Chem. Rev.* **94**, 2027–2094.
86. Kang, Y.K., Némethy, G. and Scheraga, H.A. (1987) Free Energies of Hydration of Solute Molecules. 1. Improvement of the Hydration Shell Model by Exact Computations of Overlapping Volumes, *J. Phys. Chem.* **91**, 4105–4109, and the three subsequent papers.
87. Cramer, C.J. and Truhlar, D.G. (1992) An SCF Solvation Model for the Hydrophobic Effect and Absolute Free Energies of Aqueous Solvation, *Science* **256**, 213–217.

MOLECULAR MECHANICS AND DYNAMICS SIMULATIONS OF ENZYMES

ROLAND H. STOTE[1], ANNICK DEJAEGERE[1,2]
and MARTIN KARPLUS[1,3]

[1] *Laboratire de Chimie Biophysique,*
URA 422 CNRS
Institute Le Bel
Université Louis Pasteur
67000 Strasbourg France

[2] *Ecole Supérieure de Biotechnologie de Strasbourg*
CNRS-UPR 9003
Bd Sébastien Brant
67400 Illkirch-Graffenstaden France

[3] *Department of Chemistry*
Harvard University
12 Oxford Street
Cambridge, MA 02138 USA

G. Náray-Szabó and A. Warshel (eds.), Computational Approaches to Biochemical Reactivity, 153–198.

1. Introduction

Enzymes are involved in most biological processes essential for the functioning of living organisms. For example, the enzymes choline acetyltransferase and acetylcholinesterase are essential for the transmission of signals in the nervous system; the enzyme carboxypeptidase A, found in the intestine during digestion, is involved in peptide and protein hydrolysis; the enzyme thrombin is involved in the cleavage of fibrinogen which plays a principal role in the process of blood clotting; and DNA polymerases are necessary for DNA replication. Since the first x-ray crystal structure of the enzyme lysozyme was obtained (Blake et al. 1965), the relationship between structure and function has been a central issue. A large number of experimental techniques have been used to try and understand enzyme catalysis; these methods include kinetics (Joao and Williams 1993; Fierke and Hammes 1995), x-ray crystallography (Lippard 1995; Wigley 1995) and NMR (Evans 1992; Joao and Williams 1993; Bertini and Viezzoli 1995). From structural studies, e.g., x-ray crystallography and NMR, it has been established that many enzymes undergo large conformational changes upon binding of substrate (Janin and Wodak 1983). In several cases, it has been demonstrated that these conformational changes are necessary for the proper functioning of the enzyme (Pompliano et al. 1990). With the advances in molecular biology, point mutations have been introduced into enzymes (Colman 1990; Johnson and Benkovic 1990; Plapp 1995) to test proposed enzymatic mechanisms.

A full understanding of how an enzyme functions requires a description of the reaction mechanism at the atomic level. Although kinetic measurements can establish the general mechanism of an enzyme, they do not provide information about the structural changes along the reaction path. Structural studies which examine ground-state properties and inactive complexes involving substrate analogs, are static results that do not give direct information on the transition states. Computer simulations, on the other hand, can supply both structural and kinetic information, but they are limited by inaccuracies in the results. Thus, the only way to obtain a full understanding of enzymatic reactions is to combine theoretical and experimental information.

Theoretical approaches to the study of protein structure and function draw upon concepts and techniques from theoretical chemistry, including molecular mechanics, reaction dynamics, quantum and statistical mechanics (Brooks et al. 1988). Molecular mechanics methods were first developed for the study of small organic molecules (Barton 1948; Westheimer 1956; Hendrickson 1961; Wiberg 1965; Allinger et al. 1967;

Lifson and Warshel 1968). Since the first molecular dynamics simulation of a protein with empirical energy functions (McCammon et al. 1977), the role of these methods in biology has increased over the years (Brooks et al. 1988; Karplus and Petsko 1990; Warshel 1991). Simulations have been used for the study of the structure, dynamics and thermodynamics of molecules of biological interest. The increase in speed and availablilty of computers has made it possible to address a wide range of interesting problems. The acceptance of theoretical methods in biology is based on the validation of the results by comparison with experimental measurements. The use of molecular dynamics for the refinement of X-ray crystal structures (Brünger et al. 1987; Brünger and Karplus 1991) and the determination of NMR structures (Kaptein et al. 1986) has also played a role.

Molecular mechanics and molecular dynamics simulations are based on a knowledge of the potential energy surface which is represented by an empirical function. The relative simplicity of the empirical energy function makes it appropriate for the study of enzyme systems, where one is often dealing with a large numbers of atoms (10,000 atoms and more, including the solvent environment). Although quantum mechanical calculations can yield the potential surface for small molecules, only empirical energy functions are sufficiently simple to calculate the forces and do molecular dynamics simulations for entire proteins and their environment in a reasonable time. They permit one to identify specific interactions important for the enzymatic reaction in a way that is complementary to experiment. Experimentally determined structures of enzymes complexed with analogs for substrates, intermediates, transition states and products provide a starting point for the theoretical study of these enzymes complexed with the actual molecules involved in the reactions. Many properties of enzymes can be studied by molecular mechanics and dynamics with standard empirical potentials. They include the binding of various species involved in the reaction, conformational changes and, in some cases, the nature of the reaction itself. However, for specific aspects of enzymatic mechanisms, including the energetics and dynamics of bond-breaking and bond-making events, hybrid quantum mechanical molecular mechanical (QM/MM) methods are needed (Warshel and Karplus 1972; Allinger and Sprague 1973; Field et al. 1990). This method partitions the enzyme into a quantum mechanical region and a larger molecular mechanical region (Warshel and Levitt 1976; Bash et al. 1991). The article by M. Field in this volume presents a detailed discussion of the method and its applications.

Molecular mechanics and dynamics simulations have been used to study many aspects of protein structure and function; these include the atomic motions and their correlation with experimental results, protein stability, the thermodynamics of ligand binding and more generally, protein-protein interactions. It is, therefore, necessary to limit the scope of this chapter by focusing on the applications to enzymes and, more precisely, the use of these theoretical methods to elucidate the catalytic mechanism. We begin with an overview of the general methodology and a discussion of some techniques. The importance of including long-range electrostatic interactions is pointed out and some recent developments in this area are described. An account of some applications of molecular dynamics and energy minimization to the study of enzymes is given. Although QM/MM (Field et al. 1990; Bash et al. 1991) and related methods (Warshel and Weiss 1980) are needed for a full study of enzyme catalysis, we show that MM methods have been used to examine aspects of the mechanism in a number of cases. This review is not intended to be comprehensive; instead some examples are discussed in detail to illustrate the methods and their applications. Other reviews address different aspects of computer modeling of enzyme catalysed reactions (Mulholland et al. 1993; Mulholland and Karplus 1996). Although we focus on enzymes, the techniques and potential functions are general and can be used to study other proteins, nucleic acids, lipids and oligosaccarides.

2. Potential Energy Function and Methodology

Molecular dynamics simulations require a knowledge of the energy of the system as a function of the atomic coordinates. The forces acting on the atoms are obtained from the first derivative of the potential with respect to the atomic positions. Given the forces acting on the atoms, Newton's equations of motion can be solved for all the atoms as a function of time. The potential energy functions for studying proteins are composed of terms related to bond lengths, bond angles, dihedral angles, van der Waals and electrostatic interactions. One widely used expression has the form (Brooks et al. 1983):

$$E(R)=\sum_{1,2\,pairs}\frac{1}{2}K_b(b-b_0)^2+\sum_{angles}\frac{1}{2}K_\theta(\theta-\theta_0)^2 \tag{1}$$

$$+\sum_{dihedrals}K_\phi\left[1+\cos(n\phi-\delta)\right]\quad+\sum_{i,j}\left\{4\varepsilon_{ij}\left[\left(\frac{\sigma_{ij}}{r_{ij}}\right)^{12}-\left(\frac{\sigma_{ij}}{r_{ij}}\right)^{6}\right]+\frac{q_iq_j}{\varepsilon D r_{ij}}\right\}$$

The energy, E, is a function of the Cartesian coordinate set, R, which specifies the positions of all the atoms in the system. From the coordinates, R, the internal coordinates for bond lengths (b), bond angles (θ), dihedral angles (ϕ) and the interatomic distances (r_{ij}) can be calculated. The first term in Eq. (1) is a harmonic potential representing the displacement from the ideal bond length, b_0, this is the first approximation to the energy of a bond as a function of its length. The force constant, K_b, determines the flexibility of the bond and can be evaluated from infrared stretching frequencies or quantum mechanical calculations. Ideal bond lengths can be inferred from high resolution, low temperature crystal structures or microwave spectroscopy data. The energy associated with alteration of bond angles, which is given by the second term in Eq. 1, is also represented by a harmonic potential. For rotations about bonds, torsion angle potential functions given by the third term in Eq. 1 are used. This potential is assumed to be periodic and is modeled by a cosine or sum over cosine functions. The final term in Eq. 1 represents the contribution of nonbonded interactions. This term is composed of a repulsive term which prevents atoms from interpenetrating at very short distances; an attractive term accounting for the London dispersion forces between atoms; and an electrostatic term that is attractive or repulsive depending on whether the charges q_i and q_j are opposite or equal in sign. The first two nonbonded terms combine to give the Lennard-Jones 6-12 potential, which has a minimum at an interatomic separation equal to the sum of the van der Waals radii of the atoms. The parameters σ_{ij} and ε_{ij} depend on the atoms involved and have been determined by a variety of methods, including non-bonding distances in crystals and gas-phase scattering measurements. The electrostatic interaction between a pair of atoms is represented by a Coulomb potential; D is the effective dielectric function for the medium and r_i is the distance between the two charges. The usefulness of empirical energy functions depends on the accuracy of the results obtained for macromolecules and solvent. This requires that the parameters in the equation be transferred from simple to complex systems. For evaluating the internal parameters, K_b, K_θ, K_ϕ, small model compounds are used and comparisons are made to the geometry and vibrational spectra as determined in the gas phase by IR and Raman spectroscopy, supplemented with ab initio quantum calculations. The external parameters, the Lennard-Jones parameters σ_{ij} and ε_{ij} and the atomic charges, were derived by fitting ab initio calculations on model compounds interacting with water, by comparing the calculated and experimental heats of solvation, and by intermolecular distances in crystals structures. A great deal of effort is required to determine the parameters in the potential function. The functions that are now available are adequate for macromolecular simulations, such as those described in this review. A state-of-the-art example is the CHARMM22 all atom potential function for proteins (MacKerell et al 1996), nucleic acids (MacKerell et al. 1995), lipids (Schlenkrich et al. 1996) and carbohydrates (Ha et al. 1988).

Given a potential energy function, one may use any of a variety of approaches to study enzymes and proteins, in general (Brooks et al. 1988). The most detailed information comes from molecular dynamics simulations, in which one solves Newton's equations of motion for all the atoms of the protein and the surrounding solvent. With currently available computers, one can simulate small proteins in solvent for periods of up to several nanoseconds. This is long enough to characterize the librations of small groups in the protein and to determine the dominant contribution to the atomic fluctuations.

To begin a dynamics simulation, one must have an initial set of atomic coordinates and velocities. If one is concerned with a folded protein, the coordinates are usually obtained from the x-ray or NMR structure of the protein. If using an x-ray coordinate set, hydrogen atom positions have to be introduced (Brünger and Karplus 1988). An iterative energy minimization is done to remove short van der Waals contacts and strained bond angles. Initial random velocities are assigned from a Boltzmann distribution corresponding to a low temperature and a dynamics simulation is performed for a few picoseconds. New velocities are assigned at a higher temperature followed by continued dynamics simulations; this is repeated until the temperature, which is measured in terms of the mean kinetic energy for the system of N atoms, equals the desired temperature. The equilibration period is considered finished when no systematic changes in the temperature are evident over a time period of at least 10 psec. The production simulation, which provides the coordinates and velocities for all the atoms as a function of time, is then performed by continuing to integrate the equations of motion for the desired time period. Different algorithms for integrating the equations of motion allows one to simulate proteins under well defined thermodynamic states (Andersen 1980; Nosé and Klein 1983; Berendsen et al. 1984; Hoover 1985; Allen and Tildesley 1989).

Verification of the structural and dynamic behavior of proteins predicted by simulations has been possible by comparison with experiment. Approximate agreement between the average structure obtained in the first simulation of a protein (McCammon et al. 1977), that of bovine pancreatic trypsin inhibitor (BPTI), and the structure determined by X-ray crystallography suggested that simulations were stable and meaningful results could be obtained. The fast motions sampled in simulations leave their trace in X-ray diffraction data as they cause the atom to occupy more space on average than in a rigid molecule. This spreads out the electron density, the quantity determined in crystallographic studies. If an isotropic Gaussian model is used to fit the spread in density around the average position of each atom, the width of the Gaussian can be expressed in terms of a Debye-Waller factor of the form

$$W_i(Q) = -\frac{8}{3}\pi^2 \langle \Delta r_i^2 \rangle s^2 = -B_j s^2 \qquad (2)$$

where $W_i(Q)$ is the Debye-Waller factor, Q is the scattering vector, $s = |Q|/4\pi$, $\langle \Delta r_i^2 \rangle$ is the mean-square fluctuation of atom i and B_j is usually referred to as the temperature factor. Comparisons of $\langle \Delta r_i^2 \rangle$ values from molecular dynamics simulations and measured B factors show good qualitative agreement (Ichiye et al. 1986; Arnold and Ornstein 1992), although quantitative differences are nonnegligible. These are in part due to errors in the simulation and in part due to the fact that the experimental B factors have contributions other than those arising from internal motion (Hartmann et al. 1982).

To simplify calculations, additional approximations can be introduced. In studying enzymes, for example, one is most often interested in the active site and less in the detailed behavior of the rest of the protein. The stochastic boundary molecular dynamics method (Brooks et al. 1985) allows one to simulate a small region of a protein, for example, the active site of an enzyme while including the effects of the remaining part of the protein in a more approximate way. In implementing this model, a formal partitioning of the protein/solvent system into a reaction zone and a reservoir region is done; the reaction zone is further divided into a reaction region and a buffer region (see Fig. 1) and the reservoir region is eliminated and its effects are accounted for through the mean and stochastic boundary forces (Brooks et al. 1985). Spatial criteria distinguish the reaction region atoms from buffer region atoms. These two regions differ primarily in the way the atomic dynamics are calculated, i.e., in the reaction region, atoms are treated by conventional molecular dynamics while the dynamics of the atoms in the buffer region are calculated within the framework of the Langevin model. Appropriate mean and stochastic boundary forces, which serve to maintain the structural integrity of the system and to mimic the fluctuating dynamics of the reservoir region are added. This method for molecular dynamics has been employed in simulations of

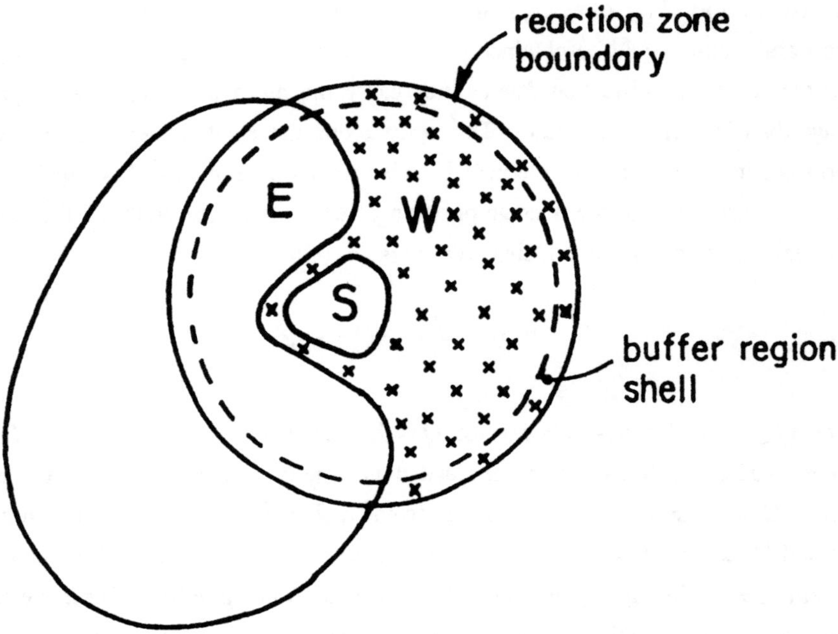

Figure 1. Schematic partitioning of an enzyme (E)-substrate (S)-water (W) system into a spherical zone and surroundings (from Brooks et al. 1988).

several solvated enzymes like RNase A (Brünger et al. 1985; Haydock et al. 1990; Straub et al. 1994) and lysozyme (Brooks and Karplus 1989). A simple extension of the original stochastic boundary method was employed in the study of the zinc binding sites in carboxypeptidase A and carbonic anhydrase (Stote and Karplus 1995), where long-range electrostatic interactions with more distant groups are important due to the presence of the highly charged zinc ion (+2). In this application of the stochastic boundary method, the Extended Electrostatics model, see below, is used to calculate the electrostatic effects from atoms in the reservoir region, which was included as a background in the system.

Another aspect of enzyme function, which can be readily studied by methods based on empirical energy functions, are large scale atomic displacements, such as domain motions and conformational transitions of loops. The energy as a function of the relative positions of domains or loops can be obtained through several techniques

(McCammon et al. 1976; Bruccoleri et al. 1986). The most direct approach is the method of adiabatic minimization (Warshel and Karplus 1974), where the coordinates involved directly in the conformational change are constrained, and the other coordinates are allowed to relax by energy minimization. A more fundamental approach is an iterative procedure to determine the normal mode or set of normal modes of the molecule that has the largest overlap with a simple model for the motion (Brooks and Karplus 1985). Alternatively, a full normal mode calculation can be preformed and the modes associated with the motion of interest can be selected (Levitt et al. 1985). For the study of loop transitions in proteins, high temperature simulations as well as more specialized algorithms which find reaction pathways and transition states can be used (Elber and Karplus 1987; Fischer and Karplus 1992). The Conjugate Peak Refinement algorithm (Fischer and Karplus 1992) serves to obtain the saddle points which are connected to one another through minimum energy paths forming a reaction coordinate between a given reactant structure and a given product structure. In the low temperature limit where atomic fluctuations are at a minimum, this reaction coordinate is the most probable path for the transition; at normal temperatures, it is likely to be a good approximation to the path obtained by averaging over dynamic pathways, particularly if the conformational change is dominated by the energy rather than by entropy. The algorithm has been successfully applied to the study of the tyrosine ring flip in BPTI (Fischer 1992) and to the mechanism of rotamase catalysis by FKBP (Fischer et al. 1993).

An important issue when using Eq. 1 especially when charged species, e.g. acidic and basic protein sidechains, are present, is the effect of the truncation that is commonly applied to the long-range interactions. Truncation is usually introduced to reduce the computational cost of simulations. This is important because for a protein like carboxypeptidase A, where there are 2439 heavy atoms, there are on the order of 6×10^6 interactions in vacuum with no truncation. In protein simulations, several different models for the long-range electrostatic truncation are available; the van der Waals interactions are sufficiently short-ranged and uniformly distributed that truncation is straightforward (Allen and Tildesley 1989). Standard truncations are the shift and switch models (Brooks et al. 1983) in which the non-bonded interactions are smoothly reduced to zero at the cutoff distance; the two models differ in the way they do this. In the shift truncation, the entire potential energy function is shifted such that the interaction at the cutoff is zero, while in the switch model, the electrostatic interactions are reduced to zero over a range of 2 to 5 Å from the cutoff distance. Standard cutoffs have been in the range of 8 to 10 Å, although longer cutoffs are being used (12 to 15 Å) as more computer time becomes available. It has been shown that the method of truncation and the cutoff distance can affect the properties of solutions (Brooks et al. 1985; Brooks 1987), of peptides (Schreiber and Steinhauser 1992) and of proteins (Whitlow and Teeter 1986; Loncharich and Brooks 1989). In the case of charged species, truncation can represent a severe approximation; e.g., at 8 to 10 Å, the unshielded Coulomb interaction between two point charges is 41.5 to 33 kcal/mole, respectively, and the force on one charge due to the other is between 5.2 and 3.0 kcal/mole-Å. A number of experimental studies have demonstrated the importance of long-range contributions in protein-ion interactions. For example, Fersht and co-workers have studied the relationship between charged surface residues of the serine protease subtilisin and the pK_a of the active site residue His-64 (Thomas et al. 1985; Russell et al. 1987; Russell and Fersht 1987; Jackson and Fersht 1993); significant contributions come from distances as large as 15 Å. The experimental pKa shifts at this distance corresponds to an interaction energy on the order of 1 kcal/mol (Russell and Fersht 1987; Jackson and Fersht 1993); this reduction is due to the shielding introduced by the solvent (Warshel and Russell 1984, Warshel 1987). Theoretical calculations of the shift in pK_a values based on continuum electrostatic models are in accord with these experimental results (Gilson and Honig 1987; Sternberg et al. 1987). Other experiments have shown that protein - Ca^{+2} affinity can be modulated by electrostatic interactions occurring over distances up to 13 Å (Pantoliano et al. 1988). In a study of azurin, a copper containing protein, the pK_a values of histidine residues 32 and 83 appear to be affected by the oxidation state of the copper ion, which lies from 12 to 17 Å from the histidine residues (Groeneveld et al. 1988). Calculations of the change in the histidine residue pK_a values induced by oxidation of the copper employing continuum models are in accord with the experimental results (Bashford et al. 1988).

In recent years, a number of models have been introduced which permit the inclusion of long-range electrostatic interactions in molecular dynamics simulation. For simulations of proteins and enzymes in a crystalline state, the Ewald summation is considered to be the correct treatment for long range electrostatic interactions (Ewald 1921; Allen and Tildesley 1989). Variations of the Ewald method for periodic systems include the particle-mesh Ewald method (York et al. 1993). To treat non-periodic systems, such as an enzyme in solution other methods are required. Kuwajima *et al.* (Kuwajima and Warshel 1988) have presented a model which extends the Ewald method to non-periodic systems. Other methods for treating explicitly long-range interactions

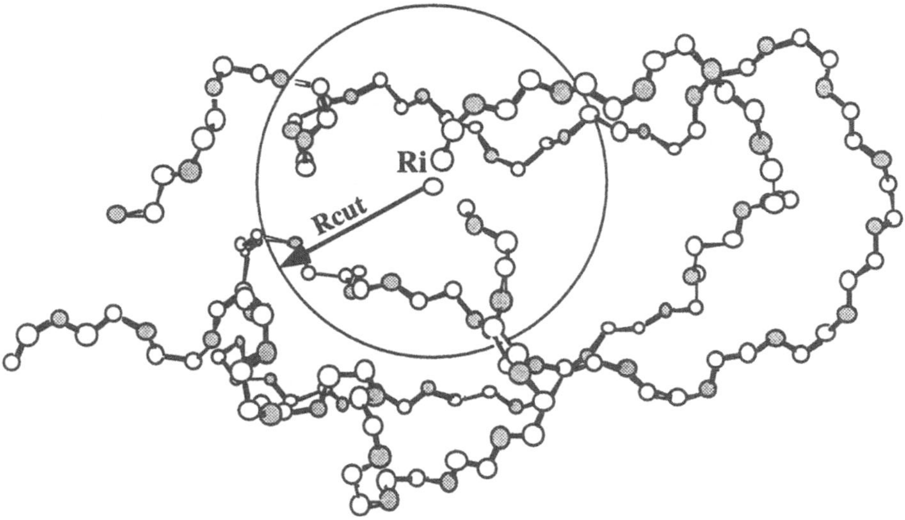

Figure 2. A schematic of the partitioning in the extended electrostatics model. The extended electrostatics model approximates the full electrostatic interaction by partitioning the electric potential and the resulting forces at atom i into a "Near" and an "Extended" contribution. The near contribution, $\phi_{Near}(r_i)$, arises from the charged particles which are spatially close to r_i while the extended contribution, $\phi_{Ext}(r_i)$, arises from the particles which are spatially distant from r_i. In calculating the near contribution, a conventional pairwise additive sum is used and the extended contribution to the potential at r_i is determined via the multipole approximation (from Stote et al. 1991).

is the twin-range cutoff method (Berendsen 1985) where shorter range interactions use a short cutoff distance while long range interactions use a longer cutoff. A class of models which appear to be particularly promising are those models which use multipole approximations for treating long-range electrostatic interactions (Brooks et al. 1983; Greengard and Rokhlin 1987; Stote et al. 1991; Lee and Warshel 1992; Shimada et al. 1994; Pérez-Jordá and Yang 1995). One model that has been applied to the study of enzymes and other proteins where charged species play an important role is the Extended Electrostatics model. The Extended Electrostatics model, described in detail elsewhere (Brooks et al. 1983; Stote et al. 1991), approximates the full electrostatic interaction by partitioning the electric potential and the resulting forces at atom i into a "near" and an "extended" contribution, see Fig. 2. Detailed tests of the Extended Electrostatics model have indicated that the model accurately approximates the full electrostatic field in a simulation of the bovine pancreatic trypsin inhibitor (Stote et al. 1991). Structural and energetic quantities obtained from the Extended Electrostatics simulation compare well to the results obtained from a simulation without a cutoff and the computational cost for using such an Extended Electrostatics algorithm is much smaller. The Extended Electrostatics model has been used in the simulation of binding interactions in the RNase A/3'-UMP enzyme-product complex (Straub et al. 1994). This study showed that inclusion of the Extended Electrostatics contribution was necessary for the uridine phosphate to remain correctly positioned in the active site. The simulations of the zinc enzymes carboxypeptidase A and carbonic anhydrase using this model show, particularly in the case of carbonic anhydrase, that long range electrostatic interactions are important for maintaining the correct geometry of the zinc binding site (Stote and Karplus 1995). The simulations of zinc enzymes are described in the applications section of this review.

In what follows we present some applications of the methodology described here to a number of different systems.

3. Applications

3.1. LYSOZYME

3.1.1. *Catalytic mechanism*

Lysozyme is a key system in the development of our understanding of the structure and function of proteins. It was the first enzyme whose x-ray structure was dètermined at high resolution (Blake et al. 1965), one of the earliest enzymes for which a detailed reaction mechanism was proposed and one of the test systems for molecular dynamics simulations. In this section, we review some of the simulation studies of lysozyme with emphasis on the reaction mechanism and on the large scale domain motion involved in the mechanism.

Lysozyme hydrolyses oligoglycosides. It is a β-glucosidase that cleaves a β-glucoside bond and releases β-glucose. Although lysozyme was the first enzyme to have its high resolution crystal structure determined (Blake et al. 1965) and the catalysis has been studied for many years, the mechanism is not fully understood and remains a source of discussion (Kirby 1987; Mooser 1992). Two features of the hydrolysis mechanism are unresolved: (1) Whether the initial bond-breaking occurs via an *endo*cyclic (Fig 3; Scheme I) or *exo*cyclic (Fig 3; Scheme II) C--O bond and

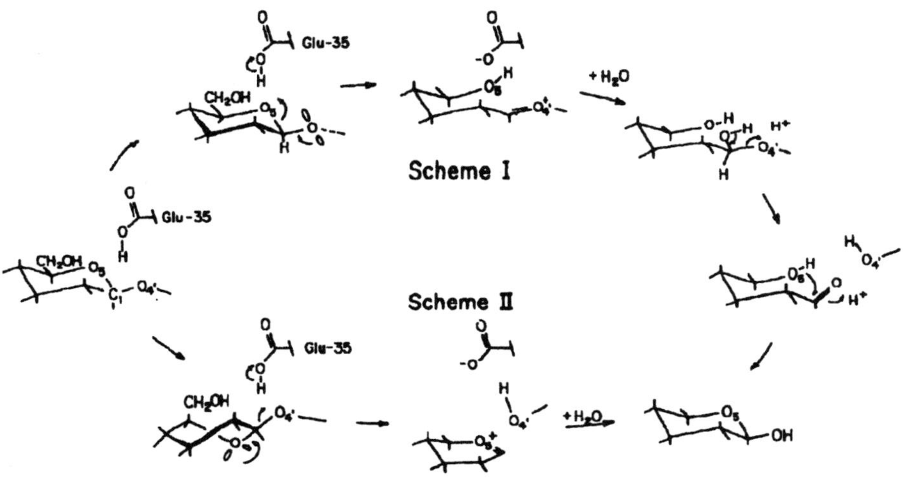

Figure 3. Scheme I and II showing proposed lysozyme mechanisms (from Post and Karplus 1986).

(2) Whether the oxocarbonium ion is stabilized by electrostatic interaction with Asp 52 or collapses to a covalent intermediate (Koshland 1953; Hadfield et al. 1994). The former question concerning the lysozyme mechanism was brought forward only relatively recently (Fleet 1985; Post and Karplus 1986), although it has a long history in glycoside chemistry (Bunton et al. 1955; Shafizadeh 1958; Miller 1967), while the latter originated with the initial proposal based on modeling a substrate into the active site cleft (Blake et al. 1967), which was in conflict with the mechanism put forward by Koshland (Koshland 1953) and recently supported by Sinnott (Sinnot 1980). It should be noted that the mechanistic features of glycoside hydrolysis need not be common to all β-glycosidases. It is possible, for example, that the active site of lysozyme is structured to promote hydrolysis by *endo*cyclic ring-opening, while that of another β-glycosidase stabilizes *exo*cyclic cleavage.

The original mechanism for lysozyme was proposed by Phillips and co-workers on the basis of model building and data for the non-enzymatic hydrolysis of glycosides. An essential element of this proposal is the distortion of the N-acetlyglucosamine residue in site D, see scheme II, Fig 3. The resulting twist-boat conformation makes it possible to take advantage of stereoelectronic assistance from the ring oxygen O5 in the transition state leading to cleavage of the exocyclic C1-O4' bond. Protonation of O4' by Glu35 yields the cyclic oxycarbonium ion which can be stabilized by the carboxylate group of Asp52. Much of Scheme II is consistent with experimental results (Blake et al. 1967; Vernon 1967; Imoto, Johnson et al. 1972; Walsh 1979), but, see below, there exists conflicting data on the required ring distortion. Strynadka and James (Strynadka and James 1991) have recently concluded that ring distortion is present in a trisaccharide, lysozyme complex. However other studies have concluded that ring distortion is not present (Schindler et al. 1977; Patt, Baldo et al. 1978; Kelly et al. 1979). The required ring distortion is not in accord with energy minimization results (Warshel and Levitt 1976; Pincus and Scheraga 1979; Pincus and Scheraga 1981).

An alternative pathway for the hydrolysis of oligoglycosides by lysozyme that differs from the original mechanism (Blake et al. 1967; Imoto, Johnson et al. 1972) was suggested based on the results of a molecular dynamics simulation of the active complex involving hen egg white lysozyme and hexa-(N-acetylglucosamine), $(GlcNAc)_6$ (Post and Karplus 1986; Post et al. 1990). In the initial step of the alternative pathway, an *endo*cyclic bond is broken in contrast to the *exo*cyclic bond cleavage in the accepted mechanism. An essential element of the accepted mechanism is distortion of the glucosidic ring occupying site D of the active site cleft away from the low-energy chair conformation (Strynadka and James 1991). The distortion is presumed to promote breaking of the *exo*cyclic C1--O4' bond following protonation by Glu35, and is intrinsic to formation of the oxocarbonium ion. The alternative mechanism does not require ring distortion (Post and Karplus 1986; Post et al. 1990), see scheme I, Fig. 3. The initial step is protonation of the endocyclic O5 by Glu 35. Cleavage of the endocyclic C1-O5 bond leads to an acyclic oxocarbonium ion intermediate. (A mechanism that includes a covalent intermediate following the endocyclic cleavage has been described by Franck (Franck 1992).) Attack by water, cleavage of the exocyclic C1-O4' bond, and ring closure lead to the final products. The accessibility to a water molecule before dissociation of the disaccharide occupying sites E and F has been questioned by Strynadka and James (Strynadka and James 1991), based on a refined x-ray structure. They have also concluded that ring distortion is present in a trisaccharide, lysozyme complex.

Both mechanisms are consistent with available mutational studies demonstrating that Glu35 and Asp52 are essential residues (Kuroki et al. 1986) and with available kinetic data (Franck 1992) that supports the formation of an oxycarbonium ion in the rate-limiting step and retention of stereochemistry at the anomeric carbon.

Efforts to trap an open-ring intermediate in the enzyme have not been successful (J. Knowles, private communication). However, evidence is accumulating that endocyclic, as well as exocyclic, cleavage contributes to non-enzymatic hydrolysis (Bunton et al. 1955; Shafizadeh 1958; Miller 1967; Franck 1992; Liras and Anslyn 1994).

Binding and structural studies have been interpreted in support of a ring distortion, but these results are complicated by ambiguous interpretations of binding affinities (Schindler et al. 1977), or by contributions irrelevant to the lysozyme mechanism since they involve unnatural substrates, see reference 20 of Post & Karplus, (Post and Karplus 1986), or inactive complexes which may not mimic conformations along the reaction coordinate. The modelling studies (carried out prior to originating the proposal for an endocyclic mechanism) have shown that positioning of a substrate in the active site with a favorable binding energy is possible without distortion from the low-energy chair conformation (Post et al. 1986; Post et al. 1990). More details concerning these calculations and their significance for the lysozyme mechanism are given in Karplus and Post (Karplus and Post 1996). Warshel and co-workers have studied the energetics of the reaction mechanism using the Empirical Valence Bond and other methods (Warshel and Levitt 1976; Warshel 1978; Warshel and Weiss 1980; Warshel 1981; Warshel 1991). QM/MM calculations of the mechanism are in progress (Mulholland, Post and Karplus, private communication).

3.1.2. *The Hinge-Bending Mode of Lysozyme*

A frequent motif in proteins (Richardson 1981) is a double-lobed structure that could make possible a large-scale, low-frequency vibration of the lobes about a hinge axis at their junction. Lysozyme (Blake, Johnson et al. 1967; Blake et al. 1967), hexokinase (Bennett and Seitz 1980), lactic acid dehydrogenase (White et al. 1976), citrate synthase (Remington et al. 1982), tomato bushy stunt virus coat protein (Harrison et al. 1978), phosphoglycerate kinase (Banks et al. 1979), and the arabinose binding protein (Gilliland and Quiocho 1981) are all examples of two domain proteins with a double-lobed structure. In many of these cases, there is experimental evidence for a hinge-bending motion associated with the lobes. Generally, the active site is located in the region between the lobes, so that their motion is likely to affect the binding and release of ligand (McCammon et al. 1976; McCammon and Northrup 1981). In the case of enzymes, the structural change may contribute directly to the catalytic process, as in hexokinase, lactic acid dehydrogenase, citrate synthase, and phosphoglycerate

TABLE 1. Summary of Hinge-Bending Calculations

Run	$\tilde{\nu}$ [a]	k[b]	Minimum[c]
Steepest descent			
Free lysozyme	4.4	3.6×10^{13}	-1.5
Lysozyme tri-NAG	5.9	6.5×10^{13}	0.8
Constrained			
Free lysozyme	3	1.7×10^{13}	-6.1
Lysozyme tri-NAG	4.3	3.5×10^{13}	4.3

[a] Wave number in units of cm^{-1}.

[b] Force constant, the quadratic coefficient of the parabola fitted to the energy vs angle function, which has units of erg rad^{-2} $mole^{-1}$.

[c] Minimum angle of the fitted parabola is in degrees.

kinase. The lobe motion in hexokinase shields the hexose substrate from the solvent (McDonald et al. 1979; Bennett and Seitz 1980). In citrate synthase, the closed form of the molecule with bound coenzyme A completely isolates the citric acid substrate from solvent whereas the open form does not (Remington et al. 1982). In addition, the two forms in the crystal are catalytically inactive, consistent with the supposition that lobe motion is essential for the complete catalytic reaction.

The first atom based calculation of a hinge bending mode was made for lysozyme where the stiffness of the hinge mode, with and without an inhibitor present, was evaluated by two types of approaches. In the first, a hinge axis was chosen by inspection of the structure and adiabatic minimizations (i.e., the hinge bending coordinate is constrained, while the other coordinates are allowed to relax by minimization) were performed to investigate the energy as a function of the relative positions of the lobes (McCammon et al. 1976; Bruccoleri et al. 1986); in the second, the normal mode corresponding to the hinge bending motion was determined (Brooks and Karplus 1985). The initial guess for the hinge-bending coordinate in the adiabatic minimization was based on a hinge axis that passes through the α carbons of Phe 38 and Lys 97 (Fig. 4). Two protocols were used for the energy minimization: steepest

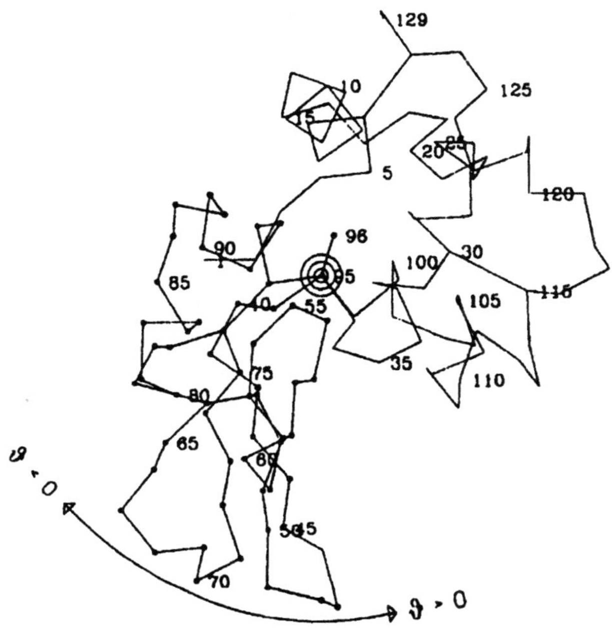

Figure 4. Hen egg white lysozyme and the hinge axis (from Bruccoleri, et al. 1986). This α-carbon drawing is oriented such that hinge axis is perpendicular to the paper, with the axis passing through the center of the bulls-eye in the middle of the picture. The α-carbons for residues in the "moving" lobe are all drawn as small circles. The arc illustrates the sense of the cleft angle where negative angles signify the opening of the cleft and positive angles signify the closing. The disulfides between residues 6 and 127, 30 and 115, 64 and 80, and 76 and 94 are drawn as lines between their respective α-carbons. Minimized coordinates were used.

descents and constrained conjugate gradient. In both protocols, one lobe is rotated rigidly by 10° at a time about the hinge bending axis, keeping the second lobe fixed, followed by energy minimization. The inhibitor, if present, was part of the fixed lobe; tri-N-acetylglucosamine (tri-NAG) was used as the inhibitor.

Figure 5 presents plots of the potential energy vs. angle for the two protocols used in the mode estimate. A least-squares fit of a parabola is drawn through the point. The fit to a parabola is good, although there are some systematic variations of the data points. The frequencies, $\nu(cm^{-1})$, and force constants, k, corresponding to each potential were calculated; Table 1 summarizes the results.

These results are consistent with measurements of Young's modulus of the native triclinic crystals and those soaked with a solution of NAG (Morozova and Morozov 1982). The estimate of the change in the force constant by these measurements, 2×10^{13} erg mole^{-1} rad^{-2}, is comparable to the calculated differences in the force constant between free enzyme and the complex (see Table 1). The presence of the inhibitor increases the frequency of the vibration in all cases and shifts the minimum energy to a more closed configuration. This was expected, as the inhibitor attracts the lobes to each other, but the magnitude of the change may be partly artifactual since no solvent effects were included in the calculations. The steepest descents protocol yields a frequency for the mode of 4.4 cm^{-1} and a minimum energy angle of -1.5° and the constrained conjugate gradient minimization yields a frequency for the mode of 3.0 cm^{-1} and a minimum energy angle of -6.1° for the free lysozyme.

Figure 6 shows the displacement of individual atoms in the free enzyme and in the complex that occur as the structure is closed from the initial structure (0°) to 30°. The 0° structure is shown with circles at each of the α-carbon whose sizes indicate the shifts that result from the minimizations performed after each 2° bend. The secondary structure elements do not shift as rigid units within each lobe. There are significant displacements of individual atoms and residues. The results with and without the inhibitor are very similar. The large displacement of Lys 96 (see figure) is due to the sidechain NH_3^+ maintaining contact with the carbonyl oxygens of His15 and Gly16 in spite of the rotation that moves the Lys 96 backbone away. The large shifts of Glu 35 and Gln 57 (see figure) are due to the rotation forcing the sidechains into contact. That for Gly 71 is due to electrostatic repulsion between its carbonyl oxygen and the carbonyl oxygens of Pro 70 and Ser 72, which are nearby. The atom shifts of the free enzyme and of the complex as the structure is opened by 12° are very similar. Closing gives considerably larger displacements than opening of the cleft; the largest shifts line the surface of the cleft.

Binding the inhibitor, GlcNAC, shifts the minimum to a more closed configuration. As the cleft closes, the accessibility to solvent of the lining of the cleft and inhibitor decreases, and this change in accessibility could play a role in the details of the enzymatic mechanism. In addition, the motion of the lobes brings Asp 52 and Glu 35 closer together. Since the putative mechanism for lysozyme involves these two residues stabilizing a carbonium ion intermediate, having two charges in proximity could increase the stabilization energy for this intermediate if they are correctly positioned.

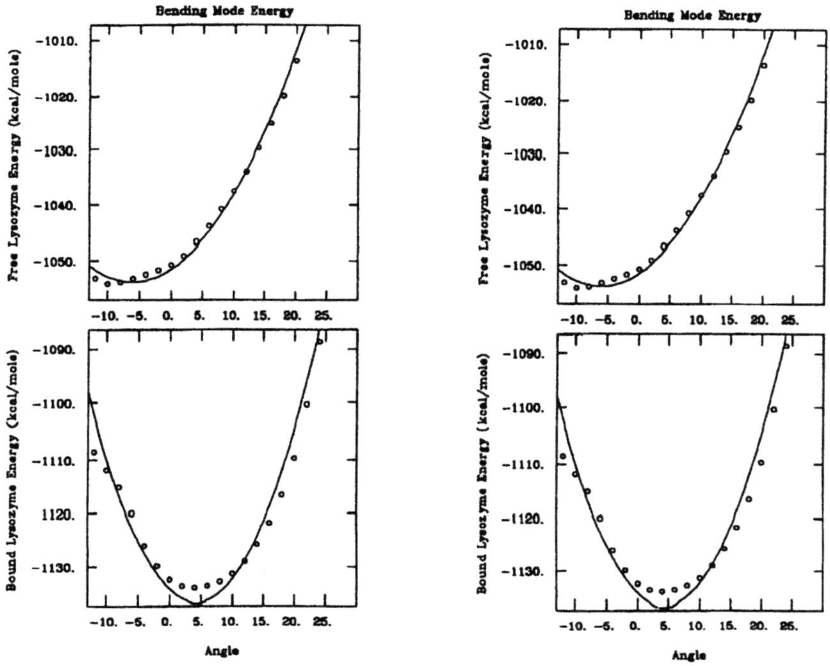

Figure 5. The hinge-bending potential for free and bound lysozyme systems (from Bruccoleri et al. 1986): (a) steepest descents protocol; (b) spatially constrained protocol. The circles are the points calculated by the protocols, the line is the least-squares fit parabola. All plots are on the same scale with energies in units of kcal/mole and angles in degrees. Positive angles correspond to closing.

The detailed energetics of the mode show that the dominant term in the hinge-bending potential is the van der Waals interaction. The shift in minimum upon binding the inhibitor is also largely due to the van der Waals term. Many residues, all clustered around the plane passing through the hinge and the center of the cleft, are involved in this interaction. The electrostatic, hydrogen bond, and bond angle interactions are also significant contributors to the mode. However, for these terms, a smaller number of residues contribute to the mode. The electrostatic interaction has the largest change in angular dependence upon binding the inhibitor.

The normal mode calculation of the hinge bending motion does not require the definition of a rotation axis. An iterative procedure (Brooks and Karplus 1985)

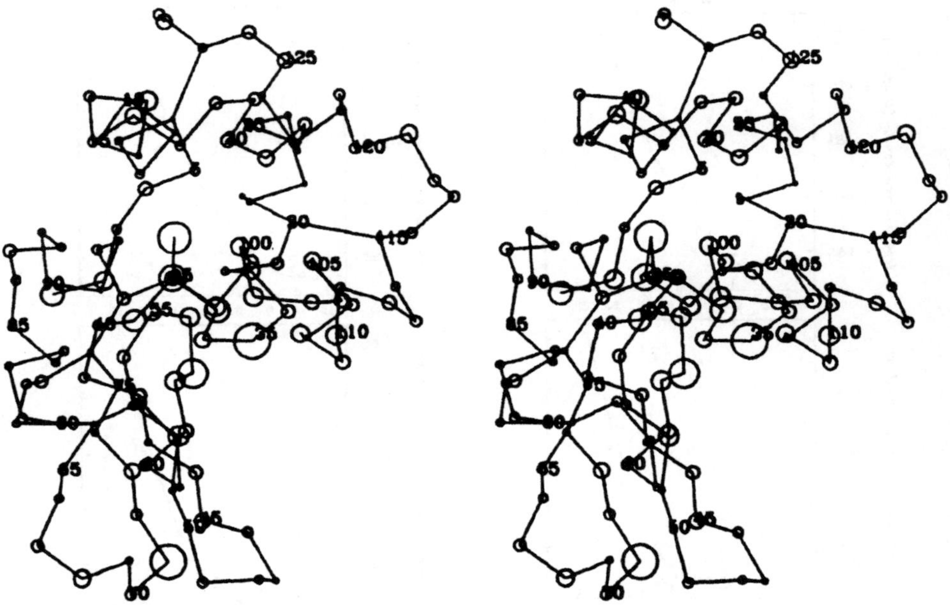

Figure 6a. A stereo α-carbon stick drawing of free lysozyme (from Bruccoleri et al. 1986), where the size of a circle associated with each α-carbon indicates the rms shift of the atoms in each residue resulting from the bending of the system from the minimized 0° structure to the 30° closed structure. The rotation of the lobes is removed from the calculation, so only relative shifts are considered here. In the drawing of the inhibitor, all nonhydrogen atoms are drawn, but no scaled cricles are used because the rms shifts for the inhibitor were quite large and they obscured the cleft. Their diameter in the drawing would be approximately the width of the cleft. In addition to plotting lines between each sequential α-carbons, four additional lines are drawn between α-carbons of residues involved in disulfide bridges. The sizes of the circles are scaled such that a rms shift of 1Å for a residue will result in a circle with diameter of 1cm.

determines the normal mode or modes of the system that overlaps most with a starting model for the motion. In the application to lysozyme, the rigid body mode in McCammon et al. (McCammon et al. 1976) was used as the initial guess. Such a procedure, once it converges, provides a description of the motion that is independent of

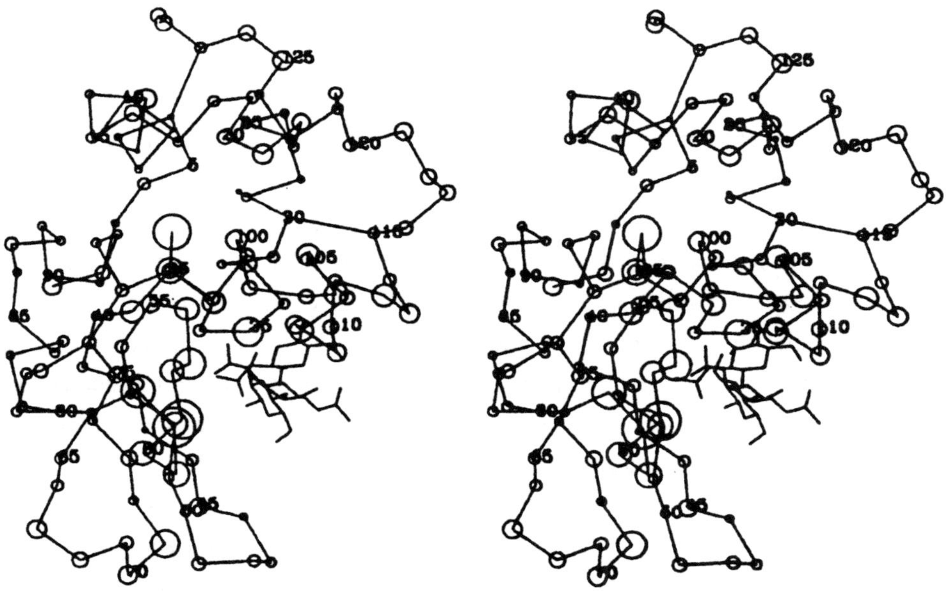

Figure 6b. As in Figure 6a, but for the complex (from Bruccoleri et al. 1986).

the details of the initial guess. Application of the method to the hinge bending mode in lysozyme yielded a frequency (3.6 cm^{-1}) very close to the value (3-5 cm^{-1}) obtained in the energy-minimization studies. However, analysis of the form of the converged mode showed that the changes associated with the hinge bending are more widely distributed over the molecule. Of particular interest is the essential role played by the tryptophan residues. Clearly this is due in part to their large size and the presence of three of them (Trp-62, Trp-63, and Trp-108) in or near the active site cleft. Tryptophans 63 and 108 rotate with a concerted motion, becoming almost parallel as the hinge closes. However, that Trp-28, which is far removed from the active site, is also important, suggests that a more general effect of tryptophans in stabilizing the structure is likely to be involved.

Most of the information on interdomain motions come from high-resolution crystal structures; several reviews are available (Janin and Wodak 1983; Bennett andHuber 1984; Gerstein et al. 1994). Calculations of hinge bending modes and domain motions in proteins other than lysozyme have been made. They include antibody molecules where the interdomain motions occur on a nanosecond time scale (McCammon and Karplus 1977; Oi et al. 1984), l-arabinose-binding protein (Mao et al. 1982), liver alcohol dehydrogenase (Colona-Cesari et al. 1986); and the mouse epidermal growth factor (Ikura and Go 1993). For the l-arabinose-binding protein, calculation and experiment suggest that the binding site is open in the unligated protein and closes by a hinge-bending motion upon ligation (Newcomer et al. 1981; Mao et al. 1982). In liver alcohol dehydrogenase, an open structure is stable in the apoenzyme (Eklund et al. 1976) and a closed structure is stable for the holoenzyme (Eklund et al. 1981). Adiabatic energy minimization calculations suggest that the apoenzyme is highly flexible and that normal thermal fluctuations could lead to a closed structure similar to that found in the holoenzyme (Colona-Cesari et al. 1986). In the mouse epidermal growth factor (EGF), the binding site for the EGF receptor is believed to be between the N-terminal and C-terminal domains; normal mode analysis indicates that the lowest frequency mode corresponds to hinge-bending motion between these two domains. Domain motion induced by ligand binding has been studied in a model of the G-protein-coupled receptor (Luo et al. 1994) using molecular dynamics simulations. This study characterized the details of ligand-induced helix motions and suggested a mechanism of activation for signal transduction. And in the normal mode calculation of citrate synthase (Marques and Sanejouand 1995), the largest amplitude normal mode compared well with the x-ray crystallographically obsevered hinge bending mode.

3.2 RIBONUCLEASE A

Ribonuclease A (RNase A) is one of the most widely studied enzymes (Richards and Wyckoff 1971; Blackburn and Moore 1982; Fersht 1985). However, despite the numerous experimental and theoretical studies devoted to this enzyme, there is still considerable debate on the details of the reaction mechanism. The enzyme catalyzes the hydrolysis of phosphoester bonds in RNA. RNA hydrolysis, whether in solution or in the active site of enzymes, is thought to proceed in two steps; the first step is transphosphorylation, in which the P-O2' bond is formed and the P-O5' bond is broken, making a 2',3'-cyclic phosphate intermediate; the second step is hydrolysis of the cyclic intermediate, in which the P-O2' bond formed in the first step is cleaved, resulting in a normal 3'-terminal phosphate group. The enzyme is thought to make use of general acid-base catalysis; i.e., an acid protonates the leaving group and a base deprotonates the attacking oxygen in each step. According to the established mechanism, when the substrate first binds to the active site His12 is in a neutral, deprotonated form and His119 is in a charged, protonated form, see Fig 7b. In the transphosphorylation step, His12 acts as the base and accepts the proton from the O2'-hydroxyl while His119 acts as the acid and protonates the O5' leaving group. In the second step, the P-O2' bond of the cyclic phosphate is hydrolyzed by an attacking water molecule. In this second step, His12 is supposed to donate a proton to O2' and act as the general acid while His119 deprotonates the attacking water molecule and acts as the general base. It is assumed that the two steps of the reaction occur sequentially in the enzyme active site before release of the products, when the enzyme is returned to its original protonation state.

A mechanism with more complex roles for His12 and 119 has been proposed by Breslow et al (Breslow et al. 1989, 1993). In the Breslow mechanism, His 12 has the same role as in the classical mechanism during transphosphorylation, i.e. it accepts a proton from the O2' hydroxyl (His12 is neutral when the substrates binds). His119 is charged when the substrates binds and it protonates a phosphoryl oxygen (as opposed to O5' in the classical mechanism). This reduces the charge on the phosphate group and facilitates nucleophilic attack by the incoming O2'. The nucleophilic attack of O2' on the phosphorus leads to the formation of a cyclic pentacovalent phosphorane intermediate. His119 then protonates O5' of the leaving group, while the positively charged His12 stabilizes the phosphorane. Cleavage of the phosphorane then occurs, making a 2',3'-cyclic phosphate intermediate. Thus in the Breslow mechanism, as in the classical one, His119 is neutral and His12 protonated after the transphosphorylation step. Cleavage of the cyclic intermediate occurs by a similar mechanism: His12 protonates a phosphoryl oxygen while His119 activates a water molecule. Nucleophilic attack by the activated water on the cyclic phosphate leads to the formation of a cyclic phosphorane intermediate. His12 regains its proton and then protonates O2' while His 119 stabilizes the phosphorane. After cleavage of the phosphorane, His12 is returned to its neutral from while His119 is protonated. The Breslow proposal is based mainly on mechanistic studies of the cleavage of dinucleotides phosphates by imidazole buffer (Breslow et al. 1989; Breslow 1993; Breslow and Xu 1993). Recently, kinetic studies (Thompson and Raines 1994) have been performed on mutants of RNAse A where either His12 or His119 are replaced by Ala. These studies have shown that the rate of hydrolysis of uridine 3'(p-nitrophenyl phosphate) is not affected by the His119→Ala mutation, while it is affected by the His12→Ala mutation. For substrates with poorer leaving groups such as polycytidylic acid or uridyl(3'→5') adenosine, the rate was affected both with the His12→Ala and the His119→Ala mutant. This suggests that the reaction can proceed without protonation of the phosphoryl oxygen by His119 in the first step, but that His119 is needed to protonate the O5' oxygen of poor leaving groups.

There are several additional active site residues which may be involved in the reaction. The importance of Lys41 is indicated by chemical modifications specific to Lys41, and also by its unusually low pKa value, 8.6-9.1 compared to pKa of 10.6-11.2 for all other lysine residues in the enzyme (Jentoft et al. 1981). Other residues of interest are Lys66 (Walter and Wold 1976), Asp121 (Umeyama et al. 1979) Gln11 and Phe120 (Gilbert et al.; Wodak et al. 1977; Wlodawer 1984). Gerlt and Gassman have proposed that the reaction proceeds through the formation of a phosphorane intermediate which is stabilized by strong hydrogens bonds between the phosphoryl oxygens and Lys41 on one hand and Gln11 or Phe120 on the other hand (Gerlt and Gassman 1993).

Crystal structures of the native enzyme and the enzyme complexed with a number of substrates and substrate analogs have provided important data to supplement the mechanistic studies (Gilbert et al.; Wodak et al. 1977; Wlodawer 1984; Fersht 1985; Wlodawer et al. 1986). In particular, Gilbert et al (Gilbert et al.) have determined the crystal structures of RNase A complexed with a substrate analog (deoxy-CpA), the cyclic phosphate (cyclic-CMP) at low temperature (-70 C), a transition-state analog (uridine vanadate), and a product (3'-UMP). Based on these structures and other data, Gilbert et al have considered specific mechanistic roles for certain residues in catalysis; their conclusions are generally in accord with the standard mechanism. The structural data support the possibility of general acid-base catalysis by His12 and His119 by an in-line mechanism; (Roberts et al. 1969; Usher et al. 1972) and stabilization of the transition state by ionic interactions with Lys41. They also concluded that Thr45 anchors His12 and helps to determine the specificity of RNase A, Asp121 anchors His119 and assists it to deprotonate the water molecule which hydrolyses the cyclic phosphate intermediate, and that Lys7 does not participate in catalysis.

Although the crystallographic studies of RNase A are very important, they do not describe the systems that are actually involved in catalysis. Other than the general difficulty that observed structures correspond to trapped species, there are more specific problems in the case of RNase A. A substrate analog (e.g. deoxy-CpA) rather than a true substrate was studied to prevent the reaction from occurring. The crystal structures have in most cases been obtained at low pH (pH ~ 5.5) rather than in the range where the enzyme is most active (pH 6 to 7); this alters the protonation state of the histidines and may affect the active site hydrogen bonding network. Since the reaction involves proton transfers, the resulting differences are likely to be significant. Finally, certain essential residues (e.g. Lys41) are disordered in most of the structures.

a

b

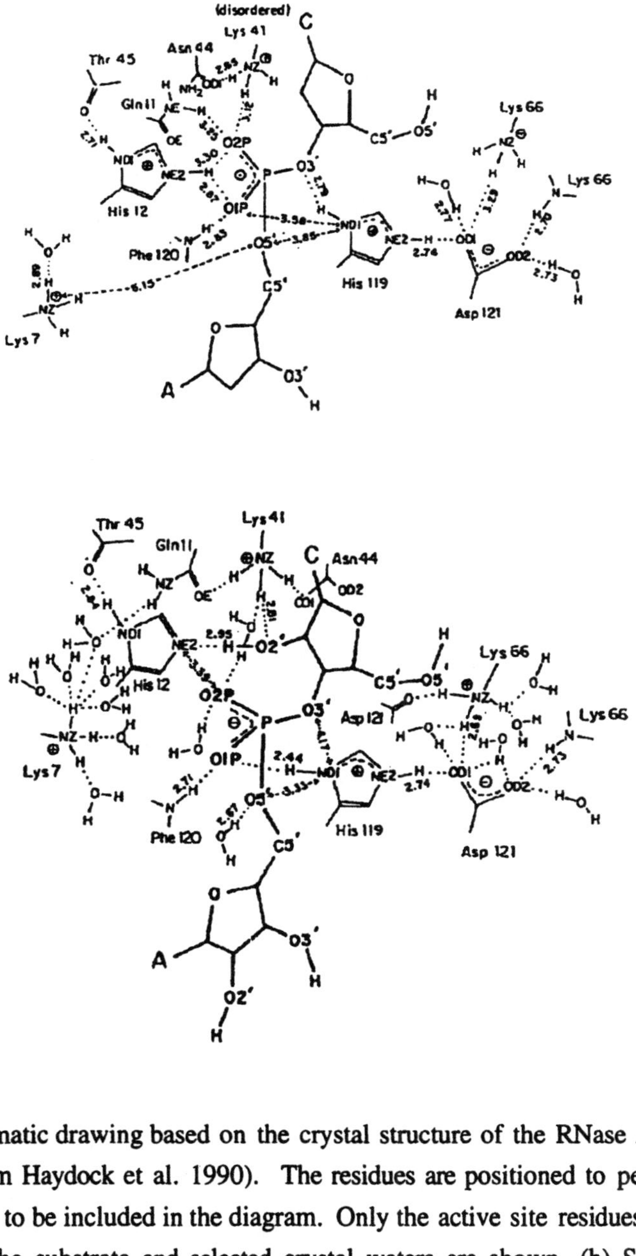

Figure 7. (a) Schematic drawing based on the crystal structure of the RNase A/deoxy-CpA complex (from Haydock et al. 1990). The residues are positioned to permit the important distances to be included in the diagram. Only the active site residues that are in the vicinity of the substrate and selected crystal waters are shown. (b) Schematic drawing based on the average structure of RNase A/CpA with His 12 deprotonated at NE2 obtained from 20 ps dynamics.

Active site, stochastic boundary simulations (Brooks and Karplus 1983; Brünger et al. 1985; Brooks and Karplus 1989) of RNase complexed with a substrate, a cyclic phosphate intermediate and a product have been made to clarify the contribution of various amino acid residues to the enzymatic reaction (Haydock et al. 1990). The structure of the enzyme complexed with a true substrate, CpA, was built from the crystal results for the enzyme complexed with deoxy-CpA, the substrate analog, by introducing the O2'H hydroxyl group; to include the deprotonated His12, the protonated His potential was replaced by that for the neutral form; His119 was protonated in the simulations.

In the RNase A/deoxy-CpA crystal structure (Fig. 7a), where there is no O2' hydroxyl, His12 makes a bifurcated hydrogen bond to the phosphoryl oxygens O1P and O2P of the substrate, His119 hydrogen bonds to O3' and Lys41 is disordered. In the simulations, the interactions of His12 and Lys41 change significantly when the His12 is deprotonated (Fig. 7b). His12 has moved into a position that is in accord with its role in deprotonating O2' during transphosphorylation by the generally accepted mechanism, as well as the Breslow proposal (Anslyn and Breslow 1989; Breslow et al. 1989). Lys41 has also moved significantly and is now positioned with its ammonium group making a strong hydrogen bond with O2'. Experimental support for the reorientation of Lys41 comes from the work of Jentoft et al (Jentoft et al. 1981) who have used NMR to measure titration curves of RNase A in which the lysines were ^{13}C-methylated. They found that the ionization of Lys41 is coupled to the ionization of a histidine residue (perhaps His12), both of which are perturbed upon substrate binding. Atom $N_{\delta 1}$ of His119, which is protonated, hydrogen bonds to O1P throughout the simulation; analysis of the dynamics trajectories shows no contribution of $N_{\delta 1}$-05' bonding. Thus, its role at the start of the transphosphorylation reaction is to make phosphorus more electropositive and facilitate attack by O2'. The position of His119 is also consistent with protonation of one of the phosphoryl oxygens in the formation of a phosphorane intermediate, as proposed by Breslow et al (Breslow et al. 1989). Interestingly, in a molecular dynamics simulation of RNase T1 (Cordes et al. 1995), it was also observed that the protonated His92, the general acid in the RNase T1 transphosphorylation step analogous to His119 in RNase A, hydrogen bonds to a phosphoryl oxygen rather than to O5' at the beginning of the reaction.

This simulation of the CpA substrate complex with a neutral His12, based on the x-ray structure with deoxy-CpA and protonated His12, demonstrated a repositioning of important residues that permits them to fulfill their catalytic role; i.e., His12 reoriented to accept the proton from the O2' hydroxyl (instead of interacting with the phosphate oxygens as in the crystal structure) and Lys41 is sufficiently close to aid in the proton transfer required for transphosphorylation (instead of being far from the substrate). The protonated His 119 interacts with a phosphoryl oxygen, in position to polarize the phosphorous or transfer a proton, as in the Breslow mechanism. These results, together with quantum mechanical calculations on organic phosphates, prompted Lim and Tole (Lim and Tole 1992; Lim and Tole 1992) to propose an alternative mechanism for the transphosphorylation catalyzed by RNase A, where His119 hydrogen bonds to a phosphoryl oxygen while His12 facilitates the intramolecular transfer of the 2'-hydroxyl hydrogen to a phosphoryl oxygen.

The mechanism of the hydrolysis step as well as that of product release from the enzyme have not been studied as thoroughly as the transphosphorylation step. It was believed from early mechanistic studies that transphosporylation and hydrolysis occured sequentially before product release (Fersht 1985). However, ^{31}P NMR was recently used to monitor the accumulation of the cyclic phosphate during the hydroysis of RNA catalyzed by RNase A, RNase T1, barnase and RNase I and it was pointed out that these enzymes released rather than hydrolyzed most of the intermediate formed by transphosphorylation of RNA (Thompson et al. 1994). It was argued that RNases have evolved primarily to catalyze the transphosphorylation of RNA and do not bind substrate water effectively. The molecular dynamics simulations of the RNase A-cyclic phosphate complex show several water molecules in the active site of the enzyme but do not point to one particular water molecule as being involved in attacking the cyclic phosphate. The average dynamics structure of the cyclic-CMP complex obtained in the simulation that started from the crystal structure showed that the residues Lys7, 41 and 66 are hydrogen bonded to water molecules, most of which are not evident in the crystal due to the high lysine mobility. In the cyclic phosphate intermediate the leaving group nucleotide is no longer present so that, in comparison with the CpA complex, additional water molecules can enter the active site. As in the crystal structure, His119 does not interact directly with the phosphate of the substrate. There is one intermediate water in the crystal structure and there are more than one in the dynamics; one of these waters is likely to be involved in attacking the phosphate. His12, which is protonated, now interacts with O2' through a bridging water molecule, instead of making a bifurcated bond to O2' and an equatorial phosphorous oxygen as seen in the crystal structure. The positions of His119 and His12 in both the crystal structure and the average dynamics structure are consistent with the conclusions of Eftink and Biltonen

(Eftink and Biltonen 1983) that the dominant interaction occurs between His12 and the cyclic phosphate, which does not interact directly with His119.

His12 does not interact with a phosphoryl oxygen of the cyclic phosphate in the simulation, as would be required in the Breslow version of the hydrolysis mechanism (Breslow 1993; Breslow and Xu 1993). In the crystal structure of the enzyme-cyclic phosphate complex, His12 makes a bifurcated hydrogen bond with O2' and a phosphoryl oxygen. Transphosphorylation and hydrolysis need not necessarily follow corresponding pathways. In the enzymatic reaction, the two processes differ in rate by 1 to 4 orders of magnitude (depending on the substrate), and thermodynamically, transphosphorylation is easily reversible while hydrolysis is not.

For the product complex with 3'-UMP, the simulation and the crystal structure have His12 in the doubly protonated form, hydrogen bonding to O1P rather than O2'. A number of experimental observations suggest that His12 is doubly protonated and interacting directly with the phosphate in the enzyme/product complex (Meadows et al. 1969; Gorenstein et al. 1976; Blackburn and Moore 1982). The experiments also imply little interaction between the phosphate and the other protonated groups at the active site, such as His119 and Lys41. The latter is supported in the average dynamics structure where Lys41 does not interact with O2' or the phosphate oxygens, and His119 is bonded to water molecules. In the crystal structure the lysine is highly mobile and His119 interacts with the hydroxyl oxygen of the phosphate. More details of the simulation results are given in Haydock et al (Haydock et al. 1990) and Straub et al. (Straub et al. 1994)

Althought there have been many experimental and theoretical investigations of RNase A, the detailed enzyme reaction mechanism remains to be determined. Even for the transphosphorylation step, which is understood best, the exact role of the catalytic residues His12 and His119 is still a matter of debate. With new mutational and other experiments, as well as QM/MM simulations, a full description may be expected in the not-to-distant future.

3.3. CARBOXYPEPTIDASE A

Carboxypeptidase A catalyzes the hydrolysis of the C-terminal amino acids in peptides and proteins; it shows a preference for aromatic and branched aliphatic side chains at the C-terminal end. The enzyme also acts as an esterase. The X-ray structure of the native form of carboxypeptidase A, was solved by Rees and Lipscomb to 1.5 Å

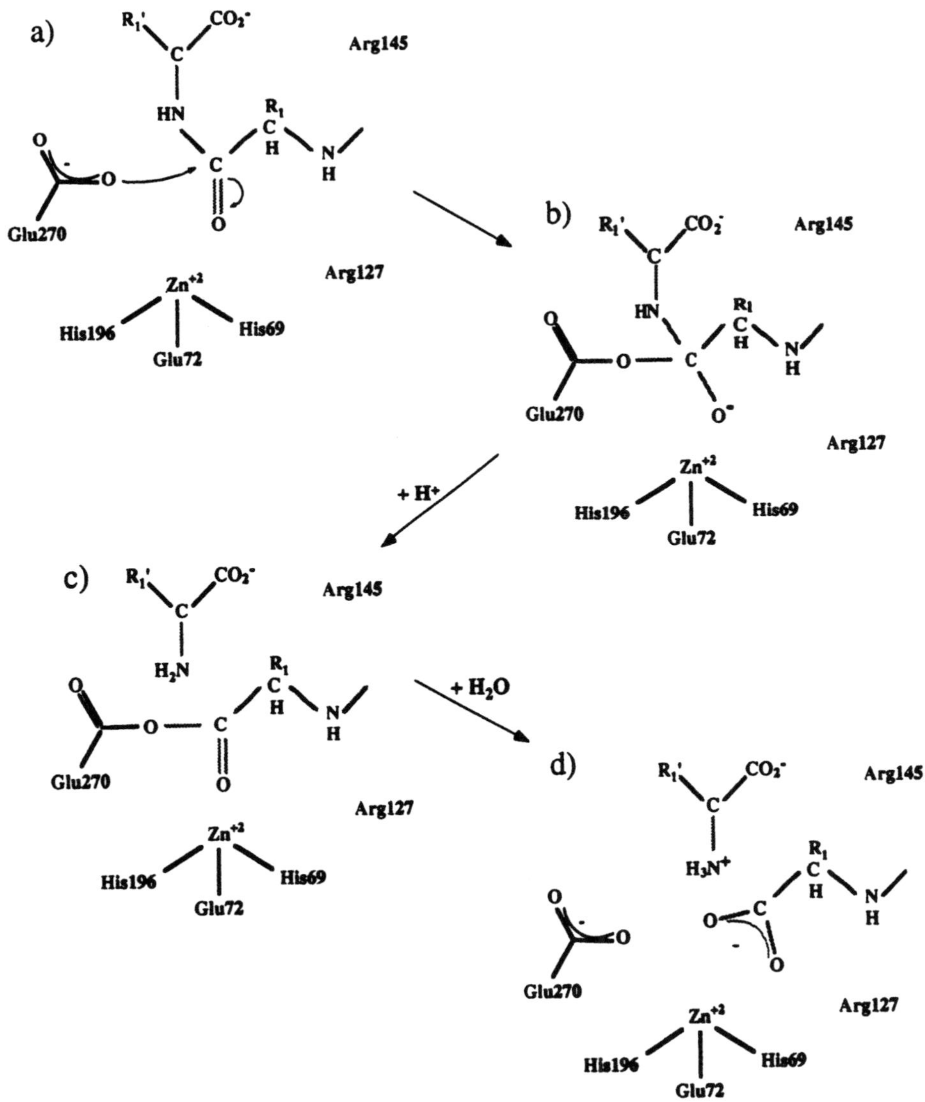

Figure 8. Proposed catalytic mechanism for carboxypeptidase A where Glu270 acts as the nucleophile a) nucleophilic attack by carboxylate oxygen of Glu270 on the carbonyl carbon of the substrate; b) forming a tetrahedral intermediate; c) the intermediate breaks down to give an anhydride species; d) hydrolysis of the acylenzyme leads to product formation and regnerates the free enzyme

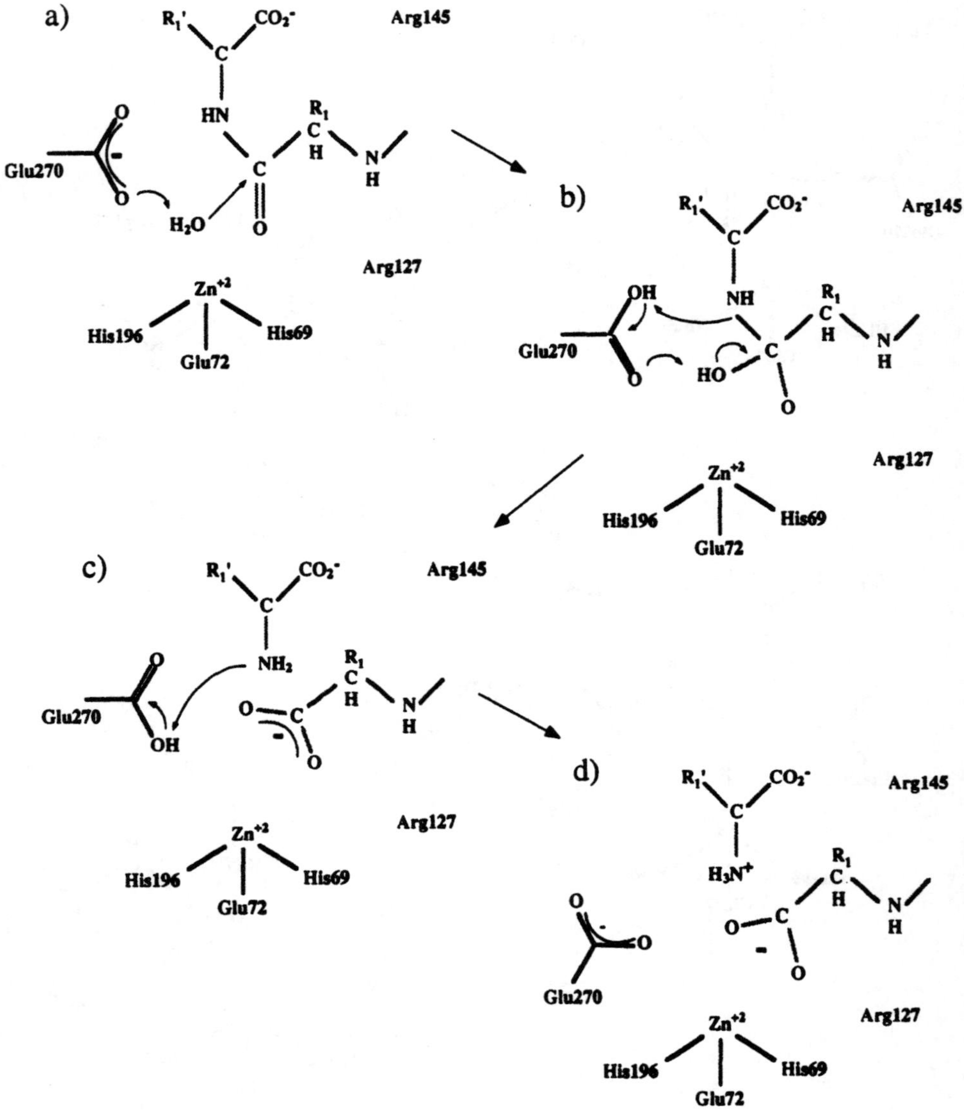

Figure 9. Proposed catalysic mechanism for carboxypeptidase A where water acts as the nucleophile. (a) nucleophilic attack by a water molecule on the carbonyl carbon of the substrate promoted by zinc and assisted by Glu270 with concommitant transfer of a proton to Glu270; (b) a tetrahedral intermediate, stabilized by interactions with Arg127 and the zinc ion, collapses with a proton donated by Glu270; (c) a second proton transfer results in product formation (d).

resolution (Rees et al. 1983); structures of the enzyme with a number of inhibitors and transition state analogs have also been determined (Christianson and Lipscomb 1985; Christianson 1991). Carboxypeptidase A has 307 protein residues and one zinc ion in the active site ligated to 2 histidines (His69 and 196), and to the carboxylate of Glu72in a bidentate fashion. A single water molecule completes the first coordination sphere of the zinc ion. Other important residues present in the active site are Glu270, Arg127, Arg145 and Tyr248. Numerous experimental studies, including x-ray crystallography (Rees et al. 1983; Kannan et al. 1984; Eriksson et al. 1988; Matthews 1988), x-ray Fine Absorption Spectroscopy (XAFS) (Zhang et al. 1992), and NMR spectroscopy of Co^{+2} analogs (Auld et al. 1992) have been done to try and determine the enzymatic mechanism of this biologically important enzyme. The experiments have established that Glu270, the zinc ion, a water molecule and the scissile peptide bond of the substrate are involved in the reaction. The structural studies have established that the carboxylate group of the substrate forms hydrogen bonds with the guanidinium group of Arg145 and that Try248 undergoes a large conformational change displacing the phenol group by approximately 12 Å.

Two mechanisms of action for the hydrolysis of peptides involving nucleophilic or general base catalysis have been proposed. While the initial binding sites for peptides and esters are different, the question of whether ester hydrolysis proceeds by a similar mechanism remains open. In the nucleophilic mechanism (see Fig 8), the zinc is believed to orient the carbonyl of the substrate in a position suitable for nucleophilic attack and to polarize the C atom so as to increase its electrophilicity and facilitate hydrolysis. Glu270, which is close to the carbonyl carbon of the C-terminal group, makes a direct nucleophilic attack on the carbon. This leads to the formation of an anhydride intermediate between the carboxyl of Glu270 and the carboxyl of the hydrolysis product. In a second step, the anhydride reacts with water to regenerate the free enzyme. It was originally thought that the phenol of Tyr248 acts as a general base in deacylation step and perhaps as a general acid in the first step by donating a proton to the leaving group. However, site specific mutagenesis studies have shown that the enzyme remains functional even when a Phe is substituted for the Tyr (Gardell et al. 1985; Hilvert et al. 1986); the experiments indicated that Tyr is important for substrate binding. In the alternative mechanism (see Fig. 9), Glu270 acts as a general base (Lipscomb 1974; Breslow and Wernick 1976; Christianson and Lipscomb 1989) to heighten the nucleophilicity of a water molecule and thus accelerate hydrolysis. Concomitant with the attack of the water molecule on the scissile carbonyl group is a transfer of a proton to Glu270. The substrate forms a tetrahedral intermediate which then collapses with a proton donated by the Glu270. The final product is realized after a second proton is transferred by Glu270.

The absence of an acyl enzyme intermediate in the hydrolysis of peptides was suggested by ^{18}O-labeling experiments (Breslow and Wernick 1976; Breslow and Wernick 1977), but these experiments did not rule out the formation of an acyl intermediate in esterolysis. Cryogenic studies of the hydrolysis of several peptides demonstrated the accumulation of intermediates. However, none of these intermediates reacted with trapping reagents, and consequently, could not be positively identified as being the acyl enzyme. Evidence has been presented for an anhydride intermediate in the hydrolysis of certain esters (Britt and Peticolas 1992; Suh 1992; Osumi et al. 1994; Lee et al. 1995). Also, cryogenic studies of cobalt substituted carboxypeptidase A suggested that peptides and esters lead to structurally distinct intermediates (Geoghegan et al. 1983). Breslow and Wernick (Breslow and Wernick 1977) proposed a unifying picture for the mechanisms of catalysis by carboxypeptidase A and suggested distinct mechanisms for esters and peptides. Peptides are hydrolyzed with the zinc atom coordinated to the peptide carbonyl and Glu270 acting as a general base for the attacking water molecule. Further, they suggested that ester carbonyl groups may be weaker ligands to the zinc than peptide carbonyl groups and on binding may not displace the water molecule coordinated to the zinc. Glu270 could act as nucleophile and form an anhydride intermediate; the aquo-zinc complex would then deliver a proton in its role as a general acid.

These experimental studies leave major aspects of the mechanism unresolved. The most important one is whether the catalysis proceeds via a general base mechanism or via a nucleophilic attack by an enzyme residue on the scissile bond. Although the experimental studies of carboxypeptidase A are providing essential information, they cannot show what species are actually involved in the reaction. For example, the suggestion that the zinc bound water acts as the nucleophile was based on the results of x-ray structures of unproductive and static complexes of carboxypeptidase A formed with pseudosubstrates and inhibitors (Christianson and Lipscomb 1985).

Recently, molecular dynamics simulations of carboxypeptidase A have been employed to further our understanding of the enzymatic mechanism of carboxypeptidase A (Makinen et al. 1989; Banci et al. 1992; Banci et al. 1993; Stote and Karplus 1995). In an early study, molecular dynamics simulations were used to characterize the secondary structure motion of enzyme as well as the motion of the different subsites involved in substrate binding (Makinen et al. 1989). The catalytically important zinc ion was not included in these simulations. The structure and dynamics of the zinc containing active site was the subject of several studies by Banci et al. (Banci et al. 1992; Banci et al. 1993; Banci et al. 1994). In one study (Banci et al. 1992), molecular dynamics simulations of the enzyme with several inhibitors were done in order to understand the structural variations produced by inhibitor binding. The effects of these variations on the catalytic mechanism were discussed. The second simulation study focused on the structural variations induced by the protonation of Glu270 (Banci et al. 1993). In the more recent study, the crystal structure of carboxypeptidase A complexed with the inhibitor, D-Phenylalanine (Christianson et al. 1989), was used to build a model of the enzyme complexed with the tetrapeptide Val-Leu-Phe-Phe. These simulations focused on the substrate and its orientation relative to the other important active site residues just prior to nucleophilic attack. Two simulation structures were generated; in one, the zinc bound water was left intact. The substrate was positioned so that the C-terminal carboxylate interacts with the Arg145, the scissile carbonyl interacts with the guanidinium group of Arg127 and the zinc bound water is nearby. In another simulation structure, the zinc bound water was replaced with the oxygen atom of the scissile carbonyl of the substrate. The enzyme was solvated with explicit water molecules and active site molecular dynamics simulations, similar in spirit to the stochastic boundary method, were done. In the simulations where the zinc bound water was retained, the water molecule moved toward the Glu270 seemingly in accord with mechanism where the water acts as the nucleophile. In the simulations where the substrate replaced the zinc bound water, the carboxylate group of Glu270 moved toward the scissile carbonyl of the substrate. These results indicate that, as far as structural changes are concerned, both mechanisms are feasible. There is evidence that both mechanisms occur depending on whether the substrate being hydrolyzed is an ester or a peptide; in the hydrolysis of esters, the anhydride mechanism is favored, while in the hydrolysis of peptides, the mechanism where an activated water acts as the nucleophile is favored.

Simulations of the native form of carboxypetidase A indicate that the zinc bound water forms a hydrogen bond with the carboxyl group of Glu270, positioning the water in a seemingly optimal orientation for nucleophilic attack on the substrate (Stote and Karplus 1995). A similar result, whereby the zinc plays a role in orienting residues important for the catalysis, was observed in the simulations of another zinc containing enzyme, carbonic anhydrase (Stote and Karplus 1995). The zinc also provides electrostatic stabilization for negatively charged intermediates formed during the course of the enzymatic reaction. These simulations used a model for zinc which permits changes in coordination geometries and ligand exchange. This model is in contrast to the model for zinc used in the simulations of Banci et al discussed above. The latter model introduces specific bonds between the zinc ion, its protein ligands as well as the oxygen of the water molecule (Banci et al. 1992). While the two models are comparable in terms of average structures and rms fluctuations of the zinc binding site, the simple nonbonded model is a more physically correct one given the flexible nature of the zinc binding site in carboxypeptidase A. Structures of the enzyme with a number of inhibitors and transition state analogs have demonstrated multiple binding modes around the ion at various stages of the enzyme reaction (Christianson and Lipscomb 1989). In addition, the bidentate coordination of Glu-72 to the zinc ion in the native state (Rees et al. 1983) tends to unidentate coordination on the binding of substrate or an inhibitor (Christianson and Lipscomb 1985). This is achieved by the movement of the ion in the direction of the Arg127 and, to a lesser degree, by the movement of Glu72. Recent XAFS studies indicate that in carboxypeptidase A, ion-protein distances differ between the solution and crystalline forms of the enzyme; they were attributed to the displacement of one oxygen of the Glu-72 from its crystallographic position (Zhang et al. 1992). It has been suggested that this flexibility may be important for enzyme function (Christianson 1991) and should be represented in the molecular models. This treatment of the zinc binding site is being used in studies of carboxypeptidase A and other zinc containing proteins.

3.4. TRIOSPHOSPHATE ISOMERASE

An enzyme that has been the subject of intensive experimental and theoretical studies is triosephosphate isomerase (TIM), which catalyses the interconversion of dihydroxyacetone phosphate (DHAP) and D-glyceraldehyde 3-phosphate (GAP), an essential step in the glycolytic pathway (Fersht 1985). The mechanism of the enzyme has been examined by QM/MM calculations which we do not describe here because it falls outside the topic of this review (Bash et al. 1991). However, an additional aspect of the overall mechanism is the conformational change of an 11-residue loop region (residues 166-176) which moves more than 7 Å and closes over the active site when substrate binds (Joseph et al. 1990; Lolis et al. 1990). Mutagenesis experiments have

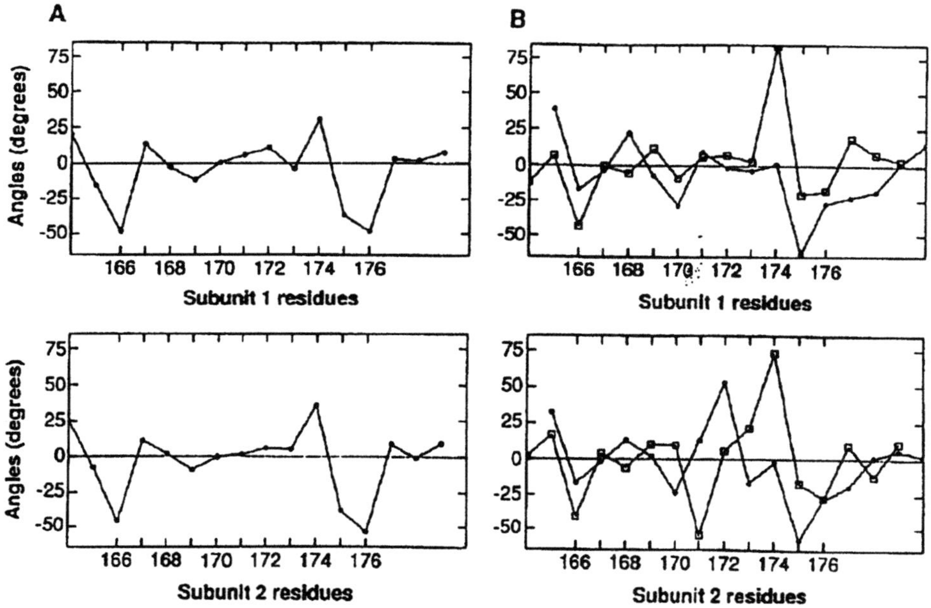

Figure 10. Plots of closed minus open loop structure dihedral angle differences for subunit 1 and for subunit 2 (from Joseph et al. 1990). (A) The α-carbon pseudodihedral angle differences and (B) main chain dihedral angle differences versus residue number. In (A) the number *i* corresponds to angle *i*, *i+1*; in (B) circles represent relative ϕ and squares relative ψ.

shown that the loop is essential for catalysis (Pompliano et al. 1990). Solid-state NMR studies have measured the motion of the loop with and without substrate and transition-state analogs (Williams and McDermott 1995). When the open and closed structures of TIM are superimposed by least-squares optimization of all α-carbons, the root-mean-square deviation of the α-carbons is only 0.42Å, whereas that for the loop residues is 4.8 Å. The α-carbon of Thr172, which is at the center of the loop, is shifted by 7.1 Å, while the ends of the loop change little; for example, the distance from the α-carbon 166 to α-carbon 176 is 7.3 Å in the open structure and 7.5 Å in the closed structure. Most important is the striking similarity between the internal structure of the loop in the open and closed forms: that is, when residues 166-176 in the initial and final states are superimposed, the rms deviation for all loop residues is only 0.73 Å.

These structural results suggest that a rigid-body, hinge-type motion takes place when the loop closes over the substrate in the active site; i.e. the loop moves more like a 'lid'. To identify the hinge regions, α-carbon pseudo-dihedral angle and main-chain dihedral angle differences between the open and closed loop were plotted against residue number. From the α-carbon plots, it is evident that there are two hinges (Fig. 10). One of these involves the angles 166-167 and 167-168, the other involves the angles 174-175, 175-176, 176-177. Changing only these angles from their values in the closed structure to their values in the open structure yields an 'open' loop with an α-carbon difference of 1.9 Å from the open structure when non-loop α-carbons are superimposed; if the dihedral 164-165 is also rotated, the rmsd difference is reduced to 1.1 Å. Analysis of the van der Waals packing and hydrogen-bonding interactions within the lid provides an explanation for its rigid body motion and the presence of two hinges. To obtain information about the lid motion, molecular dynamics simulations were performed. High temperatures were used to decrease the time required for a conformational change. All simulations began with the closed form since an open structure with increased entropy was expected to be favored at high temperature. At 298 K, the loop oscillated about the closed position, whereas during the 1000 K simulation, the loop opened and closed repeatedly. In the simulation at 500K; the loop oscillated about the closed position until 29 ps when it flipped to a more open position. An examination of the dihedral angle differences between the MD structures at 29 ps and the closed and open structures show the largest changes between residues 174 and 175 suggesting that the loop starts to open in the region of residues 174 and 175. The structural and simulation analysis demonstrates that the conformational change observed in TIM involves a rigid lid rather than a flexible loop and suggests that similar behavior will be found in other enzymes that have loop regions involved in substrate binding. A lid type of transition has has been identified recently in lactate dehydrogenase. Similar simulation studies have been used to better understand the nature of loop motion in HIV protease (Collins et al. 1995).

4. Conclusions

Since the first work describing a molecular dynamics simulation of a small protein was published (McCammon et al. 1977), there has been explosive growth in research concerned with theoretical studies of proteins and enzymes (McCammon and Harvey 1987; Brooks, et al 1988; Warshel 1991). Most of the studies have used empirical energy functions. This chapter describes the nature of the empirical energy function, its use in molecular mechanics calculations and molecular dynamics simulations and presents several applications to enzymes. Most important is the result that starting with x-ray structures of unproductive and static complexes of the enzyme formed with pseudosubstrates and inhibitors, simulations of the enzyme complexed with true substrates can provide direct information on the positions and dynamics of catalytically important residues. This makes it possible to explore the contribution of various amino acids to the enzyme reaction, even without a complete calculation of the reaction path. The latter requires extended potential surfaces of the QM/MM type that include bond rupture and bond formation.

Large scale conformational changes play an important role in enzyme function; several examples are reviewed in this chapter. Molecular mechanics calculations, in particular, energy minimization and normal mode calculations, as well as molecular dynamics simulations have been employed to provide an understanding of the mechanisms of such motions and their role in the enzyme.

Acknowledgements

Much of the material in this review was taken from previously published articles involving at least one of the authors. For lysozyme, references (Brooks and Karplus 1985; Post and Karplus 1986; Karplus and Post 1996) were used; for RNase A reference (Haydock et al. 1990) was used; for carboxypeptidase A (Stote and Karplus 1995) was used and for triosephosphate isomerase, (Joseph et al. 1990) was used. We thank Arieh Warshel for a careful reading of the manuscript. The work was supported in part by a grant from the National Science Foundation to Harvard University. MK is the Theodore William Richards Professor of Chemistry at Harvard University and a Professeur Conventioné at the Université Louis Pasteur in Strasbourg, France. RS is a Chargé de Recherche in the CNRS and AD is a Maitre de Conférences at the Université Louis Pasteur in Strasbourg France.

References

Allen, M. P. and Tildesley, D. J. (1989) *Computer Simulation of Liquids*. Oxford University Press, Oxford.

Allinger, N. L., Miller, M. A., Van Catledge, F. A. and Hirsch, J. A. (1967) Conformational Analysis LVII. The Calculation of the Conformational Structures of Hydrocarbons by the Westheimer-Hendrickson-Wiberg Method, *J. Am. Chem. Soc.* **89**, 4345-4357.

Allinger, N. L. and Sprague, J. T. (1973) Calculation of the Structures of Hydrocarbons Containing Delocalized Electronic Systems by the Molecular Mechanics Method., *J. Am. Chem. Soc.* **95**, 3893-3907.

Andersen, H. C. (1980) Molecular dynamics simulations at constant pressure and/or temperature, *J. Chem. Phys.* **72**, 2384-2393.

Anslyn, E. and Breslow, R. (1989) On the Mechanism of Catalysis by Ribonuclease: Cleavage and Isomerization of the Dinucletide UpU Catalyzed by Imidazole Buffers, *J. Am. Chem. Soc.* **111**, 4473-4480.

Arnold, G. E. and Ornstein, R. L. (1992) A molecular dynamics simulation of bacteriophage T4 lysozyme, *Prot. Eng.* **5**, 703-714.

Auld, D. S., Bertini, I., Donaire, A., Messori, L. and Moratal, J. M. (1992) pH-Dependent Properties of Cobalt(II) Carboxypeptidase A - Inhibitor Complexes, *Biochemistry* **31**, 3840 - 3846.

Banci, L., Bertini, I. and La Penna, G. (1993) A Molecular Dynamics Study of Carboxypeptidase A: Effect of Protonation of Glu 270, *Inorg. Chem.* **32**, 2207-2211.

Banci, L., Bertini, I. and La Penna, G. (1994) The Enzymatic Mechanism of Carboxypeptidase: A Molecular Dynamics Study, *Proteins: Struct. Func. Gen.* **18**, 186-197.

Banci, L., Schroder, S. and Kollman, P. A. (1992) Molecular Dynamics Characterization of the Active Cavity of Carboxypeptidase A and Some of Its Inhibitor Adducts, *Proteins: Struct. Func. Gen* **13**, 288 - 305.

Banks, R. D., Blake, C. C. F., Evans, P. R., Haser, R., Rice, D. W., Hardy, G. W., Merrett, M. and Phillips, A. W. (1979) Sequence, structure and activity of phosphoglycerate kionase: a possible hinge-bending enzyme, *Nature* **279**, 773-777.

Barton, D. H. R. (1948) Interactions between Non-bonded Atoms, and the Structure of cis-Decalin, *J. Chem. Soc.* 340-342.

Bash, P. A., Field, M. J., Davenport, R. C., Petsko, G. A., Ringe, D. and Karplus, M. (1991) Computer Simulation and Analysis of the Reaction Pathway of Triosephosphate Isomerase, *Biochemistry* **30**, 5826-5832.

Bashford, D., Karplus, M. and Canters, G. W. (1988) Electrostatic Effects of Charge Perturbations Introduced by Metal Oxidation in Proteins. A Theoretical Analysis, *J. Mol. Biol.* **203**, 507-510.

Bennett, W. C. and Seitz, T. A. (1980) Structure of a complex between yeast hexokinase A and glucose, *J. Mol. Biol.* **140**, 211-388.

Bennett, W. S. and Huber, R. (1984) Structural and Functional Aspects of Domain Motions in Proteins, *Crit. Rev. Biochem.* **15**, 291-384.

Berendsen, H. (1985). in *Molecular Dynamics and Protein Structure*. Western Springs, Il., Polycrystal Book Service.

Berendsen, H. J. C., Postma, J. P. M., van Gunsteren, W. F., DiNola, A. and Haak, J. R. (1984) Molecular dynamics with coupling to an external bath, *J. Chem. Phys.* **81**, 3684-3690.

Bertini, I. and Viezzoli, M. S. (1995). NMR of paramagnetic molecules: A contribution to the understanding of enzymic mechanisms. *NATO ASI Ser., Ser. C459(Bioinorganic Chemistry)*. 93-104.

Blackburn, P. and Moore, S. (1982). Pancreatic ribonuclease. *The Enzymes.* New York, Academic. 317-433.

Blake, C. C. F., Johnson, L. N., Mair, G. A., North, A. C. T., Phillips, D. C. and Sarma, V. R. (1967) Crystallographic Studies of the activity of hen egg-white lysozyme., *Proc. Royal Soc. London (Ser. B)* **167**, 378-388.

Blake, C. C. F., Koenig, D. F., Mair, G. A., North, A. C. T., Phillips, D. C. and Sarma, V. R. (1965) Structure of Hen Egg-White Lysozyme. A Three-dimensional Fourier Synthesis at 2 Å Resolution, *Nature* **206**, 757-761.

Blake, C. C. F., Mair, G. A., North, A. C. T., Phillips, D. C. and Sarma, V. R. (1967) On the Conformation of the Hen egg-white lysozyme molecule., *Proc. R. Soc. London, Ser. B* **167**, 365-377.

Breslow, R. (1993) Kinetics and mechanism in RNA cleavage, *Proc. Natl. Acad. Sci. USA* **90**, 1208-1211.

Breslow, R., Huang, D.-L. and Anslyn, E. (1989) On the mechanism of action of ribonucleases: Dinucleotide cleavage catalyzed by imidazole and Zn2+, *Proc. Natl. Acad. Sci. USA* **86**, 1746-1750.

Breslow, R. and Wernick, D. (1976) On the Mechanism of Catalysis by Carboxypeptidase A, *J. Am. Chem. Soc.* **98**, 259-261.

Breslow, R. and Wernick, D. L. (1977) Unified picture of mechanisms of catalysis by carboxypeptidase A, *Proc. Natl. Acad. Sci. U. S.A.* **74**, 1303-1307.

Breslow, R. and Xu, R. (1993) Recognition and çatalysis in nucleic acid chemistry, *Proc. Natl. Acad. Sci. USA* **90**, 1201-1207.

Britt, B. M. and Peticolas, W. L. (1992) Raman Spectral Evidence for an Anhydride Intermediate in the Catalysis of Ester Hydrolysis by Carboxypeptidase A, *J. Am. Chem. Soc.* **114**, 5295-5303.

Brooks, B. R., Bruccoleri, R. E., Olafson, B. D., States, D. J., Swaminathan, S. and M., K. (1983) CHARMM: A Program for Macromolecular Energy Minimization and Dynamics Calculations, *J. Comp. Chem* **4**, 187-217.

Brooks, B. R. and Karplus, M. (1985) Normal Modes for Specific Motions of Macromolecules: Application to the Hinge-Bending Mode of Lysozyme, *Proc. Natl. Acad. Sci. USA* **82**, 4995-4999.

Brooks, C., L. III (1987) The influence of long-range force truncation on the thermodynamics of aqueous ionic solutions, *J. Chem. Phys.* **86**, 5156-5162.

Brooks, C. L. III and Karplus, M. (1983) Deformable Stochastic Boundaries in Molecular Dynamics, *J. Chem. Phys.* **79**, 6312-6325.

Brooks, C. L. III, Karplus, M. and Pettitt, B. M. (1988) *Proteins: A Theoretical Perspective of Dynamics, Structure, and Thermodynamics.* John Wiley & Sons,

Brooks, C. L., III Pettitt, B. M. and Karplus, M. (1985) Structural and energetic effects of truncating long range interactions in ionic and polar fluids, *J. Chem. Phys.* **83**, 5897-5908.

Brooks, C. L. III Brünger, A. T. and Karplus, M. (1985) Active Site Dynamics in Protein Molecules: A Stochastic Boundary Molecular-Dynamics Approach, *Biopolymers* **24**, 843-865.

Brooks, C. L. III and Karplus, M. (1989) Solvent Effects on Protein Motion and Protein Effects on Solvent Motion, *J. Mol. Biol.* **208**, 159 - 181.

Bruccoleri, R. E., Karplus, M. and McCammon, J. A. (1986) The Hinge-Bending Mode of a Lysozyme-Inhibitor Complex, *Biopolymers* **25**, 1767-1802.

Brünger, A. T., Brooks, C. L. III and Karplus, M. (1985) Active Site dynamics of ribonuclease, *Proc. Natl. Acad. Sci. USA* **82**, 8458-8462.

Brünger, A. T. and Karplus, M. (1988) Polar Hydrogen Positions in Proteins: Empirical Energy Placement and Neutron Diffraction Comparison, *Proteins: Struc. Func. Genet.* **4**, 148-156.

Brünger, A. T. and Karplus, M. (1991) Molecular Dynamics Simulations with Experimental Restraints, *Acc. Chem. Res.* **24**, 54-61.

Brünger, A. T., Kuriyan, J. and Karplus, M. (1987) Crystallographic *R* Factor Refinement by Molecular Dynamics, *Science* **235**, 458-460.

Bunton, C. A., Lewis, T. A., Llewellyn, D. R. and Vernon, C. I. (1955) Mechanisms of Reaction in the Sugar Series. Part I. The Acid-catalysed Hydrolysis of α- and β-Methyl and α- and β-Phenyl D-Glucopyranosides., *J. Chem. Soc.* 4419-4423.

Christianson, D. W. (1991) Structural Biology of Zinc, *Adv. Protein Chem.* **42**, 281-355.

Christianson, D. W. and Lipscomb, W. N. (1985) Binding of a possible transition state analog to the active site of carboxypeptidase A, *Proc. Natl. Acad. Sci. USA* **82**, 6840-6844.

Christianson, D. W. and Lipscomb, W. N. (1989) Carboxypeptidase A, *Acc. Chem. Res* **22**, 62 - 69.

Christianson, D. W., Mangani, S., Shoham, G. and Lipscomb, W. N. (1989) Binding of D-Phenylalanine and D-Tyrosine to Carboxypeptidase A, *J. Biol. Chem.* **264**, 12849-12853.

Collins, J. R., Burt, S., K. and Erickson, J. W. (1995) Flap opening in HIV-1 protease simulated by 'activated' molecular dynamics, *Struc. Biol.* **2**, 334-338.

Colman, R. F. (1990). Site-Specific Modification of Enzymes Sites. *The Enzymes*. 19, San Diego, Academic Press. 283-321.

Colona-Cesari, F., Perahia, D., Karplus, M., Ecklund, H., Brandem, C. and Tapia, O. (1986) Interdomain Motion in Liver Alcohol Dehydrogenase: Structural and Energetic Analysis of the Hinge Bending Mode, *J. Biol. Chem.* **261**, 15273-15289.

Cordes, F., Starikov, E. B. and Saenger, W. (1995) Initial State of an Enzymatic Reaction. Theoretical Prediction of Complex Formation in the Active Site of RNase T1, *J. Am. Chem. Soc.* **117**, 10365-10372.

Eftink, M. R. and Biltonen, R. L. (1983) Energetics of Ribonuclease A Catalysis. 1. pH, Ionic Strength, and Solvent Isotope Dependence of the Hydrolysis of Cytidine Cyclic 2',3' Phosphate, *Biochemistry* **22**, 5123-5134.

Eklund, H., Nordström, B., Zeppezauer, E., Söderlund, G., Ohlsson, I., Boiwe, T., Söderberg, B.-O., Tapia, O. and Brändén, C.-I. (1976) Three-dimensional Structure of Horse Liver Alcohol Dehydrogenase at 2.4 Å Resolution, *J. Mol. Biol.* **102**, 27-59.

Eklund, H., Samama, J.-P., Wallén, L., Brändén, C.-I., Åkeson, Å. and Jones, T. A. (1981) Structure of a Triclinic Ternary Complex of Horse Liver Alcohol Dehydrogenase at 2.9 Å Resolution, *J. Mol. Biol.* **146**, 561-587.

Elber, R. and Karplus, M. (1987) A Method for Determining Reaction Paths in Large Molecules: Application to Myoglobin, *Chem. Phys. Lett.* **139**, 375-380.

Eriksson, E. A., Jones, T. A. and Liljas, A. (1988) Refined Structure of Human Carbonic Anhydrase II at 2.0 Å Resolution, *Proteins: Struct. Funct. Genet.* **4**, 274-282.

Evans, J. N. S. (1992). NMR and enzymes. *Pulsed Magn. Reson.: NMR, ESR, Opt.* Oxford, UK., Oxford Univ. Press. 123-73.

Fersht, A. (1985). Enzyme Structure and Mechanism. New York, N. Y., W.H. Freeman.

Field, M. J., Bash, P. A. and Karplus, M. (1990) A Combined Quantum Mechanical and Molecular Mechanical Potential for Molecular Dynamics Simulations, *J. Comp. Chem.* **11**, 700-733.

Fierke, C. A. and Hammes, G. G. (1995) Transient kinetic approaches to enzyme mechanisms, *Methods Enzymol. (Enzyme Kinetics and Mechanism, Part D)* **249**, 3-37.

Fischer, S. (1992) Curvilinear reaction coordinates of conformational changes in macromolecules. Application to rotamase catalysis., Thesis, Harvard University.

Fischer, S. and Karplus, M. (1992) Conjugate Peak Refinement: An Algorithm for Finding Reaction Paths and Accurate Transition-States in Systems with Many Degrees of Freedom, *Chem. Phys. Lett.* **194**, 252-261.

Fischer, S., Michnick, S. and Karplus, M. (1993) A Mechanism for Rotamase Catalysis by the FK506 Binding Protein (FKBP), *Biochemistry* **32**, 13830-13837.

Fleet, G. W. J. (1985) An Alternative Mechanism for the Mode of Inhibition of Glycosidase Activity by Polyhydroxylated Piperidines, Pyrrolidines, and Indolizidines: The Mechanism of Action of Some Glycosidases, *Tetrahedron Lett.* **26**, 5073-5076.

Franck, R. W. (1992) The Mechanism of β-Glycosidases: A Reassessment of Some Seminal Papers, *Biorg. Chem.* **20**, 77-88.

Gardell, S. J., Craik, C. S., Hilvert, D., Urdea, M. S. and Rutter, W. J. (1985) Site directed mutagenesis shows that tyrosine 248 of carboxypeptidase A does not play a crucial role in catalysis, *Nature* **317**, 551-555.

Geoghegan, K. F., Galdes, A., Martinelli, R. A., Holmquist, B., Auld, D. S. and Vallee, B. L. (1983) Cryospectroscopy of Intermediates in the Mechanism of Carboxypeptidase A, *Biochemistry* **22**, 2255-2262.

Gerlt, J. A. and Gassman, P. G. (1993) Understanding the Rates of Certain Enzyme-Catalyzed Reactions: Proton Abstraction from Carbon Acids, Acyl-Transfer Reactions, and Displacement Reactions of Phosphodiesters, *Biochemistry* **32**, 1194311952.

Gerstein, M., Lesk, A. M. and Chothia, C. (1994) Structural Mechanisms for Domain Movements in Proteins, *Biochemistry* **22**, 6739-6749.

Gilbert, W. A., Fink, A. L. and Petsko, G. A. unpublished results,

Gilliland, G. L. and Quiocho, F. Q. (1981) Structure of the L-Arabinose-binding Protein From Escherichia coli at 2.4 Å Resolution, *J. Mol. Biol.* **146**, 341-362.

Gilson, M. K. and Honig, B. H. (1987) Calculation of electrostatic potentials in an enzyme active site, *Nature* **330**, 84-86.

Gorenstein, D. G., Wywicz, A. M. and Bode, J. (1976) Interaction of Uridine and Cytidine Monophosphates with Ribonuclease A. IV Phosphorus-31 Nuclear Magnetic Resonance Studies, *J. Am. Chem. Soc.* **98**, 2308.

Greengard, L. and Rokhlin, V. (1987) A Fast Algorithm for Particle Simulations, *J. Comp. Phys.* **73**, 325-348.

Groeneveld, C. M., Ouwerling, M. C., Erkelens, C. and Canters, G. W. (1988) ^{1}H Nuclear Magnetic Resonance Study of the Portonation Behaviour of the Histidine Residues and the Electron Self-exchange Reactions of Azurin from Alcaligenes denitrificans, *J. Mol. Biol.* **200**, 189-199.

Ha, S. N., Giammona, A., Field, M. and Brady, J. W. (1988) A revised potential-energy surface for molecular mechanics studies of carbohydrates, *Carbohydr. Res.* **180**, 207-221.

Hadfield, A. T., Harvey, J. D., Archer, D. B., MacKenzie, D. A., Jeenes, D. J., Radford, S. E., Lowe, G., Dobson, C. M. and Johnson, L. N. (1994) Crystal Structure of the Mutant D52S Hen Egg White Lysozyme with an Oligosaccharide Product, *J. Mol. Biol.* **243**, 856-872.

Harrison, S. C., Olson, A. J., Schutt, C. E., Winkler, F. K. and Bricogne, G. (1978) Tomato bushy stunt virus at 2.9 Å Resolution, *Nature* **276**, 368-373.

Hartmann, H., Parak, F., Steigemann, W., Petsko, G. A., Ringe Ponzi, D. and Frauenfelder, H. (1982) Conformational substates in a protein: structure and dynamics of metmyoglobin at 80 K, *Proc. Natl. Acad. Sci. USA* **79**, 4967-4971.

Haydock, K., Lim, C., Brünger, A. T. and Karplus, M. (1990) Simulation Analysis of Structures on the Reaction Pathway of RNAse A, *J. Am. Chem. Soc.* **112**, 3826-3831.

Hendrickson, J. B. (1961) Molecular Geometry. I. Machine Computation of the Common Rings, *J. Am. Chem. Soc.* **83**, 4537.

Hilvert, D., Gardell, S. J., Rutter, W. J. and Kaiser, E. T. (1986) Evidence against a Crucial Role for the Phenolic Hydroxyl of Tyr-248 in Peptide and Ester Hydrolyses Catalyzed by Carboxypeptidase A: Comparative Studies of the pH Dependencies of the Native and Phe-248-Mutant Forms, *J. Am. Chem. Soc.* **108**, 5298-5304.

Hoover, W. G. (1985) Canonical dynamics: equilibrium phase-space distributions, *Phys. Rev. A* **31**, 1695-1697.

Ichiye, T., Olafson, B., Swaminathan, S. and Karplus, M. (1986) Structure and Internal Mobility of Proteins: A Molecular Dynamics Study of Hen Egg White Lysozyme, *Biopolymers* **25**, 1909-1939.

Ikura, T. and Gō, N. (1993) Normal Mode Analysis of Mouse Epidermal Growth Factor: Characterization of the Harmonic Motion, *Proteins: Struct. Func. Gen.* **16**, 423-436.

Imoto, T., Johnson, L. M., North, A. C. T., Phillips, D. C. and Rupley, J. A. (1972) Vertebrate Lysozymes. *The Enzymes*. New York, Academic. 665-868.

Jackson, S. E., Fersht, A. R. (1993) Contribution of Long-Range Electrostatic Interactions to the Stabilization of the Catalytic Transition State of the Serine Protease Subtilisin BPN', *Biochemistry* **32**, 13909-13916.

Janin, J. and Wodak, S. Y. (1983) Structural Domains in Proteins and Their Role in the Dynamics of Protein Function, *Prog. Biophys. Mol. Biol.* **42**, 21-78.

Jentoft, J. E., Gerken, T. A., Jentoft, N. and Dearborn, D. G. (1981) [^{13}C]Methylated Ribonuclease A. ^{13}C NMR Studies of the Interaction of Lysine 41 with Active Site Ligands, *J. Biol. Chem.* **256**, 231-236.

Joao, H. C. and Williams, R. J. P. (1993) The anatomy of a kinase and the control of phosphate transfer, *Eur. J. Biochem.* **216**, 1-18.

Johnson, K. A. and Benkovic, S. J. (1990). Analysis of Protein Function by Mutagenesis. *The Enzymes*. 19, San Diego, Academic Press. 159-211.

Joseph, D., Petsko, G. A. and Karplus, M. (1990) Anatomy of a Protein Conformational Change: Hinged "Lid" Motion of the Triosephosphate Isomerase Loop, *Science* **249**, 1425-1428.

Kannan, K. K., Ramanadham, M. and Jones, T. A. (1984) Structure, refinement, and function of carbonic anhydrase isozymes: Refinement of human carbonic anhydrase I., *Ann. N.Y. Acad. Sci.* **429**, 49-60.

Kaptein, R., Zuiderweg, E. R. P., Scheek, R. M., Boelens, R. and van Gunsteren, W. F. (1986) A Protein Structure from Nuclear Magnetic Resonance Data. lac Repressor Headpiece, *J. Mol. Biol.* **182**, 179-182.

Karplus, M. and Petsko, G. A. (1990) Molecular dynamics simulations in biology, *Nature* **347**, 631-639.

Karplus, M. and Post, C. B. (1996). Simulations of lysozyme: Internal motions and the reaction mechanism. *Lysozymes: Model Enzymes in Biochemistry and Biology*. Basel, Switzerland, Birkhäuser.

Kelly, J. A., Sielecki, A. R., Sykes, B. D., James, M. N. G. and Phillips, D. C. (1979) X-ray crystallography of the binding of the bacterial cell wall trisaccharide NAM-NAG-NAM to lysozyme, *Nature (London)* **282**, 875-878.

Kirby, A. J. (1987) Mechanism and Stereoelectronic Effects in the Lysozyme Reaction, *Crit. Rev. Biochemistry* **22**, 283-315.

Koshland, D. E., Jr. (1953) Stereochemistry and the Mechanism of Enzymatic Reactions, *Biol. Rev.* **23**, 416-436.

Kuroki, R., Yamada, H., Moriyama, T. and Imoto, T. (1986) Chemical Mutations of the Catalytic Carboxyl Groups in Lysozyme to the Corresponding Amides, *J. Biol. Chem.* **261**, 13571-13574.

Kuwajima, S. and Warshel, A. (1988) The extended Ewald method: a general treatment of long-range electrostatic interactions in microscopic simulations, *J. Chem. Phys.* **89**, 3751-3759.

Lee, F. S. and Warshel, A. (1992) A local-reaction-field method for fast evaluation of long-range electrostatic interactions in molecular simulations, *J. Chem. Phys.* **97**, 3100-3107.

Lee, S., Hwang, B. K., Myoung, Y. C. and Suh, J. (1995) Carboxypeptidase A-Catalyzed Hydrolysis of O-(α-Acylamino-2-styrylacryloyl)-L-β-phenyllactates: Search of Specific Ester Substrates Hydrolyzed through Accumulation of Intermediates, *Bioorg. Chem.* **23**, 183-192.

Levitt, M., Sander, C. and Stern, P. S. (1985) Protein Normal-mode Dynamics: Trypsin Inhibitor, Crambin, Ribonuclease and Lysozyme, *J. Mol. Biol.* **181**, 423-447.

Lifson, S. and Warshel, A. (1968) Consistent Force Field for Calculations of Conformations, Vibrational Spectra, and Enthalpies of Cycloalkane and n-Alkane Molecules, *J. Chem. Phys.* **49**, 5116-5129.

Lim, C. and Tole, P. (1992) Concerted Hydroxl Ion Attack and Pseudorotation in the Base-Catalyzed Hydrolysis of Methyl Ethylene Phosphate, *J. Phys. Chem.* **96**, 5217-5219.

Lim, C. and Tole, P. (1992) Endocyclic and Exocyclic Cleavage of Phosphorane Monoanion: A Detailed Mechanism of the RNase A Transphosphorylation Step, *J. Am. Chem. Soc.* **114**, 7245-7254.

Lippard, S. J. (1995). Methane monooxygenase. NATO ASI Ser., Ser. C (Bioinorganic Chemistry). 1-12.

Lipscomb, W. N. (1974) Relationship of the Three Dimensional Structure of Carboxypeptidase A to Catalysis. Possible Intermediates and pH Effects, *Tetrahedron* **30**, 1725-1732.

Liras, J. L. and Anslyn, E. V. (1994) Exocyclic and Endocyclic Cleavage of Pyranosides in Both Methanol and Water Detected by a Novel Probe, *J. Am. Chem. Soc.* **116**, 2645-2646.

Lolis, E., Alber, T., Davenport, R. C., Rose, D., Hartman, F. C. and Petsko, G. A. (1990) Structure of Yeast Triosphosphate Isomerase at 1.9-Å Resolution, *Biochemisty* **29**, 6609-6618.

Loncharich, R. J. and Brooks, B. R. (1989) The Effects of Truncating Long-Range Forces on Protein Dynamics, *Proteins: Struct. Funct. Genet.* **6**, 32-45.

Luo, X., Zhang, D. and Weinstein, H. (1994) Ligand-induced domain motion in the activation mechanism of a G-protein-coupled receptor, *Protein Engineering* **7**, 1441-1448.

MacKerell, A. D., Karplus, M., et al (1996) *to be published*

MacKerell, A. D., Wiókiewicz-Kuczera, J. and Karplus, M. (1995) An All-Atom Empirical Energy Function for the Simulation of Nucleic Acids, *J. Am. Chem. Soc.* **117**, 11946-11975.

Makinen, M. W., Troyer, J. N., van der Werff, H., Berendsen, J. C. and van Gunsteren, W. F. (1989) Dynamical Structure of Carboxypeptidase A, *J. Mol. Biol.* **207**, 210-216.

Mao, B., Pear, M. R., McCammon, J. A. and Quiocho, F. A. (1982) Hinge-bending in L-Arabinose-binding Protein, *J. Biol. Chem.* **257**, 1131-1133.

Marques, O. and Sanejouand, Y.-H. (1995) Hinge-Bending Motion in Citrate Synthase Arising From Normal Mode Calculations, *Proteins: Struct. Funct. Genet.* **23**, 557-560.

Matthews, B. W. (1988) Structural Basis of the Action of Thermolysin and Related Zinc Peptidases, *Acc. Chem. Res.* **21**, 333 - 340.

McCammon, J. A., Gelin, B. R. and Karplus, M. (1977) Dynamics of Folded Proteins, *Nature* **267**, 585-590.

McCammon, J. A., Gelin, B. R., Karplus, M. and Wolynes, P. G. (1976) The Hinge-Bending Mode in Lysozyme, *Nature* **262**, 325-326.

McCammon, J. A. and Harvey, S. (1987) Dynamics of Proteins and Nucleic Acids, CambridgeUniversity Press, Cambridge.

McCammon, J. A. and Karplus, M. (1977) Internal Motions of Antibody Molecules, *Nature* **268**, 765-766.

McCammon, J. A. and Northrup, S. H. (1981) Gates binding of ligands to proteins, *Nature* **293**, 316-317.

McDonald, R. C., Steitz, T. A. and Engelman, D. M. (1979) Yeast Hexokinase in Solution Enhibits a Large Conformational Change upon Binding Glucose or Glucose 6-Phosphate, *Biochemistry* **18**, 338-342.

Meadows, D. H., Roberts, G. C. K. and Jardetsky, O. (1969) Nuclear magnetic resonance studies of the structure and binding sites of enzymes. VIII. Inhibitor binding to ribonuclease., *J. Mol. Biol.* **45**, 491-511.

Mooser, G. (1992) Glycosidases and Glycosyltransferases, *The Enzymes* **20**, 187-233.

Morozova, T. Y. and Morozov, V. N. (1982) Viscoelasticity of Protein Crytal as a Probe of the Mechanical Properties of a Protein Molecule, *J. Mol. Biol.* **157**, 173-179.

Mulholland, A., J., Grant, G. H. and Richards, W. G. (1993) Computer modelling of enzyme catalysed reaction mechanisms, *Protein Engineering* **6**, 133-147.

Mulholland, A., J. and Karplus, M. (1996) in press,

Newcomer, M. E., Lewis, B. A. and Quiocho, F. A. (1981) The Radius of Gyration of L-Arabinose-binding Protein Decreases upon Binding of Ligand, *J. Biol. Chem.* **256**, 13218-13222.

Nosé, S. and Klein, M. L. (1983) Constant pressure molecular dynamics for molecular systems, *Mol. Phys.* **50**, 1055-1076.

Oi, V. T., Vuong, T. M., Hardy, R., Reidler, J., Dangl, J., Herzenberg, L. A. and Stryer, L. (1984) Correlation between segmental flexibility and effector function antibodies, *Nature (Lond.)* **307**, 136-140.

Osumi, A., Rahmo, A., King, S. W., Przystas, T. and Fife, T. H. (1994) Substituent Effects in the Carboxypeptidase A catalyzed Hydrolysis of Substituted L,β-Phenyllactate Esters, *Biochemistry* **33**, 14750-14757.

Pantoliano, M. W., Whitlow, M., Wood, J. F., Rollence, M. L., Finzel, B. C., Gilliland, G. L., Poulos, T. L. and Bryan, P. N. (1988) The Engineering of Binding Affinity at Metal Ion Binding Sites for the Stabilization of Proteins: Subtilisin as a Test Case, *Biochemistry* **27**, 8311-8317.

Patt, S. L., Baldo, J. H., Boekelheidi, K., Weisz, G. and Sykes, B. D. (1978) The nuclear magnetic resonance determination of the conformation of saccharides bound in subsite D of lysozyme, *Can. J. Biochem.* **56**, 624-629.

Pérez-Jordá, J. and Yang, W. (1995) A simple O(NlogN) algorithm for the rapid evaluation of particle-particle interactions, *Chem. Phys. Lett.* **247**, 484-490.

Pincus, M. R. and Scheraga, H. A. (1979) Conformational energy calculations of enzyme-substrate and enzyme-inhibitor complexes of lysozyme. 2. Calculation of the structures of complexes with a flexible enzyme., *Macromolecules* **12**, 633-644.

Pincus, M. R. and Scheraga, H. A. (1981) Prediction of the Three-Dimensional Structures of Complexes of Lysozyme with Cell Wall Substrates, *Biochemistry* **20**, 3960-3965.

Plapp, B. V. (1995) Site-directed mutagenesis: A tool for studying enzyme catalysis, *Methods Enzymol. (Enzyme Kinetics and Mechanism, Part D)* **249**, 91-199.

Pompliano, D. L., Peyman, A. and Knowles, J. R. (1990) Stabilization of a Reaction Intermediate as a Catalytic Device: Definition of the Functional Role of the Flexible Loop in Triosphosphate Isomerase, *Biochemistry* **29**, 3186-3194.

Post, C. B., Brooks, B. R., Karplus, M., Dobson, C. M., Artymiuk, P. J., Cheetham, J. C. and Phillips, D. C. (1986) Molecular Dynamics Simulations of Native and Substrate-bound Lysozyme, *J. Mol. Biol.* **190**, 455-479.

Post, C. B., Dobson, C. M. and Karplus, M. (1990). Lysozyme hydrolysis of β-glycosides: a consensus between binding interactions and mechanisms. *ACS Symposium Series*. chapter 23.

Post, C. B. and Karplus, M. (1986) Does Lysozyme Follow the Lysozyme Pathway? An Alternative Based on Dynamic, Structural, and Stereoelectronic Considerations, *J. Am. Chem. Soc.* **108**, 1317-1319.

Rees, D. C., Lewis, M. and Lipscomb, W. N. (1983) Refined Crystal Structure of Carboxypeptidase A at 1.54 Å Resolution, *J. Mol. Biol.* **168**, 367 - 387.

Remington, S., Wiegand, G. and Huber, R. (1982) Crystollographic Refinement and Atomic Models of two Different Forms of a Citrate Synthase at 2.7 and 1.7 Å Resolution, *J. Mol. Biol.* **158**, 111-152.

Richards, F. M. and Wyckoff, H. W. (1971) *The Enzymes* **4**, (P.D. Boyer, Ed.) pp. 647-806, Academic, New York.

Richardson, J. S. (1981) The Anatomoy and Taxonomy of Protein Structure, *Adv. Protein Chem.* **34**, 167-339.

Roberts, G. C. K., Dennis, E. A., Meadows, D. H., Cohen, J. S. and Jardetzky, O. (1969) Mechanism of action of ribonuclease, *Proc. Natl. Acad. Sci. USA* **62**, 1151-1158.

Russell, A., Thomas, P. G. and Fersht, A. R. (1987) Electrostatic Effects on the Modification of Charged Groups in the Active Site Cleft of Subtilisin by Protein Engineering, *J. Mol. Biol.* **193**, 803-813.

Russell, A. J. and Fersht, A. R. (1987) Rational modification of enzyme catalysis by engineering surface charge, *Nature* **328**, 496-500.

Schindler, M., Assaf, Y., Sharon, N. and Chipman, D. M. (1977) Mechanism of Lysozyme Catalysis: Role of Ground-State Strain in Subsite D in Hen Egg-White and Human Lysozymes, *Biochemistry* **16**, 423-431.

Schlenkrich, M., Brickmann, J., MacKerell, A. D. J. and Karplus, M., Ed. (1996). An Empirical Potential Energy Function for Phospholipids: Criteria for Parameter Optimization and Applications. *Membrane Structure and Dynamics*. Birkhauser.

Schreiber, H. and Steinhauser, O. (1992) Molecular dynamics studies of solvated polypeptides: why the cut-off scheme does not work, *Chem. Phys.* **168**, 75-89.

Shafizadeh, F. (1958) Formation and cleavage of the oxygen ring in sugars, *Adv. Carbohydr. Chem.* **13**, 9-61.

Shimada, J., Kaneko, H. and Takada, T. (1994) Performance of fast multipole methods for calculating electrostatic, *J. Comp. Chem* **15**, 28-43.

Sinnot, M. L. (1980). Glycosyl Group Transfer. Enzyme Mechanisms. London, Royal Society of Chemistry. 259-291.

Sternberg, M. J. E., Hayes, F. R. F., Russell, A. J., Thomas, P. G. and Fersht, A. R. (1987) Prediction of electrostatic effects of engineering of protein charges, *Nature* **330**, 86-88.

Stote, R. H. and Karplus, M. (1995) Zinc Binding in Proteins and Solution: A Simple but Accurate Nonbonded Representation, *Proteins: Struct. Func. Gen.* **23**, 12-31.

Stote, R. H., States, D. J. and Karplus, M. (1991) On the treatment of electrostatic interactions in biomolecular simulation, *J. Chim. Phys.* **88**, 2419-2433.

Straub, J. E., Lim, C. and Karplus, M. (1994) Simulation Analysis of the Binding Interactions in the RNase A/3'-UMP Enzyme-Product Complex as a function of pH, *J. Am. Chem. Soc.* **116**, 2591-2599.

Strynadka, N. C. J. and James, M. N. G. (1991) Lysozyme Revisited: Crystallographic Evidence for Distortion of an N-Acetylmuramic Acid Residue Bound in Site D, *J. Mol. Biol.* **220**, 401-424.

Suh, J. (1992) Model studies of Metalloenzymes Involving Metal Ions as Lewis Acid Catalysts, *Acc. Chem. Res.* **25**, 273-279.

Thomas, P. G., Russel, A. J. and Fersht, A. R. (1985) Tailoring the pH dependence of enzyme catalysis using protein engineering, *Nature* **318**, 375-376.

Thompson, J. E. and Raines, R. T. (1994) Value of General Acid-Base Catalysis to Ribonuclease A, *J. Am. Chem. Soc.* **116**, 5467-5468.

Thompson, J. E., Venegas, F. D. and Raines, R. T. (1994) Energetics of Catalysis by Ribonucleases: Fate of the 2',3'-Cyclic Phosphodiester Intermediate, *Biochemistry* **33**, 7408-7414.

Umeyama, H., Nakayawa, S. and Fujii, T. (1979) Simulation of the charge relay structure in ribonuclease A, *Chem. Pharm. Bull.* **27**, 974-980.

Usher, D. A., Erenrich, E. S. and Eckstein, F. (1972) Geometry of the First Step in the Action of Ribonuclease A, *Proc. Natl. Acad. Sci. USA* **69**, 115-118.

Vernon, C. A. (1967) The mechanisms of hydrolysis of glycosides and their relevance to enzyme-catalyzed reactions, *Proc. Roy. Soc., Ser. B* **167**, 389-401.

Walsh, C. (1979) *Enzymatic Reaction Mechanisms*. W. H. Freeman, San Francisco.

Walter, B. and Wold, F. (1976) The Role of Lysine in the Action of Bovine Pancreatic Ribonuclease A, *Biochemistry* **15**, 304-310.

Warshel, A. (1978) Energetics of enzyme catalysis, *Proc. Natl. Acad. Sci. USA* **75**, 5250-5254.

Warshel, A. (1981) Calculations of Enzymatic Reactions: Calculations of pKa, Proton Transfer Reactions, and General Acid Catalysis Reactions in Enzymes, *Biochemistry* **20**, 3167-3177.

Warshel, A. (1991) *Computer Simulation of Chemical Reactions in Enzymes and Solutions*. John Wiley & Sons, New York.

Warshel, A. and Karplus, M. (1972) Calculation of Ground and Excited State Potential Surfaces of Conjugated Molecules. I. Formulation and Parametrization, *J. Am. Chem. Soc.* **94**, 5612-5625.

Warshel, A. and Karplus, M. (1974) Calculation of $\pi\pi^*$ Excited State Conformations and Vibronic Structure of Retinal and Related, *J. Am. Chem. Soc.* **96**, 5677-5689.

Warshel, A. and Levitt, M. (1976) Theoretical Studies of Enzymic Reactions: Dielectric, Electrostatic and Steric Stabilization of the Carbonium Ion in the Reaction of Lysozyme, *J. Mol. Biol.* **103**, 227-249.

Warshel, A. and Russell, S. T. (1984) Calculations of electrostatic interactions in biological systems and in solutions, *Q. Rev. Biophys.* **17**, 283-422.

Warshel, A. and Weiss, R, M. (1980) An Empirical Valence Bond Approach for Computing Reactions in Solutions and in Enzymes, *J. Am. Chem. Soc.* **102**, 6218-6226.

Westheimer, F. H. (1956) in *Steric Effects in Organic Chemistry*. New York, N. Y., John Wiley and Sons, Inc. 523-555.

White, J. L., Hackert, M. L., Buehner, M., Adams, M. J., Ford, G. C., Lentz, P. J., Jr., Smiley, I. E., Steindel, S. J. and Rossman, M. G. (1976) A Comparison of the Structures of Apo Dogfish M4 lactate Dehydrogenase and its Ternary Complexes, *J. Mol. Biol.* **102**, 759-779.

Whitlow, M. and Teeter, M. M. (1986) An Empirical Examination of Potential Energy Minimization Using the Well-Determined Structure of the Protein Crambin, *J. Am. Chem. Soc.* **108**, 7163-7172.

Wiberg, K. B. (1965) A Scheme for Strain Energy Minimization. Application to the Cyclo-alkanes, *J. Am. Chem. Soc.* **87**, 1070-1078.

Wigley, D. B. (1995) Structure and mechanism of DNA topoisomerases, *Annu. Rev. Biophys. Biomol. Struct.* **24**, 185-208.

Williams, J. C. and McDermott, A. E. (1995) Dynamics of the Flexible Loop of Triosephosphate Isomerase: The Loop Motion is Not Ligand Gated, *Biochemistry* **34**, 8309-8319.

Wlodawer, A. (1984) Structure of bovine pancreatic ribonuclease by x-ray and neutron diffraction, *Biol. Macromol. Assem.* **2**, 393-439.

Wlodawer, A., Borkakoti, N., Moss, D. S. and Howlin, B. (1986) Comparison of Two Independently Refined Models of Ribonuclease-A, *Acta. Cryst. B* **42**, 379-387.

Wodak, S. Y., Liu, M. Y. and Wyckoff, H. W. (1977) The Structure of Cytidilyl(2',5')Adenosine When Bound to Pancreatic Ribonuclease S, *J. Mol. Biol.* **116**, 855-875.

York, D. M., Darden, T. A. and Pedersen, L. G. (1993) The effect of long-range electrostatic interactions in simulations of macromolecular crystals: a comparison of the Ewald and truncated list methods, *J. Chem. Phys.* **99**, 8345-8348.

Zhang, K., Chance, B., Auld, D. S., Larsen, K. S. and Vallee, B. L. (1992) X-ray Absorption Fine Structure Study of the Active Site of Zinc and Cobalt Carboxypeptidase A in Their Solution and Crystalline Forms, *Biochemistry* **31**, 1159 - 1168.

ELECTROSTATIC INTERACTIONS IN PROTEINS

KIM A. SHARP

Department of Biochemistry and Biophysics

University of Pennsylvania

Philadelphia, PA 19104-6059

1. Introduction

Electrostatics is one of the fundamental interactions that determine the structure, stability, binding affinity, chemical properties, and hence the biological reactivity, of proteins. Electrostatic interactions may contribute to the biological reactivity of proteins in a number of ways, which may be divided into the following general areas:

i) Association rates. A prerequisite for most biological interactions involving proteins is that they must form a close interaction with one or more molecules. The maximal rate at which this can happen is governed by general physical properties such as molecular size and shape, solvent viscosity, temperature, and these put a fundamental limit on the maximal rates of biological reactions. Electrostatics, however, being the longest range fundamental interaction, can considerable enhance association rates, either by facilitating translation diffusion, correct orientation, or by reduction in the dimensionality of the space through which the protein or its ligand must diffuse.

G. Náray-Szabó and A. Warshel (eds.), Computational Approaches to Biochemical Reactivity, 199–235.

ii) Binding constants. Tight binding between molecules is often a prerequisite for biological activity, particularly for catalysis, where the the protein must bind the transition state tightly for effective rate enhancement, and for achieving specificity, which depends on differences in binding constants between different targets that are greater than thermal energy (kT).

iii) Modification of chemical and physical properties of functional groups. Proteins use a repertoire of functional groups including the amino acid side chains and various cofactors to achieve a wide range of functions. The same functional groups are often used in quite different ways; For example the heme group is used for both oxygen binding and electron transfer reactions. This is achieved by modifying the properties of these functional groups via interaction with other groups, and through other structural features of the protein, such as accessibility to solvent and ligands. Very often this modification involves electrostatic interactions. Notable examples are the alteration of the ionization energy (pKa) of a group, shifting of redox midpoints, and the creation of potentials or fields at certain points in the active site to stabilize functionally important charged or dipolar intermediates in enzyme catalysis.

In this review I will only consider contributions to protein energetics and reactivity that can be treated using classical electrostatics. Nevertheless, many of the approaches described here can be combined with quantum mechanical treatments, so that one part of the system (often the active site and/or substrate) is treated quantum mechanically, while the effect of more distant (not directly bonded) parts is treated classically. Broadly speaking, the scope of a classical treatment excludes such properties as the energies of breaking or forming of bonds, determination of intrinsic pKa's and redox midpoints, electronic energy levels, and electronic coupling between electron donors and acceptors. It does however cover *changes* in intrinsic pKa's and redox midpoints, changes in electronic energy levels (due for example to classical Stark type effects), treatment of nuclear Frank-Condon factor contributions to electron transfer rates, solvation effects. It may also include electrostatic stabilization of an intermediate or transition state in which significant bond rearrangement has occurred if the charge distribution is obtained by other methods.

The general problem of relating structural and dynamic properties of a protein to its function requires, in the context of classical electrostatics, the calculation of the electrostatic potential distribution throughout space produced in response to some given charge distribution. The charge distribution may represent either the instantaneous or average state and structure of the protein and its surrounding environment and solvent. From the potential distribution one may calculate electrostatic energies and forces, and thence experimental properties. In most situations of interest using a complete description of the charge state and motions of all atoms is neither possible or desirable so some charges are represented explicitly while the others are either determined within the calculation itself (so called induced or mobile charges), or their electrostatic *effect* is included implicitly.

The theoretical background to classical electrostatics is first reviewed, beginning with the physical basis for the electrostatic response of a protein/solvent system to a charge distribution, and ways of modelling this response. Consistent classical electrostatic frameworks for describing a protein/solvent system are then described. In addition the methods must be able to model temperature, pH and ionic strength effects if these affect the property of interest. Of particular importance is the way one may extract experimentally observable properties from such models.

Since one goal of classical electrostatic applications to protein reactivity is to incorporate available detailed structural information, the system is usually quite complicated and analytical solutions to the various electrostatic equations are rarely available. Numerical methods for solving these equations rapidly and accurately are therefore a non-trivial requirement. Various methods of calculation are briefly discussed. A key component of such calculations is to have reliable input parameters and data. These typically include the position, size, and charge distribution of all the atoms or groups being explicitly treated, and parameters describing the electrostatic response properties, such as the polarizability and dielectric constant, of the parts of the system not treated explicitly. Throughout the chapter examples of successful applications of classical electrostatics to problems in protein function and reactivity are mentioned to illustrate the application of different theoretical concepts, although many interesting studies cannot be covered in the scope of this chapter. More specific application of electrostatic calculations to catalysis are described in the following chapter.

2. Theory

2.1. THE RESPONSE OF THE SYSTEM TO ELECTROSTATIC FORCES

The goal of electrostatic calculations on proteins is to determine from a given structure with a certain explicit charge distribution what the electrostatic potential distribution is. This requires a physical description of how the system (protein plus surrounding solvent) responds to the electrostatic potential arising from the charge distribution. The explicit charge distribution is commonly represented as a collection of point charges and dipoles, usually distributed on the atom centers. Although this representation is somewhat simplistic, and more complicated charge distributions such as bond dipoles or higher order multipoles may be used, in practice this representation has proved more than adequate, more complicated effects being incorporated through suitable parameterization. The response to this charge distribution arises from three physical processes: Electronic polarization, reorientation of permanent dipolar groups, and movement of whole charges. The latter usually involves redistribution of mobile ions in the solvent, although it may include movement of ionized side-chains if significant changes in protein structure occur. A number of phenomenological models are used to model these responses.

2.1.1. Electronic Polarizability

Electronic polarizability can be represented by either point inducible dipoles or by a dielectric constant. In the first approach, the distortion of the electron cloud around an atom by an applied field **E** results in an induced dipole given by

$$\mathbf{m} = \alpha \mathbf{E} \tag{1}$$

where α is the polarizability. Equation (1) applies to a linear, isotropic polarizability, i.e. the induced dipole is aligned with and proportional to field. The inducible dipole approach may be extended to non-isotropic and non-linear polarizabilities but in practice Eqn. (1) has proved sufficiently realistic for proteins. In addition, not enough experimental or theoretical information is available on protein polarizabilities to justify the increased complexity, and larger redistributions of electron density should be treated by quantum mechanical methods anyway. Similar considerations generally restrict the placing of induced dipoles at the atom centers, although in principle they may be positioned at bond centers as well.

The second approach to modelling the electronic polarizability is via the susceptibility, χ,

$$\mathbf{P} = <\mathbf{m}>/\mathrm{V} = \chi \mathbf{E} \tag{2}$$

where **P** is the mean dipole moment induced in some volume V by the Maxwell (total) field **E** in this case. In this model, in contrast to the point dipole model, the response is distributed over some volume V. In practice the polarization is not calculated directly using Eqn. (2), which is presented primarily to show the connection to the point inducible dipole model (Sharp et al. 1992). Instead the polarization is calculated using the Poisson equation (described below) using the dielectric constant which is related to the susceptibility by

$$\varepsilon = 1 + 4\pi\chi \tag{3}$$

The contribution of electronic polarizability to the dielectric constant of most organic material, evaluated e.g. by high frequency dielectric (ε_∞) measurements or the refractive index is usually in the range 2-2.5.

2.1.2. Dipole Reorientation

The reorientation of groups such as the peptide bond or surrounding water molecules which have large permanent dipoles is an important part of the overall response. If these are not treated explicitly using molecular mechanics (MM) or Monte-Carlo (MC) methods these too may be treated using a dielectric constant, i.e. via Eqn. (2). Four factors determine the degree of response from permanent dipoles: i) The dipole moment magnitude. ii) The density of such groups in the protein or solvent iii) The freedom of such groups to reorient. iv) The degree of cooperativity between dipole motions. A treatment of dielectric that incorporates all these factors is the Kirkwood-Frohlich equation for the dielectric constant ε (Kirkwood 1939)

$$\frac{\varepsilon_o(\varepsilon-\varepsilon_\infty)(2\varepsilon+\varepsilon_\infty)}{\varepsilon(\varepsilon_\infty+2)^2} = \frac{<\mathrm{m}^2> - <\mathrm{m}>^2}{\mathrm{VkT}} \tag{4}$$

where k is Boltzmann's constant, T is temperature, ε_0 is the permittivity of vacuum. The numerator on the righthand side is the mean squared fluctuation in dipole moment observed in a volume V due to thermal fluctuations. This quantity reflects the contributions of the four factors above, and an increase in any of these factors will increase the dielectric constant. The left hand side accounts for both the electronic polarizability contribution and the effect of the surrounding material on the volume V in a self consistent manner. Physically, this equation says that the larger the fluctuations induced by thermal motions, the more easily the material can be polarized by an applied field. Equation (4) has been applied to the calculation of isotropic dielectric constants for amorphous protein solids (Gilson and Honig 1986; Nakamura et al. 1988) and to the interior of proteins in solution (Simonson and Perahia 1995; Smith et al. 1993). These calculations and experimental measurements (Takashima and Schwan 1965) provide an estimate of ε=2.5-4 for the contribution of dipolar groups.

For electrostatic models based on dielectric theory the experimental solvent dielectric constant, reflecting the contribution of electronic polarizability and dipole reorientation, is usually used throughout (e.g. for water at 25°C ε=78.6). In principle however, Eqn. (4) is equally applicable to water and could be used to model how the dielectric constant of water might be perturbed at the protein surface, although the contribution of cooperative motions is particularly hard to treat accurately. In addition the accuracy of explicit water molecules, and the difficulties of obtaining dielectric behavior from MM and MC simulations have restricted this approach.

An alternative approach to treating the dipole reorientation contribution that is particularly suited to the water surrounding the protein is the Langevin dipole model which describes the polarization due to a permanent solvent dipole of magnitude m^L in terms of the applied field

$$\mathbf{m} = \frac{\mathbf{E}}{E} m^L \left(\coth X - \frac{1}{X} \right) \tag{5}$$

where X is a function of the dipole moment and applied field. In the original development of the Langevin dipole model $X = m^L E/kT$ However use of this function in Eqn. (5) as it stands is not sufficiently accurate for a highly polar associated liquid like water. The Langevin model can however be extended by making X a more sophisticated function of the local applied field, which can be parameterized to accurately reproduce solvation energies of ionic species of various radii and charges (Lee et al. 1993; Warshel and Åqvist 1991; Warshel and Russell 1984). The Langevin model, unlike the linear dielectric model (Eqns. (2-3)) can account for the saturation of the response at high fields that occurs if the dipoles become highly aligned with the field. The dielectric model can also be extended to incorporate saturation effects (Warwicker 1994), although there is a compensating effect of electrostriction, which increases the local dipole density (Jayaram et al. 1989a), so that the linear solvent dielectric model has proven sufficiently accurate for most protein applications to date.

2.1.3. Charge Rearrangement

Charge rearrangement can occur in any circumstance where there are formally charged groups which have some degree of mobility. The most mobile, and easiest to treat are salt ions in the solvent. The commonest way of treating this is via the Boltzmann model, where the local concentration c(**r**) of an ion of type i, valence z_i, and bulk concentration c_i^o at a position **r** in the solvent is given by

$$c_i(\mathbf{r}) = c_i^o \exp(-z_i e \phi(\mathbf{r})/kT) \tag{6}$$

where $\phi(\mathbf{r})$ is the average potential at **r**. This equation says that cations will be found preferentially in regions of negative potential around a protein, and depleted in regions of positive potential, and *vice versa* for anions. The net charge density of mobile ions is then given by

$$\rho^m(\mathbf{r}) = \sum_i z_i e c_i(\mathbf{r}) \tag{7}$$

with the electroneutrality condition $\sum_i z_i c_i^{\,o} = 0$. An assumption made in the use of Eqn. (6) is that the potential of mean force can be approximated by the mean electrostatic potential in the exponent of the Boltzmann factor. This neglects the effect of ion size and correlations between ion positions. Other models for the mobile ion behavior which account for these effects are integral equation models and MC models (Bacquet and Rossky 1984; Murthy et al. 1985; Olmsted et al. 1989; Olmsted et al. 1991; Record et al. 1990). While these can be combined with different dielectric treatments and models that incorporate detailed molecular shape (Jayaram et al. 1990), they have generally been applied to simpler systems because of their greater computational requirements. Comparisons between different models have been used to check the different simplifications made in each (Murthy et al. 1985; Sharp 1995). These studies show that ion size and correlation effects do not significantly compromise the Boltzmann model for monovalent (1-1) salts at mid-range concentrations 0.001-0.5M, it can significantly underestimate ion concentrations near charge molecules for divalent ions. Thus because of the simplicity of the Boltzmann model, and the ease of combining it with the dielectric treatment of electronic and dipole polarization, Eqn. (6) is widely used for modelling the effects of 1-1 salts on proteins.

Formally charged groups on proteins, particularly the longer side-chains on the surface of proteins, Arginine, Lysine, and to a lesser extent Glutamate and Aspartate, have the ability to alter their conformation in response to electrostatic fields. In addition information about fluctuations about their mean position may need to be included in calculating average properties. Three approaches to modelling protein formal charge movements can be taken. The first is to treat the motions within the dielectric response, e.g. via the Kirkwood-Frohlich model, Eqn. (4). In this case a method of evaluating the dipole fluctuations arising from charge movements is required, e.g. using molecular dynamics (MD). With this approach, the protein may be viewed as having a higher dielectric than 2.5-4 in the regions of these charged groups, particularly at the surface, where the concentration and mobility of these groups may give an effective dielectric of 30 or more (Simonson and Perahia 1995; Smith et al. 1993). A second approach is to model the effect of charge motions on the electrostatic quantity of interest explicitly with MM simulations, without formulating these motions as a dielectric at all (Langsetmo et al. 1991; Wendoloski and Matthew 1989). This involves generating an ensemble of structures with different explicit charge distributions. The third approach is based on the fact that often one is interested in a specific biological process A->B, in which one can evalute the structure of the protein in states A and B, (experimentally or by modelling) and any change in average charge positions is incorporated at the level of different average explicit charge distribution inputs for the calculation, modelling only the electronic, dipolar, and salt contributions as the response.

2.1.4. Concepts of Polarizability , Polarity and Dielectric constants in Proteins

All of the models described above for the electrostatic response are based on well defined physical treatments that have been used successfully for many years in applications to homogeneous solid state and liquid state systems. Their use in protein-solvent systems is more recent, but has been driven by the increasing amount of atomic level structural information available on proteins. Nevertheless, the use of such phenomonological approaches for these complex, non-homogeneous systems raises a number of questions about how realistic they are, and when different treatments are applicable. To some extent this is best determined by using them on particular problems in protein structure and function and seeing whether they have explanatory and predictive power. There have been many such applications (e.g. see (Gilson 1995; Sharp and Honig 1990b; Sharp 1994; Warshel and Åqvist 1991) for reviews) which have raised a number of conceptual issues.

The first issue is the distinction between polarity and polarizability. Briefly, polarity may be thought of as describing the density of charged and dipolar groups in a particular region. Polarizability, by contrast, refers to the *potential* for reorganizing charges, orienting dipoles and inducing dipoles. Thus polarizability depends both on the polarity and the freedom to reorganize in response to an applied electric field, as described say by Eqn. (4). This ability to reorganize may depend on the environment in which the charges and dipoles find themselves (e.g. bulk water vs. water near the protein surface), or on what part of a process is of interest. Thus when a protein is folding, or undergoing a large conformational rearrangement, the peptide groups may be quite free to reorient. In the folded protein these may become spatially organized so as to stabilize another charge or dipole, creating a region with high polarity. Within the folded protein, however, there will usually be much less ability to reorient the dipolar groups in response to a new charge or dipole without significant disruption of the structure. Thus a protein may contain regions that are both highly polar and have a low polarizability.

The second issue is what is meant by a dielectric constant. Referring back to Eqns. (2-3), it is just a particular model for describing electronic and dipolar polarizability in which the response is averaged over some region V. If V is large on the atomic scale then the dielectric is being using in a macroscopic sense. If V is on the order of an atom or a group then it is being used in a more microscopic sense, and different dielectric responses may be used for different atoms or groups (Sharp et al. 1992). The concern about the use of a dielectric constant is less whether it is macroscopic or microscopic and more whether it describes well the average polarization produced in some region by an applied field. A considerable number of applications show that for treating electronic polarization, small protein dipole reorientations, and water polarization the dielectric model is appropriate (For example see reviews (Gilson 1995; Sharp and Honig 1990b; Sharp 1994)). For large protein dipole motions, due for example, to protein conformational changes which involve specific and very nonhomogeneous motions, it is inappropriate to treat the response as a dielectric or indeed with any phenomenological response function. Such motions should be treated in terms of changes in the explicit charge distribution

A third issue is the use of 'effective' dielectric constants to describe the strength of interaction between two charges, q_1 and q_2. This is defined as the ratio of the observed or calculated interaction strength, U, to that expected between the same two charges in vacuum:

$$\varepsilon^{eff} = \frac{q_1 q_2}{r_{12}} \Big/ U \tag{8}$$

where r_{12} is the distance between the charges. If the system were completely homogeneous in terms of its electrostatic response and involved no charge rearrangement then ε^{eff} would describe the dielectric constant of the medium containing the charges. This is generally never the case: The strength of interaction in a protein system is determined by the net contribution from protein, solvent and ions, so ε^{eff} does not give information about the dielectric property of any particular region of space. In fact in the same system ε^{eff} will general have quite different values for different charge-charge interactions. In this case ε^{eff} is no more than its definition- a measure of the strength of interaction. While this is an important quantity, it cannot be used directly to answer questions about the protein dielectric constant for example. Rather, it is one of quantities that one aims to extract from theoretical models to compare to experiment.

The final issue concerns a question that is often raised: What is the dielectric constant of a protein? Conflation of the concepts of polarity, polarizability, dielectric constant and effective dielectric constant has resulted in a confusing range of answers, ranging from high (greater than or equal to that of water), to low (similar to organic solids). As the preceding sections and discussion should have made clear, in fact this is not an entirely meaningful or useful question. The more pertinent questions are what are the different contributions to the electrostatic response, what is the best way to model them (using dielectric response models for some contributions if appropriate), and how can one combine them into a consistent electrostatic model to extract experimentally important quantities.

2.2. DEPENDENCE OF POTENTIAL ON THE CHARGE DISTRIBUTION

All the models described above for the electrostatic response provide either the charge or dipole distribution due to a potential or field. To complete the description of an electrostatic model these equations must be combined with the appropriate equations that provide the dependence of the potential and field on the charge and dipole distribution.

2.2.1. The Potential from Point Charges and Dipoles

The potential at a point in space $\mathbf{r}_j$ arising from a collection of n point charges, q_i, and m point dipoles (permanent and/or induced), $\mathbf{m}_j$ is given by

$$\phi(\mathbf{r}_j) = \sum_i^n \frac{q_i}{r_{ij}} + \sum_i^m \frac{\mathbf{m}_i . \mathbf{r}_{ij}}{r_{ij}^3} \tag{9}$$

where $\mathbf{r}_{ij} = \mathbf{r}_j - \mathbf{r}_i$ is the position vector with respect to the i'th charge or dipole positions respectively. The field is obtained from Eqn. (9) by taking the gradient with respect to the position vector,

$$\mathbf{E}(\mathbf{r}_j) = -\nabla\phi(\mathbf{r}_j) = \sum_i^n \frac{q_i \mathbf{r}_{ij}}{r_{ij}^3} + \sum_i^m \frac{\mathbf{m}_i (3\mathbf{r}_{ij}\mathbf{r}_{ij} - r_{ij}^2)}{r_{ij}^5} \tag{10}$$

Equations (9-10), which are derived from Coulomb's law, are valid in vacuum (ε=1) or for applications where a dielectric response model is not being used. For the special case of a uniform dielectric ε the potential and field are just given by (9) and (10) scaled, or screened, by the factor 1/ε.

2.2.2. The Potential from a Charge or Dipole Distribution

Equations analogous to (9) and (10) can be written for when the charge and dipole density distributions, ρ and **P** respectively, are described as a function of position $\mathbf{r}_i$

$$\phi(\mathbf{r}_j) = \int_V \left\{ \frac{\rho(\mathbf{r}_i)}{\mathbf{r}_{ij}} + \frac{\mathbf{P}(\mathbf{r}_i).\mathbf{r}_{ij}}{\mathbf{r}_{ij}^3} \right\} d\mathbf{r}_i \tag{11}$$

$$\mathbf{E}(\mathbf{r}_j) = \int_V \left\{ \frac{\rho(\mathbf{r}_i)\mathbf{r}_{ij}}{\mathbf{r}_{ij}^3} + \frac{\mathbf{P}(\mathbf{r}_i)(3\mathbf{r}_{ij}\mathbf{r}_{ij} - \mathbf{r}_{ij}^2)}{\mathbf{r}_{ij}^5} \right\} d\mathbf{r}_i \tag{12}$$

In contrast to Eqns. (9) and (10), Eqns. (11), and (12) can be used with dielectric models where **P** is determined using the susceptibility or dielectric constant (Eqn. 2).

2.2.3. Interaction Potentials and Reaction Potentials

The concepts of interaction and reaction potentials are important for a physical understanding of the results of electrostatic calculations and their relationship to experimental quantities. The potential at a charge or dipole *i* created by another charge or dipole *j* is referred to as an interaction potential, since it quantitates the strength of the electrostatic interaction between those two groups. In a system in which all the charges and dipoles are represented explicitly then the only contribution is the direct or Coulombic potential of *j* at *i*. If the electrostatic model represents part of the system by some polarizability function(s), then the potential at *i* includes the potential from the polarization induced by *j*. This is often referred to as the screening potential, since it opposes the direct, Coulombic potential. A third potential contribution at *i* arises from the polarization induced by *i* itself. This is often referred to as the reaction or self potential, and if solvent is involved as the solvation potential. This potential is of opposite sign to the charge on *i*, thus preferentially stabilizing, or solvating the charge in a more polarizable environment. Equations (9-12) include the direct, screening, and reaction potentials and these individual contributions may be calculated separately if required by appropriate use of these equations (Sharp and Honig 1990b). For example the self and interaction potentials would be obtained from calculations in which charge was assigned only to groups *i* or *j* respectively, while the screening contribution to the *i-j* interaction can be obtained by taking the difference between values obtained *in vacuo* and with polarizability with only charge *j* assigned.

2.3. MODELS FOR PROTEIN ELECTROSTATICS

2.3.1. Self Consistent Field Models

Combining Equations (1) and (5) for the protein electronic polarizability and the solvent polarizability, respectively, with Eqn. (10) for the field Warshel and co-workers (Lee et al. 1993; Warshel and Åqvist 1991; Warshel and Russell 1984) developed the Protein Dipole Langevin Dipole (PDLD) model which was the first consistent model for treating protein/solvent polarizabilities in protein electrostatic applications. The electrostatic field distribution in this model is given by

$$\mathbf{E}(\mathbf{r}_j) = \sum_i^n \frac{q_i^e \mathbf{r}_{ij}}{r_{ij}} + \sum_i^n \frac{\mathbf{m}_i^e (3\mathbf{r}_{ij}\mathbf{r}_{ij} - r_{ij}^2)}{r_{ij}^5} + \sum_i^n \frac{\alpha_i \mathbf{E}_i (3\mathbf{r}_{ij}\mathbf{r}_{ij} - r_{ij}^2)}{r_{ij}^5} + \sum_k^l \frac{m_k^L \frac{\mathbf{E}_k}{E_k}(\coth X - \frac{1}{X})(3\mathbf{r}_{kj}\mathbf{r}_{kj} - r_{kj}^2)}{r_{kj}^5} \tag{13}$$

where $q^e{}_i$, $\mathbf{m}^e{}_i$, and α_i are the charge, permanent dipole moment and electronic polarizability assigned to the i'th atom, and $m^L{}_k$ is the k'th solvent Langevin dipole. The Langevin dipoles representing the solvent are placed on lattice points suitably distributed within the solvent volume. Values for atomic polarizabilities, charges and Langevin parameters suitable for PDLD electrostatic calculations are available in the literature (Russell and Warshel 1985) . Notice that the field appears on both sides of Enq. (13). The most straightforward way to solve this equation is by setting the initial field on the righthand side to be that from the permanent charges and dipoles, calculating the induced dipoles, recalculating the field, and so on until the whole procedure has converged to a self consistent set of fields and dipoles. To extend this model to account for dipole reorientation, Warshel et al have combined the PDLD model with molecular dynamics (MD) simulations (Warshel and Creighton 1989), in which MD is used to generate different explicit charge distributions.

2.3.2. Poisson Boltzmann Models

Equations (2) and (12) may be combined and re-expressed, using Gauss's law, as a partial differential equation, the Poisson Equation

$$\nabla.\varepsilon(\mathbf{r})\nabla\phi(\mathbf{r})+4\pi\rho(\mathbf{r})=0 \qquad (14)$$

which relates the potential, charge and dielectric distributions, $\phi(\mathbf{r})$, $\rho(\mathbf{r})$, $\varepsilon(\mathbf{r})$ respectively. Contributions to the polarizability from electrons, protein permanent dipoles, and solvent dipoles are incorporated into this model by using an appropriate value for the dielectric for each region of protein and solvent. Recognizing that the charge distribution can be split up into the explicit charge distribution on the protein plus the mobile solvent ion distribution, $\rho=\rho^e+\rho^m$, one can include salt effects in Eqn. (14) by combining it with Eqns. (6-7) to yield the Poisson-Boltzmann (PB) equation

$$\nabla.\varepsilon(\mathbf{r})\nabla\phi(\mathbf{r}) + 4\pi\sum_i z_i e c_i^o e^{-z_i e\phi(\mathbf{r})/kT} + 4\pi\rho^e(\mathbf{r}) = 0 \tag{15}$$

The Poisson-Boltzmann model is the other major model for protein electrostatics, which has been widely used, especially because of its ability to treat salt effects. In applications, where the protein is not highly charged the Boltzmann term in Eqn. (15) can be linearized since $e\phi/kT <\approx 1$, giving

$$\nabla.\varepsilon(\mathbf{r})\nabla\phi(\mathbf{r}) - 8\pi I e^2\phi(\mathbf{r})/kT + 4\pi\rho^e(\mathbf{r}) = 0 \tag{16}$$

where $I = \frac{1}{2}\sum_i z_i^2 c_i^o$ is the ionic strength. The linearized equation is not however applicable to many applications involving highly charged species, such as membrane systems, nucleic acids, nucleic acid binding proteins.

Values for protein atomic charges, radii, and dielectric constants suitable for use with the Poisson-Boltzmann equation are available in the literature (Jean-Charles et al. 1990; Mohan et al. 1992; Simonson and Brunger 1994; Sitkoff et al. 1994b)

2.4. CALCULATION OF ENERGIES AND FORCES

2.4.1. Energies

Once the electrostatic potential distribution has been obtained, calculation of experimental properties almost always requires evaluation of the electrostatic energy or force. The most general expression for the electrostatic energy is obtained via the charging integral approach, a thermodynamic coupling process in which one takes an initial state where all the explicit charges and dipoles, q_i^e, $\mathbf{m}_i^e$, are discharged and converts it into the fully charged state by simultaneously charging each charge and dipole, using a charging parameter λ. The explicit charge and dipole values are λq_i^e and

$\lambda \mathbf{m}_i^e$ at any point in the charging procedure. The electrostatic free energy is given by the integrals

$$\Delta G^{el} = \sum_i \int_0^1 \phi_i(\lambda) q_i^e \delta\lambda + \sum_i \int_0^1 \mathbf{E}_i(\lambda).\mathbf{m}_i^e \delta\lambda \tag{17}$$

where the sum is over all explicit charges and dipoles, and $\phi_i(\lambda)$ and $\mathbf{E}_i(\lambda)$ are the corresponding potential and field at each charge or dipole respectively. Note that the second summation does *not* include the induced dipoles, since the creation of these dipoles is part of the reorganization energy, and their inclusion would mean double counting. If all the response functions are linear, then ϕ_i and $\mathbf{E}_i$ are linear in λ and Eqn. (17) becomes

$$\Delta G^{el} = \frac{1}{2} \sum_i \phi_i q_i^e + \frac{1}{2} \sum_i \mathbf{E}_i.\mathbf{m}_i^e \tag{18}$$

where ϕ_i and $\mathbf{E}_i$ are the final potential and field, given for example by one of the electrostatic models described above. Equation (18) is widely used in the literature to compute electrostatic energies. It is sometimes stated that the factor of one half appears in order to avoid double counting, but it is in fact a deeper consequence of the linearity of the system. In non-linear systems, the integration factors will not be equal to one half. When applicable, a linear Eqn. (18) easily allows one to decompose the energy into different interactions.

A common source of non-linearity is the Boltzmann term in the PB equation (15). In this case Eqn. (18) is invalid, and the integration in Eqn. (17) become tedious since the electrostatic equations have to be re-solved many times. Variational expressions for the total electrostatic energy for the non-linear PB model are available (Reiner and Radke 1990; Sharp and Honig 1990a; Zhou 1994)

$$\Delta G^{el} = \int_V \rho^e \phi - \frac{\varepsilon E^2}{8\pi} - kT \sum_i c_i^o (e^{-z_i e\phi / kT} - 1) d\mathbf{r} \tag{19}$$

where the integration is now over all space, not over the charging process.

Regarding the equations (17-19) for electrostatic energy, the questions sometimes arises as to whether they describe energies or free energies, and if the latter, whether they are Gibbs or Helmholtz free energies. From the origin of Eqn. (17) it should be clear that the resulting quantity is a free energy contribution since it is obtained as work during a thermodynamic process. Furthermore, consideration of the physical behavior behind the dielectric constant of, for example, water shows that the degree to which the dipoles are aligned by an applied field depends on a balance between electrostatic forces and the unfavorable entropy of dipole ordering. Consequently the dielectric constant is temperature dependent, which will impart a temperature dependence, and hence an entropic component, to the free energy. This also allows one to analyze entropic and enthalpic components of electrostatic free energies if this are of interest (Rashin and Honig 1985; Sharp 1995). As to whether the free energy is a Gibbs or a Helmholtz free energy, neither constant pressure nor constant volume conditions are explicitly imposed in the charging process, but since the change in volume at 1 Atm. in most processes is very small PdV is negligible and one need not distinguish them. Gibbs free energy, G, is used here.

2.4.2. Forces

The general expression for the electrostatic force on a charge q is given by the gradient of the total free energy with respect to that charge's position

$$\mathbf{f}_q = -\nabla_{\mathbf{r}_q} (G^{el}) \tag{20}$$

If the movement of that charge does not affect the potential distribution due to the other charges and dipoles, then Eqn. (20) can be evaluated using the 'test charge' approach, in which case the force depends only on the gradient of the potential, or the field at the charge

$$\mathbf{f} = q\mathbf{E} \tag{21}$$

Similar considerations lead to expressions for the translational and rotational forces, **f** and **T** respectively, on a dipole

$$\mathbf{f} = -\frac{\mathbf{m}.\mathbf{E}}{\mathrm{E}}\nabla\mathrm{E} \qquad \mathbf{T} = \mathbf{m}\mathrm{x}\mathbf{E} \tag{22}$$

Equations (21) and (22) are well suited for calculating forces on atoms within the self-consistent field electrostatic models, and also in combination with molecular mechanics, since the latter use potentials whose electrostatic component is expressed in terms of point charges, as given by equations like Eqn. (9). For models that use the dielectric formulism, however, they are not always applicable. For example forces arise between a single charge and a dielectric boundary due to image charge (reaction potential) effects. Also movement of the dielectric boundary between the solute and solvent will change the energy of the system. This results in a force acting on the atoms that are at the solvent-solute surface. Ways to account for this effect have been used to combine the PB description of solvent with MD (Niedermeier and Schulten 1992; Sharp 1991; Zauhar 1991) A similar effect to the 'dielectric pressure' force arises from solvent ion pressure at the solute-solvent boundary. An elegant general expression for the force density that includes both these effect has been derived within the PB model using variational methods (Gilson et al. 1993)

$$\mathbf{f} = \rho^e \mathbf{E} - \frac{1}{2}\mathrm{E}^2 \nabla \varepsilon - \frac{4\pi \mathrm{I} e^2 \phi^2}{kT} \nabla \mathrm{A} \tag{23}$$

where A is a function describing the accessibility to solvent ions, which is 0 inside the protein, and 1 in the solvent, and whose gradient is non-zero only at the solute solvent surface. Similarly, in a two dielectric model (solvent plus protein) the gradient of ε is non-zero only at the molecular surface. The first term accounts for the force acting on a charge due to a field, as in Eqn. (21), while the second and third terms account for the dielectric surface pressure and ionic atmosphere pressure terms respectively. Equation (23) has been used to combine the PB equation and molecular mechanics (Gilson et al. 1994).

3. Numerical Methods

Numerical solution of self consistent field type electrostatic equations is generally straightforward. The protein is represented by the positions, charges and polarizabilities of all its atoms. The solvent is represented, in the PDLD method, by a lattice of Langevin dipoles (Lee et al. 1993). An initial guess is made for the field, usually by setting it equal to that arising from the charges and permanent dipoles. The dipoles induced by this field are calculated, the field recalaculated with the induced dipole contribution, and so on, until convergence. A possible pitfall that is easily avoided is the polarization catastrophe, where two or more groups are too close and too polarizable so their polarizations mutually reinforces each other, leading to infinite polarization. This should not happen if the distribution and size of polarizabilities is realistic.

Numerical solution of the Poisson and Poisson-Boltzmann equations is more complicated since these are three dimensional partial differential equations, which in the latter case can be non-linear. Solutions in planar, cylindrical and spherical geometry, are well known but are of limited use in applications where one wishes to account for the detailed three dimensional structure of a protein in a realistic way. In these applications the protein is represented by the positions, charges and radii of all the atoms which describe its shape and polarity. The region inside the protein is assigned dielectric(s) values for the protein to account for electronic polarizability, and if appropriate dipolar polarizability. The region outside the protein's molecular surface is assigned the solvent dielectric value(s), and the required ionic strength. To obtain the electrostatic potential requires the ability to solve the equations for this complex, assymmetric distribution of charge, dielectric and salt concentration. Inevitably, this involves discretizing space and using finite methods.

Many numerical techniques, some developed in engineering fields to solve differential equations, have been applied to the PB equation. These include finite difference methods (Gilson et al. 1988; Nicholls and Honig 1991; Warwicker and Watson 1982), finite element methods (Rashin 1990; Yoon and Lenhoff 1992; Zauhar and Morgan 1985), multigridding (Holst and Saied 1993; Oberoi and Allewell 1993), conjugate gradient methods (Davis and McCammon 1989) and fast multipole methods (Bharadwaj et al. 1994; Davis 1994). Methods for treating the non-linear PB include under-relaxation (Jayaram et al. 1989b), and powerful inexact Newton methods (Holst et al. 1994a). The non-linear PB equation can also be solved via a self consistent field approach, in which one calculates the potential using Eqn (11), then the mobile charge density is calculated using Eqn. (7), and the procedure is repeated until convergence (Pack and Klein 1984; Pack et al. 1986). The method allows one to put in more elaborate models for the ion distribution, for example incorporating the finite size of the ions (Pack et al. 1993). Approximate methods based on spherical approximations (Born-type models) have also been used (Schaeffer and Frommel 1990; Still et al. 1990). Considerable numerical progress has been made in finite methods, and accurate rapid algorithms are available. The reader is referred to the original references for numerical details.

4. Application to Experiment

4.1. INTERACTION POTENTIALS

An important reason for calculating the electrostatic potential distribution is to obtain experimentally measurable properties. In some cases the electrostatic potential or field itself is directly measurable. In these cases no further calculation is necessary to obtain the appropriate experimental quantity. Examples of applications like this include the measurement and calculation of α-helix potentials and fields (Lockhart and Kim 1993; Lockhart and Kim 1992; Sitkoff et al. 1994a) , the measurment and calculation of potentials around DNA (Hecht and Honig 1995; Shin and Hubbell 1992) and at membrane surfaces (McDaniel et al. 1986). Another application is when the stabilization potential at a particular site in a protein is required. An example is the evaluation of the protein potential at the oxyanion hole in a serine protease (Soman et al. 1989). Most experimental parameters of interest, however, are obtained by combining the results of an electrostatic calculation with further theoretical analysis or calculations.

4.2 SOLVATION ENERGIES

The solvation energy is defined as the change in free energy upon transfer of a molecule or group from one solvent (or environment) to another. As such it can be an important component of shifts in charge transfer equilibria, and of the total free energy of binding. Free energies of transfer of small molecules between solvents are also an important source of data for parameterizing and testing electrostatic models. The electrostatic component of the solvation free energy change is obtained as the difference in reaction field or self energy of a group or molecule in the two environments, A and B

$$\Delta G^{rf} = \frac{1}{2}\sum_i q_i(\phi_i^{rf}(B) - \phi_i^{rf}(A)) \tag{24}$$

where the summation is over all charges q_i of the group or molecule, and the potential is that due to the reaction field of those charges. To calculate total solvation energy changes Eqn. (24) must be combined with a method for calculating the non-electrostatic solvation contribution, e.g. by surface area methods. Applications to solvation include comparisions of electrostatic and free energy perturbation studies (Ewing and Lybrand 1993; Jean-Charles et al. 1990) and analysis of small molecule and ion solvation (Mohan et al. 1992; Rashin and Namboodiri 1987; Sharp et al. 1992; Simonson and Brunger 1994; Sitkoff et al. 1994b)

4.3. CHARGE TRANSFER EQUILIBRIA

Acid-base equilibria, electron transfer and ion binding are examples of charge transfer equilibria important in protein function, in which the transferred species is a proton, an electron or a salt ion respectively. The theory of the dependence of these three equilibria on classical electrostatic effects can be treated in an identical manner.

Beginning with acid-base equilibria, a titratable group will have an intrinsic ionization equilibrium, expressed in terms of a known intrinsic pK^oa (where $pK^oa = -\log_{10}(K^oa)$, and K^oa is the dissociation constant for the reaction $H^+A = H^+ + A$ and A can be an acid or a base) which is determined by all the quantum chemical, electrostatic and environmental effects operating on that group in some reference state. For example a reference state for the aspartic acid side-chain ionization might be the isolated amino acid in water, for which $pK^oa = 3.85$. In the environment of the protein the pKa will be altered by three electrostatic effects. The first is by moving the group to an environment with a different polarizability, the second is due to interaction with permanent dipoles in the protein, the third is due to charged, perhaps titratable, groups. The effective pKa is given by

$$pKa = pK^{o}a + (\Delta\Delta G^{rf} + \Delta\Delta G^{perm} + \Delta\Delta G^{tit})/2.303kT \qquad (25)$$

where the factor of 1/2.303kT converts units of energy to units of pKa. The first contribution, $\Delta\Delta G^{rf}$, arises because the completely solvated group induces a strong favourable reaction field in the high dielectric water, which stabilizes the charged form of the group (The neutral form is also stabilized by the solvent reaction field induced by any dipolar groups, but to a lesser extent). Desolvating the group to any degree by moving it into a less polarizable environment will preferentially destabilize the charged form of that group, shifting the pKa by an amount

$$\Delta\Delta G^{rf} = \frac{1}{2}\sum_{i}\left(q_i^d \Delta\phi_i^{rf,d} - q_i^p \Delta\phi_i^{rf,p}\right) \qquad (26)$$

where q_i^p and q_i^d are the charge distributions on the group, $\Delta\phi_i^{rf,p}$ and $\Delta\phi_i^{rf,d}$ are the changes in the group's reaction potential upon moving it from its reference state into the protein, in the protonated (superscript p) and deprotonated (superscript p) forms respectively, and the sum is over the group's charges. The contribution of the permanent dipoles is given by

$$\Delta\Delta G^{tit} = \sum_{i}\left(q_i^d - q_i^p\right)\phi_i^{perm} \qquad (27)$$

where ϕ_i^{perm} is the interaction potential at the *i*'th charge due to all the permanent dipoles in the protein, including the effect of screening. At this stage it is worth noting that intrinsic pKa's of groups in proteins are rarely shifted by more than 1 pKa unit indicating that the effects of desolvation are often compensated to a large degree by the $\Delta\Delta G^{perm}$ term.

The final term accounts for the contribution of all the other charge groups:

$$\Delta\Delta G^{titr} = \sum_i \left(q_i^d < \phi_i >^d_{pH,c,\Delta V} - q_i^p < \phi_i >^p_{pH,c,\Delta V} \right) \tag{28}$$

where $< \phi_i >$ is the mean potential at group charge *i* from all the other titratable goups. The charge state of the other groups in the protein depend in turn on their intrinsic "pKa's", on the external pH if they are acid-base groups, the external redox potential ΔV if they are redox groups, and the concentration of ions, c, if they are ion binding sites, as indicated by the subscript on $<\phi_i>$. Moreover, the charge state of the group itself will affect the equilibrium at the other sites. Because of this linkage, exact determination of the complete charged state of a protein is a complex procedure. If there are N such groups, the rigorous way to do that is to compute the titration state partition function by evaluating the relative electrostatic free energies of all 2^N ionization states for a given set of pH, c, ΔV. From this one may calculate the mean ionization state of any group as a function of pH, ΔV etc. For large N this becomes impractical, but various approximate schemes work well, including a Monte-Carlo procedure (Beroza et al. 1991; Yang et al. 1993) or partial evaluation of the titration partition function by clustering the groups into strongly interacting sub-domains (Bashford and Karplus 1990; Gilson 1993; Yang et al. 1993)

Calculation of ion binding equilibria in proteins proceeds exactly as for calculation of acid-base equilibria, the results usually being expressed in terms of changes in an association constant, Ka for the reaction $Y^z + A = Y^zA$, where z is the ion valence. Similarly for shifts in the redox midpoint potential, defined by [oxdised]/[reduced] = $\Delta V - E_m$, where the midpoint potential E_m is the external reducing potential at which the group is half oxidised and half reduced. Shifts in E_m can be calculated using the equivalent of Eqn. (24), where now the conversion factor of 25.7/kT converts the electrostatic energies to the customary units of mV. E_m's are customarily measured and tabulated at pH 7 (E_{m7}), since redox midpoints, like pKa's (and ion binding equilibria) are dependent on pH, ion concentration and the protein's charged state.

Using the proton transfer event as an example, shifts in pKa's have a number of consequences. Consider some event or conformational change X->Y that shifts the pKa of a group. If the pKa is shifted to or from a value near the pH, this event will result in release or uptake of protons. Calculation of this release is straightforward once all the pKa's are known. A reciprocal consequence is that the process X->Y will have a pH dependence. Furthermore, removal of that group, or replacement with another group of a different pKa will affect the energetics of X->Y. For example consider a protein in which the group A has values of pKa^x and pKa^y in the two states X and Y respectively. The release of protons is given by

$$\Delta n_{H^+} = \frac{1}{(1+10^{(pH-pK_a^x)})} - \frac{1}{(1+10^{(pH-pK_a^y)})} \tag{29}$$

which has a maximum at a pH midway between the two pKa's. The equilibrium between X and Y given by K = [X]/[Y] is

$$K/K_o = \frac{(1+10^{-(pH-pK_a^x)})}{(1+10^{-(pH-pK_a^y)})} \tag{30}$$

where K_o is the high pH limiting value of the equilibrium constant (Alternatively the low pH state may be used as a reference). The change in free energy for the process X->Y is given by $-kT\ln(K/K_o)$. Equation (30) can also be used to obtain the change in energy upon mutating group A, by substituting the appropriate values of pKa. Equations (29-30) can be generalized to deal with a collection of linked groups. Analogous considerations apply to release of bound ions or electrons, and the dependence on salt concentration or redox potential respectively.

There have been many studies of the electrostatics of charge transfer equilibria, which include the effect of mutations on active site histidine pKa's and enzyme activity in subtilisin (Gilson and Honig 1987; Sternberg et al. 1987), the linkage between redox potentials and pKa's in azurin (Bashford et al. 1988), the relationship between pKa's, salt bridges and stability in thioredoxin (Langsetmo et al. 1991) and pKa's and stability in myoglobin (Yang and Honig 1994), a detailed analysis of redox potential shifts in ion-sulfur proteins (Langen et al. 1992) and cytochromes (Gunner and Honig 1990; Gunner and Honig 1991), proton relay events in bacteriorhodopsin (Bashford and Gerwert 1992; Sampogna and Honig 1994).

4.4. ELECTROSTATIC CONTRIBUTIONS TO BINDING ENERGY

The electrostatic contribution to the binding energy of two molecules is obtained by taking the difference in total electrostatic energies in the bound (AB) and unbound A+B states. For the linear case

$$\Delta\Delta G_{\text{bind}}^{\text{elec}} = \frac{1}{2}\sum_{i}^{N_A} q_i(\phi_i^{AB} - \phi_i^{A}) + \frac{1}{2}\sum_{j}^{N_B} q_j(\phi_j^{AB} - \phi_j^{B}) \tag{31}$$

where the first and second sums are over all charges in molecule A and B respectively, and ϕ^X is the total potential produced by x=A, B or AB. Equation (31) allows for the possiblity that the conformation may change upon binding, since different charge distributions may be used for the complexed and uncomplexed forms of A, and similarly for B. However other energetic terms, including those involved in any conformational change have to be added to Eqn. (31) to obtain net binding free energy changes. Nevertheless, changes in binding free energy due to charge modifications or changes in external factors such as pH and salt concentration may be estimated using Eqn. (31) alone. For the latter, salt effects are usually only significant in highly charged molecules, in which the non-linear form for the total electrostatic energy, Eqn. (19) must be used. The salt dependence of binding of drugs and proteins binding to DNA has been studied using this approach (Misra et al. 1994a; Misra et al. 1994b; Sharp et al. 1995), including the pH dependence of drug binding (Misra and Honig 1995). Other applications include the binding of sulfate to the sulfate binding protein (Aqvist et al. 1991) and antibody and antigen interactions (Lee et al. 1992; Slagle et al. 1994)

4.5. ASSOCIATION RATES

One of the most successful applications of electrostatic calculations is to the calculation of association rates. Quite often these rates are greatly affected by the long range forces (i.e. just the electrostatics) acting between protein and ligand/substrate, which make this a suitable place to use the results of electrostatic calculations. The general approach to calculating the bimolecular first encounter association rate constant for the reaction A+B -> B is based on the following formula (Allison et al. 1985; Northrup et al. 1984)

$$k_a = 4\pi D_{ab} R_{init} P_{ab} \tag{32}$$

where D_{ab} is the mutual translational diffusion constant of A and B ($D_{ab}=D_a+D_b$), R_{init} is some initial encounter distance between A and B that is large enough so that the electrostatic forces between A and B are negligible (So that the term $4\pi D_{ab}R_{init}$ is just the pure diffusion determined rate constant for arrival of the substrate to a distance R_{init} from the enzyme obtained using the usual Smoluchowski model). P_{ab} is the probability that the substrate, having approached to distance R_{init}, will collide productively with the protein. P_{ab} can be estimated through Brownian dynamics (BD) simulations as follows. The substrate is positioned randomly in some orientation and location a distance R_{init} from the target. A substrate trajectory is propagated using a combination of random walk and drift steps, where the size of the random walk step is determined by the mutual diffusion constant, D_{ab}, and the drift is determined by the translational frictional coefficient kT/D_{ab}, and the electrostatic rigid body force on it, **f** (Ermak and McCammon 1978)

$$\Delta\mathbf{r} = \mathbf{n}\sqrt{(2D_{ab}\tau)} + \mathbf{f}D_{ab}\tau/kT \tag{33}$$

where $\Delta\mathbf{r}$ is the change in position over a simulation time step τ, typically 0.1-1 nsec, and **n** is normally distributed random vector. A trajectory is propagated until either a productive collision occurs or the substrate-enzyme separation exceeds some cutoff $R_{cut}>R_{init}$. Many such trajectories are run until an estimate of the fraction of productive trajectories, P_h is estimated to the desired statistical precision. P_{ab} is then given by (Allison et al. 1985; Northrup et al. 1984)

$$P_{ab} = \frac{P_h}{\left[1-(1-P_h)\dfrac{R_{init}}{R_{cut}}\right]} \tag{34}$$

Euqation (34) corrects for the probability that the substrate will re-enter and diffucse to R_{init} again, and alows one to simulate the effects of an infinite solution in a finite simulation. Electrostatic body forces are usually calculated using the test charge approach (Eqns. (20-21)), from precomputed electrostatic potential distributions obtained from the PB equation (Sharp et al. 1987a; Sharp et al. 1987b). This is rapid, which is necessary given the many hundreds of thousands of force evaluations required by BD simulations. It neglects, however, the effect one molecule may have on the potential from the other (Davis et al. 1991; Davis and McCammon 1990). Although current computer resources are insufficient to implement the exact force calculation method (e.g. using Eqn. (23)), many useful results can be obtained from such BD simulations. Equation (33) describes only simple translational diffusion. More sophisticated simulations can incorporate adaptive step sizes, rotational diffusion (Andrew et al. 1993; Northrup et al. 1987), hydrodynamic effects (Allison et al. 1986), and the flexibility of substrate or enzyme (Luty et al. 1993; Wade et al. 1993). The combined PB/BD approach has been used to study the effects of ionic strength and chemical modification on SOD turnover (Davis et al. 1991; Sharp et al. 1987a), to predict the effect of charged residue modifications (Getzoff et al. 1992; Sharp et al. 1987a), to study orientational and ionic strength effects in cytochrome-c/cytochrome-c peroxidase interactions (Andrew et al. 1993; Northrup et al. 1993), to study antibody-antigen interactions (Holst et al. 1994b), and to look at acetyl choline-acetyl choline esterase interactions (Antosiewicz et al. 1995).

4.6. CATALYSIS

All of the above electrostatic effects have general implications for enzyme catalysis. The protein by creating specific interaction potentials at certain sites can preferentially stabilize desired intermediates (or destabilize unwanted intermediates) e.g. the oxyanioin hole in trypsin stabilises the serine ion (Soman et al. 1989). Electrostatic contributions to binding that favor the transition state charge distribution will accelerate reactions. Acidic and basic groups are often intimately involved in catalysis, and shifts in pKa's can alter enzyme reactivities. An example is the pH and charged residue dependence of reactivity in subtilisin (Gilson and Honig 1987). Finally, the turnover rate of an enzyme at low substrate concentration ($<K_m$) is, in simple Michaelis-Menton kinetics, proportional to the association rate times the reaction probability ($k_{turnover}$ =

$k_aP_r = k_ak_{cat}/(k_{cat}+k_{diss})$), so electrostatic rate enhancement will affect the turnover rate. A good example is superoxide dismutase (Sharp et al. 1987a). More specific treatments of electrostatic effects in catalysis are covered in the next chapter.

5. References

Allison, S. A., S. H. Northrup and J. A. McCammon. (1985). "Extended Brownian dynamics of diffusion controlled reactions." J Chem. Phys. **83**: 2894.

Allison, S. A., S. H. Northrup and J. A. McCammon. (1986). "Simulation of biomolecular diffusion and complex formation." Biophysical J. **49**: 167.

Andrew, S., K. Thomasson and S. Northrup. (1993). "Simulation of electron-transfer self-exchange in cytochrome-c and cytochrome-b(5)." Journal of the American Chemical Society. **115**: 5516-5521.

Antosiewicz, J., M. K. Gilson, I. H. Lee and J. A. McCammon. (1995). "Acetylcholinesterase: diffusional encounter rate constants for dumbbell models of ligand." Biophys. J. **68**: 62-68.

Aqvist, J., H. Luecke, F. A. Quiocho and A. Warshel. (1991). "Dipoles localized at helix termini of proteins stabilize charges." : 2026-30.

Bacquet, R. and P. Rossky. (1984). "Ionic Atmosphere of Rodlike Polyelectrolytes. A Hypernetted Chain Study." J. Phys. Chem. **88**: 2660.

Bashford, D. and K. Gerwert. (1992). "Electrostatic calculations of the pka values of ionizable groups in bacteriorhodopsin." J. Mol. Biol. **224**: 473-86.

Bashford, D. and M. Karplus. (1990). "pka's of ionizable groups in proteins: atomic detail from a continuum electrostatic model." Biochemistry. **29**: 10219-25.

Bashford, D., M. Karplus and G. W. Canters. (1988). "Electrostatic effects of charge perturbations introduced by metal oxidation in proteins: A theoretical analysis." J Mol. Biol. **203**: 507.

Beroza, P., D. Fredkin, M. Okamura and G. Feher. (1991). "Protonation of interacting residues in a protein by a Monte-Carlo method." Proc. Natl. Acad. Sci. USA. **88**: 5804-8.

Bharadwaj, R., A. Windemuth, S. Sridharan, B. Honig and A. Nicholls. (1994). "The fast multipole boundary-element method for molecular electrostatics--an optimal approach for large systems." J. Comp. Chem. **16**: 898-913.

Davis, M. E. (1994). "The inducible multipole solvation model-a new model for solvation effects on solute electrostatics." J. Chem. Phys. **100**: 5149-59.

Davis, M. E., J. D. Madura, J. Sines, B. A. Luty, S. A. Allison and J. A. McCammon. (1991). "Diffusion-Controlled Enzymatic Reactions." : 473-497.

Davis, M. E. and J. A. McCammon. (1989). "Solving the finite difference linear Poisson Boltzmann equation: Comparison of relaxational and conjugate gradient methods." J Comp. Chem. **10**: 386.

Davis, M. E. and J. A. McCammon. (1990). "Calculating Electrostatic Forces from Grid-Calculated Potentials." Journal of Computational Chemistry. **11**: 401-409.

Ermak, D. L. and J. A. McCammon. (1978). J Chem. Phys. **69**: 1532.

Ewing, P. J. and T. P. Lybrand. (1993). "A comparison of perturbation method and PB electrostatic calculations for estimation of relative solvation free energies." J. Phys. Chem. **98**: 1748-52.

Getzoff, E., D. Cabelli, C. Fisher, H. Parge, M. Viezzoli, L. Banci and R. Hallewell. (1992). "Faster superoxide dismutase mutants designed by enhancing electrostatic guidance." Nature. **358**: 347-351.

Gilson, M. (1993). " Multiple-site Titration and Molecular Modeling: 2 Rapid Methods for Computing Energies and Forces For Ionizable Groups in Proteins." Proteins-structure Function and Genetics. **15**: 266-282.

Gilson, M., M. Davis, B. Luty and J. McCammon. (1993). "Computation of Electrostatic Forces on Solvated Molecules Using the Poisson-Boltzmann Equation." Journal of Physical Chemistry. **97**: 3591-3600.

Gilson, M. and B. Honig. (1986). "The Dielectric Constant of a Folded Protein." Biopolymer. **25**: 2097-2119.

Gilson, M. and B. Honig. (1987). "Calculation of Electrostatic Potentials in an Enzyme Active Site." Nature. **330**: 84.

Gilson, M., J. McCammon and J. Madura. (1994). "Molecular dynamics simulation with continuum solvent." J. Comp. Chem. :

Gilson, M., K. A. Sharp and B. Honig. (1988). "Calculating the Electrostatic Potential of Molecules in Solution: Method and Error Assessment." J. Comp. Chem. **9**: 327-335.

Gilson, M. K. (1995). "Theory of electrostatic interactions in macromolecules." Curr. Op. Struct. Biol. **5**: 216-223.

Gunner, M. and B. Honig. (1990). Electrostatic Analysis of the Midpoints of the Four Hemes in the Bound Cytochrome of the Reaction Center of Rp. Viridis. Perspectives in Photosynthesis. Netherlands, Kluwer Academic Publishers.

Gunner, M. R. and B. Honig. (1991). "Electrostatic Control of Midpoint Potentials in the Cytochrome Subunit of the Rhodopseudomonas Viridis Reaction Center." Proc. Natl. Acad. Sci. USA. **88**: 9151-9155.

Hecht, J. L. and B. Honig. (1995). "Electrostatic potentials near the surface of DNA-Comparing theory and experiment." J. Phys. Chem. **99**: 7782-7786.

Holst, M., R. Kozack, F. Saied and S. Subramaniam. (1994a). "Protein electrostatics-Rapid multigrid-based Newton algorithm for solution of the full nonlinear Poisson-Boltzmann equation." J. Biomol. Struct. Dynam. **11**: 1437-1445.

Holst, M., R. E. Kozack, F. Saied and S. Subramaniam. (1994b). "Treatment of electrostatic effects in proteins: multigrid-based Newton iterative method for solution of the full nonlinear Poisson-Boltzmann equation." Proteins. **18**: 231-45.

Holst, M. and F. Saied. (1993). " Multigrid Solution of the Poisson-Boltzmann Equation." J. Comp. Chem. **14**: 105-113.

Jayaram, B., R. Fine, K. A. Sharp and B. Honig. (1989a). "Free Energy Calculations of Ion Hydration: An Analysis of the Born Model in Terms of Microscopic Simulations." J Phys. Chem. **93**: 4320-4327.

Jayaram, B., K. A. Sharp and B. Honig. (1989b). "The Electrostatic Potential of B-DNA." Biopolymers. **28**: 975-93.

Jayaram, B., S. Swaminathan, D. L. Beveridge, K. Sharp and B. Honig. (1990). "Monte Carlo Simulation Studies on the Structure of the Counterion Atmosphere of B-DNA. Variations on the Primitive Dielectric Model." Macromolecules. **23**: 3156-3165.

Jean-Charles, A., A. Nicholls, K. Sharp, B. Honig, A. Tempczyk, T. Hendrickson and C. Still. (1990). "Electrostatic contributions to solvation energies: Comparison of free energy perturbation and continuum calculations." J. Am. Chem. Soc. **113**: 1454-1455.

Kirkwood, J. G. (1939). J. Chem. Phys. **7**: 911.

Langen, R., G. Jensen, U. Jacob, P. Stephens and A. Warshel. (1992). " Protein Control of Iron-sulfur Cluster Redox Potentials." Journal of Biological Chemistry. **267**: 25625-25627.

Langsetmo, K., J. A. Fuchs, C. Woodward and K. A. Sharp. (1991). "Linkage of Thioredoxin stability to titration of ionizable groups with perturbed pKa." Biochemistry. **30**: 7609-7614.

Lee, F., Z. Chu and A. Warshel. (1993). "Microscopic and Semimicroscopic Calculations of Electrostatic Energies in Proteins by the Polaris and Enzymix Programs." Journal of Computational Chemistry. **14**: 161-185.

Lee, F. S., Z. T. Chu, M. B. Bolger and A. Warshel. (1992). "Calculations of antibody-antigen interactions: microscopic and semi- microscopic evaluation of the free energies of binding of phosphorylcholine analogs to McP603." Protein Engineering. **5**: 215-28.

Lockhart, D. and P. Kim. (1993). "Electrostatic Screening of Charge and Dipole Interactions with the Helix Backbone." Science. **260**: 198-202.

Lockhart, D. J. and P. S. Kim. (1992). "Internal stark effect measurement of the electric field at the amino terminus of an alpha helix." Science. **257**: 947-51.

Luty, B., R. Wade, J. Madura, M. Davis, J. Briggs and J. McCammon. (1993). " Brownian Dynamics Simulations of Diffusional Encounters Between Triose Phosphate Isomerase and Glyceraldehyde Phosphate: Electrostatic Steering of Glyceraldehyde Phosphate." Journal of Physical Chemistry. **97**: 233-237.

McDaniel, R., K. A. Sharp, D. E. Brooks, A. C. McLaughlin, A. P. Winiski, D. Cafiso and S. McLaughlin. (1986). "Electrokinetic and Electrostatic Properties of Bilayers Containing Gangliosides Gm1, Gd1a, of Gt1." Biophys. J. **49**: 741-752.

Misra, V., J. Hecht, K. Sharp, R. Friedman and B. Honig. (1994a). "Salt effects on Protein-DNA interactions: The Lambda cI repressor and Eco R1 endonuclease." J. Mol. Biol. **238**: 264-280.

Misra, V. and B. Honig. (1995). "On the magnitude of the electrostatic contribution to ligand-DNA interactions." Proc. Natl. Acad. Sci. USA. **92**: 4691-5.

Misra, V., K. Sharp, R. Friedman and B. Honig. (1994b). "Salt effects on ligand-DNA binding: Minor groove antibiotics." J. Mol. Biol. **238**: 245-263.

Mohan, V., M. E. Davis, J. A. McCammon and B. M. Pettitt. (1992). " Continuum Model Calculations of Solvation Free Energies - Accurate Evaluation of Electrostatic Contributions." Journal of Physical Chemistry. **96**: 6428-6431.

Murthy, C. S., R. J. Bacquet and P. J. Rossky. (1985). "Ionic Distributions near Polyelectrolytes. A comparison of Theoretical Approaches." J Phys. Chem. **89**: 701.

Nakamura, H., T. Sakamoto and A. Wada. (1988). "A theoretical study of the dielectric constant of a protein." Protein Engineering. **2**: 177-183.

Nicholls, A. and B. Honig. (1991). "A rapid finite difference algorithm utilizing successive over-relaxation to solve the Poisson-Boltzmann equation." J. Comp. Chem. **12**: 435-445.

Niedermeier, C. and K. Schulten. (1992). "Molecular-dynamics simulations in Heterogeneous Dielectric and Debye-Huckel Media: Application to the Protein Bovine Pancreatic Trypsin-inhibitor." Molecular Simulation. **8**: 361-387.

Northrup, S., K. Thomasson, C. Miller, P. Barker, L. Eltis, J. Guillemette, S. Inglis and A. Mauk. (1993). "Effects of charged amino-acid mutations on the bimolecular kinetics of reduction of yeast iso-1-ferricytochrome-c by bovine ferrocytochrome-b(5)." Biochemistry. **32**: 6613-6623.

Northrup, S. H., S. A. Allison and J. A. McCammon. (1984). "Brownian dynamics simulations of diffusion influenced bimolecular reactions." J Chem. Phys. **80**: 1517.

Northrup, S. H., J. O. Boles and J. C. Reynolds. (1987). "Electrostatic effects in the Brownian dynamics association and orientation of heme proteins." J Phys. Chem. **91**: 5991.

Oberoi, H. and N. Allewell. (1993). "Multigrid solution of the nonlinear Poisson-Boltzmann equation and calculation of titration curves." Biophysical Journal. **65**: 48-55.

Olmsted, M. C., C. F. Anderson and M. T. Record. (1989). "Monte Carlo description of oligoelectrolyte properties of DNA oligomers." Proc. Natl. Acad. Sci. USA. **86**: 7766-7770.

Olmsted, M. C., C. F. Anderson and M. T. Record. (1991). "Importance of oligoelectrolyte end effects for the thermodynamics of conformational transitions of nucleic acid oligomers." Biopolymers. **31**: 1593-1604.

Pack, G., G. Garrett, L. Wong and G. Lamm. (1993). "The Effect of a Variable Dielectric Coefficient and Finite Ion Size on Poisson-boltzmann Calculations of Dna-electrolyte Systems." Biophysical Journal. **65**: 1363-1370.

Pack, G. R. and B. J. Klein. (1984). "The Distribution of Electrolyte Ions Around the B- and Z-Conformers of DNA." Biopolymers. **23**: 2801.

Pack, G. R., L. Wong and C. V. Prasad. (1986). "Counterion Distribution around DNA." Nucl. Acid Res. **14**: 1479.

Rashin, A. A. (1990). "Hydration phenomena, classical electrostatics and the boundary element method." J. Phys. Chem. **94**: 725-733.

Rashin, A. A. and B. Honig. (1985). "Reevaluation of the Born Model of Ion Hydration." J Phys. Chem. **89**: 5588.

Rashin, A. A. and K. Namboodiri. (1987). "A simple Method for the Calculation of Hydration Enthalpies of Polar Molecules with Arbitrary Shapes." J Phys. Chem. **91**: 6003.

Record, T., M. Olmsted and C. Anderson. (1990). Theoretical studies of the thermodynamics of ion interaction with DNA. Theoretical Biochemistry and Molecular Biophysics. Adenine Press.

Reiner, E. S. and C. J. Radke. (1990). "Variational approach to the electrostatic free energy in charged colloidal suspensions." J. Chem. Soc. Faraday Trans. **86**: 3901.

Russell, S. T. and A. Warshel. (1985). "Calculations of electrostatic energies in proteins." J Mol. Biol. **185**: 389-404.

Sampogna, R. and B. Honig. (1994). "Environmental effects on the protonation states of active site residues in bacteriorhodopsin." Biophysical J. **66**: 1341-52.

Schaeffer, M. and C. Frommel. (1990). "A precise analytical method for calculating the electrostatic energy of macromolecules in aqueous solution." J. Mol. Biol. **216**: 1045-1066.

Sharp, K., R. Fine and B. Honig. (1987a). "Computer Simulations of the Diffusion of a Substrate to an Enzyme Active Site." Science. **236**: 1460.

Sharp, K., R. Fine, K. Schulten and B. Honig. (1987b). "Brownian Dynamics Simulations of Diffusion to Irregular Bodies." J. Phys. Chem. **91**: 3624.

Sharp, K. and B. Honig. (1990a). "Calculating total electrostatic energies with the non-linear Poisson-Boltzmann Equation." J. Phys. Chem. **94**: 7684-7692.

Sharp, K. and B. Honig. (1990b). "Electrostatic Interactions in Macromolecules: Theory and Applications." Ann. Rev. Biophys. Biophys. Chem. **19**: 301-332.

Sharp, K. A. (1991). "Incorporating Solvent and Ion Screening into Molecular Dynamics Using the Finite-Difference Poisson-Boltzmann Method." J. Comp. Chem. **12**: 454-468.

Sharp, K. A. (1994). "Electrostatic Interactions in Macromolecules." Current Opinion in Structural Biology. **4**: 234-239.

Sharp, K. A. (1995). "Polyelectrolyte electrostatics: Salt dependence, entropic and enthalpic contributions to free energy in the nonlinear Poisson-Boltzmann model." Biopolymers. **36**: 227-243.

Sharp, K. A., R. Friedman, V. Misra, J. Hecht and B. Honig. (1995). "Salt effects on polyelectrolyte-ligand binding: Comparison of Poisson-Boltzmann and limiting law counterion binding models." Biopolymers. **36**: 245-262.

Sharp, K. A., A. Jean-Charles and B. Honig. (1992). "A local dielectric constant model for solvation free energies which accounts for solute polarizability." J. Phys. Chem. **96**: 3822-3828.

Shin, Y. K. and W. L. Hubbell. (1992). " Determination of electrostatic potentials at biological interfaces using electron spin double resonance." Biophys J. **61**: 1443-53.

Simonson, T. and A. Brunger. (1994). "Solvation free energies estimated from a macroscopic continuum theory." J. Phys. Chem. **98**: 4683-4694.

Simonson, T. and D. Perahia. (1995). "Internal and interfacial dielectric properties of cytochrome c from molecular dynamics in aqueous solution." Proc. Natl. Acad. Sci. USA. **92**: 1082-6.

Sitkoff, D., D. J. Lockhart, K. A. Sharp and B. Honig. (1994a). "Calculation of electrostatic effects at the amino terminus of an alpha helix." Biopysical Journal. **64**: 1-10.

Sitkoff, D., K. Sharp and B. Honig. (1994b). "Accurate calculation of hydration free energies using macroscopic solvent models." J. Phys. Chem. **98**: 1978-1988.

Slagle, S., R. E. Kozack and S. Subramaniam. (1994). "Role of electrostatics in antibody-antigen association: anti-hen egg lysozyme/lysozyme complex (HyHEL-5/HEL)." J. Biomol. Struct. Dyn. **12**: 439-56.

Smith, P., R. Brunne, A. Mark and W. vanGunsteren. (1993). "Dielectric properties of trypsin inhibitor and lysozyme calculated from molecular dynamics simulations." Journal of Physical Chemistry. **97**: 2009-2014.

Soman, K., A.-S. Yang, B. Honig and R. Fletterick. (1989). "Electrical Potentials in Trypsin Isozymes." Biochemistry. **28**: 9918-9926.

Sternberg, M., F. Hayes, A. Russell, P. Thomas and A. Fersht. (1987). "Prediction of electrostatic effects of engineering of protein charges." Nature. **330**: 86.

Still, C., A. Tempczyk, R. Hawley and T. Hendrickson. (1990). "Semianalytical treatment of solvation for molecular mechanics and dynamics." J. Am. Chem. Soc. **112**: 6127-6129.

Takashima, S. and H. P. Schwan. (1965). "Dielectric constant measurements on dried proteins." J Phys. Chem. **69**: 4176.

Wade, R., M. Davis, B. Luty, J. Madura and J. McCammon. (1993). "Gating of the Active-site Of Triose Phosphate Isomerase: Brownian Dynamics Simulations of Flexible Peptide Loops in The Enzyme." Biophysical Journal. **64**: 9-15.

Warshel, A. and J. Åqvist. (1991). "Electrostatic energy and macromolecular function." Ann. Rev. Biophys. Biophys. Chem. **20**: 267-298.

Warshel, A. and S. Creighton. (1989). In: Computer Simulation of Biomolecular Systems. Leiden, ESCOM.

Warshel, A. and S. Russell. (1984). "Calculations of electrostatic interactions in biological systems and in solutions." Quart. Rev. Biophys. **17**: 283.

Warwicker, J. (1994). "Improved continuum electrostatic modelling in proteins, with comparison to experiment." J. Mol. Biol. **236**: 887-903.

Warwicker, J. and H. C. Watson. (1982). "Calculation of the Electric Potential in the Active Site Cleft due to Alpha-Helix Dipoles." J. Mol. Biol. **157**: 671.

Wendoloski, J. J. and J. B. Matthew. (1989). "Molecular dynamics effects on protein electrostatics." Proteins. **5**: 313.

Yang, A., M. Gunner, R. Sampogna, K. Sharp and B. Honig. (1993). "On the Calculation of Pk(a)s in Proteins." Proteins-Structure Function and Genetics. **15**: 252-265.

Yang, A. and B. Honig. (1994). "Structural origins of pH and ionic strength effects on protein stability. Acid denaturation of sperm whale apomyoglobin." J. Mol. Biol. **237**: 602-14.

Yoon, L. and A. Lenhoff. (1992). "Computation of the electrostatic interaction energy between a protein and a charged suface." J. Phys. Chem. **96**: 3130-3134.

Zauhar, R. (1991). "The incorporation of hydration forces determined by continuum electrostatics into molecular mechanics simulations." J. Comp. Chem. **12**: 575-583.
Zauhar, R. and R. J. Morgan. (1985). "A new method for computing the macromolecular electric potential." J Mol. Biol. **186**: 815.
Zhou, H. X. (1994). "Macromolecular electrostatic energy within the nonlinear Poisson-Boltzmann equation." J. Phys. Chem. **100**: 3152-3162.

ELECTROSTATIC BASIS OF ENZYME CATALYSIS

G. NÁRAY-SZABÓ and M. FUXREITER
Department of Theoretical Chemistry, Eötvös Loránd University
H-1117 Budapest, Pázmány Péter st. 2, Hungary
A. WARSHEL
Department of Chemistry, University of Southern California
University Park, Los Angeles, CA 90082-1062, USA

1. Introduction

Since the proposition of the lock-and-key analogy by Emil Fischer (1894) several attempts have been done to understand the basis of enzymatic action. The key factor of rate acceleration by enzymes with a factor of 10^{10} or more (Kraut, 1977) has been thought to be the fit of transition state to the enzyme active site (Haldane, 1930; Pauling, 1946), however, it is still somewhat unclear what do we mean by the term *fit*. What are the real factors leading to the stabilisation of the transition state?

Catalysis is generally defined as a phenomenon where the rate of a chemical reaction increases through interaction of the reacting partners with a substance (the catalyst) without changing the equilibrium constant. This means that the catalyst must find a way to modify the original potential surface and reduce the activation barrier. The same is true in the case of enzyme catalysis: the enzyme must find a way to interact with the substrate(s) and to change the potential surface so that the resulting activation barrier is lower than that of the original process in water. In the following we make an attempt to discuss most important factors leading to the switch from the reaction path of the uncatalysed process to that of the catalytic reaction.

While there is a growing tendency to attribute enzyme catalysis to transition-state stabilisation (Schowen, 1978), one can argue that this effect is the result of ground-state destabilisation (e.g. Menger, 1992; Tapia et al., 1994b). Some ground-state destabilisation effects such as strain can be excluded by computer simulation studies (e.g. Warshel and Levitt, 1976) while entropic factors should still be examined carefully (see below). Perhaps the most important problem with ground-state destabilisation effects is that they cannot change k_{cat}/K_M. That is, as explained by Warshel (1991), this crucial parameter is determined by the difference between the free energy of the transition state and the free energy of the state where the substrate is not bound to the enzyme and therefore is *independent* of the energy of the Michaelis complex (the ground state). The increase in k_{cat}/K_M may become ineffective when the ground state is overstabilised (i.e. K_M is lower than the substrate concentration) but in a

G. Náray-Szabó and A. Warshel (eds.), Computational Approaches to Biochemical Reactivity, 237–293.

large range of K_M there is no evolutional advantage of ground-state destabilisation. Thus, although one can argue that enzymes could exploit any possible catalytic effect, most of these cannot be used effectively due to fundamental limitations (see discussion in Chapter 9 of Warshel, 1991). Thus, we focus here on the main factors that are likely to contribute to enzyme catalysis: chemical factors (e.g. general acid-base catalysis), entropic factors, and most importantly, electrostatic effects. Since this chapter is devoted to electrostatics, we treat in Section 3 more than a dozen enzymatic reactions where electrostatic effects play a crucial role in catalysis, either in rate acceleration or in specificity.

2. Sources of Enzymatic Rate Acceleration

2.1. ENTROPY

The hypothesis that enzymes should be capable of efficiently catalysing reactions with unfavourable entropies of activation by acting as "entropy traps", i.e. that the binding energy of the enzyme is used to freeze out the rotational and translational degrees of freedom, necessary to form the activated complex, is quite old (Bruice and Sturtevant, 1959; Westheimer, 1962; Storm and Koshland, 1970; Jencks, 1975). Page and Jencks (1971) have suggested that translational and rotational entropy provides an important driving force for enzymatic rate accelerations. Their analysis was based on gas-phase considerations and was therefore criticised by Larsen (1973) who suggested that the loss of translational degrees of freedom upon combining the two reactants is accompanied by the destruction of the solvent cage as the reactants combine and the corresponding entropy effects cancel each other. Kraut (1977) pointed out that the entropic effects that may be associated with the acylation step in serine proteases cannot help in accelerating the deacylation step. Unfortunately, an actual demonstration that entropic effects are important is far from simple. Some of the original proposals suffer from significant problems in terms of their assumption and even the definition of the given problem (see e.g. the discussion of the "orbital steering" mechanism in Warshel, 1991). More importantly, while entropic effects can lead to ground-state destabilisation they cannot help in increasing k_{cat}/K_M. Nevertheless, it is instructive to see if entropy can help in increasing k_{cat}.

Exact determination of entropy effects in enzymatic reactions is not an easy task even nowadays when sophisticated Monte Carlo and molecular dynamics methods are available for calculations (Warshel, 1991; Åqvist and Warshel, 1993). One way to examine the importance of entropy is to analyse the configuration space available to the system in its ground and transition states, both in the enzyme and in solution. The entropic contribution to the catalytic effect, relative to the uncatalysed solution can be expressed as

$$-T\Delta\Delta S^{\ddagger} \approx -\beta^{-1}\ln[(v_p^{\ddagger}/v_p^{0})/(v_s^{\ddagger}/v_s^{0}) \qquad (1)$$

where $v^{\ddagger}$ and v^{0} denote the configurational volumes in the transition and ground states, respectively, while p and s refer to the protein and solution, respectively. The resulting entropic contribution reflects the differences in the motion of the reacting fragments in the enzyme active site and in the reference solvent cage (Åqvist and Warshel, 1993). These contributions seem to be smaller than previously thought since enzyme molecules are quite flexible and a qualitative examination of the entropic contribution to the rate acceleration of serine proteases indicated that this contribution is small (Warshel, 1991).

Several organic model compounds have been designed to examine the possible importance of proximity effects. These model compounds seem to indicate that proximity effects can account for rate enhancement by a factor of 10^{10} or more (Fersht and Kirby, 1968). For example, in the model compound of Figure 1 the relative rate of intermolecular *vs*. intramolecular acetyl migration is 10^{8} thus proximity of the reacting

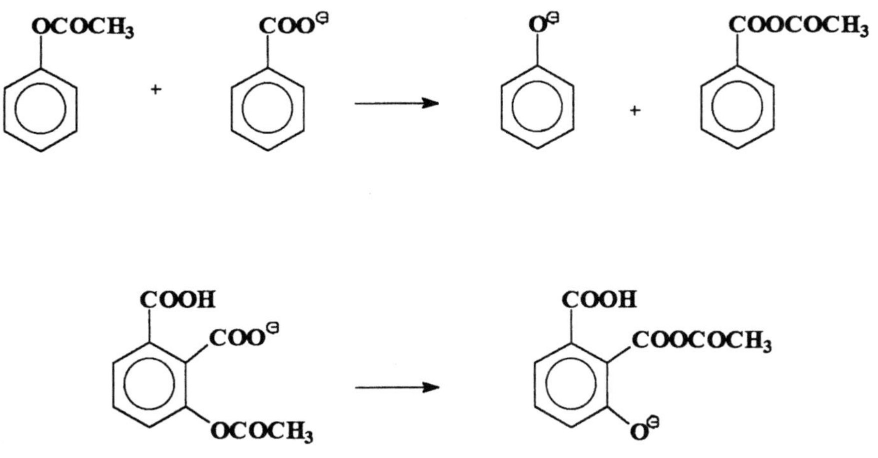

Figure 1. Comparison of relative rates of intermolecular (top) and intramolecular (bottom) acetyl migration (after Fersht and Kirby, 1968). k_2 for the intermolecular process is 10^{-10} $s^{-1}M^{-1}$ while k_1 for the intramolecular process is 0.02 s^{-1}.

groups can be an important determinant of rate acceleration. This does not prove, however, that enzymes provide significant rate acceleration by proximity effects. That is, enzymes are quite flexible and the model compound used for demonstrating proximity effects freeze more degrees of freedom than what is done in a typical enzymatic reaction, except possible in ring closure reactions (see a detailed analysis in Warshel, 1991). Furthermore, as argued above, the ground state destabilisation associated with proximity effects does not help in increasing of k_{cat}/K_M which corresponds to the height of the transition state relative to the state where the substrate is not bound to the enzyme.

Catalytic antibodies have been used recently as an impressive proof that enzymes catalyse reactions by providing strong binding to the corresponding transition states (Lerner et al., 1991). These are proteins with loops (often called hypervariable segments) that display a high degree of sequence variability and provide the basis for the ability of the antibody molecule to bind a wide variety of molecules penetrating into the body. Small molecules (haptens) are typically bound in a cleft but for large ones the binding site can extend to large surfaces. The specificity of antibodies for their ligands can exceed that of enzymes for substrates. Several antibodies have been generated that catalyse a wide array of chemical reactions. Typically, the antibody-catalysed reactions proceed with rates faster by a factor of 10^3 to 10^6 than the uncatalysed ones, but in some cases rates may approach to those of enzymes.

Some attempts have been made to account for the action of catalytic antibodies in terms of entropic effects. For example, an antibody, induced by a a bicyclic hapten that resembles the conformationally restricted chair-like transition state for the Claisen rearrangement of chorismic acid to prephenic acid, accelerated the rearrangement 10^4-fold over the uncatalysed rate (Jackson et al., 1988). For comparison, the enzyme chorismate mutase accelerates the reaction approximately 3×10^6-fold over the background reaction. It was found that the entropy of activation of the antibody-catalysed reaction is close to zero, compared with a $\Delta S^\dagger$ of -13 entropy units for the uncatalysed reaction (Schultz, 1989). This results in about 17 kJ/mol catalytic advantage for k_{cat} which is not inconsistent with our argument that entropy can be important in ring closure reactions (the transition state of this reaction involves a cyclic arrangement). However, a major point that was overlooked in the analysis of the above reaction and in the entropy-trap hypothesis is the fact that there is no entropic advantage in k_{cat}/K_M relative to the uncatalysed reaction, since in both cases we start with the substrate in water.

It seems to us that the question about the actual advantage of entropic contributions is still unresolved. This is partially due to the difficulties in obtaining reliable estimates of entropic effects by molecular modelling methods and it presents a major challenge for the future.

2.2. GENERAL ACID-BASE CATALYSIS

Many enzymatic processes involve general acid or base catalysis (cf. e.g. Jencks, 1969). Typically, a step of the given reaction that involves a proton transfer process is modified and the proton transfer is mediated by an acidic or basic residue of the protein. In this way the free energy of the proton transfer (PT) step is reduced and the reaction is accelerated. It is tempting to attribute this rate acceleration to the distance between the proton donor and acceptor, which is known to determine the reaction barrier for PT reactions (cf. e.g. Scheiner and Kar, 1995), but as argued in (Warshel, 1991) both in enzymes and in reference reactions in water we have similar distances thus the enzyme can only slow down the reaction by preventing an optimal distance and it cannot accelerate the reaction by providing an optimal distance. General acid/base catalysis comes into play in enzymes by determining the free energy of the

PT step, ΔG_{PT}. This quantity is determined by the difference between the pK_a's of the donor and acceptor (ΔpK_a). The value of this ΔpK_a in water is the "chemical part" of the general acid-base catalysis and is independent of the specific enzyme active site. In fact, this effect can be simply considered as the result of using different reaction mechanisms with different reactants rather than an actual catalytic effect. The change of the given ΔpK_a from its value in water to the corresponding value in the enzyme active site is a true catalytic effect. This change reflects the electrostatic effect of the enzyme active site which is the subject of the next section.

A typical example of a general acid-base assistance is provided by the reaction catalysed by triosephosphate isomerase (cf. Figure 2.) The proton on C1 of the substrate (dihydroxyacetone phosphate DHAP) moves to the carbonyl carbon while the

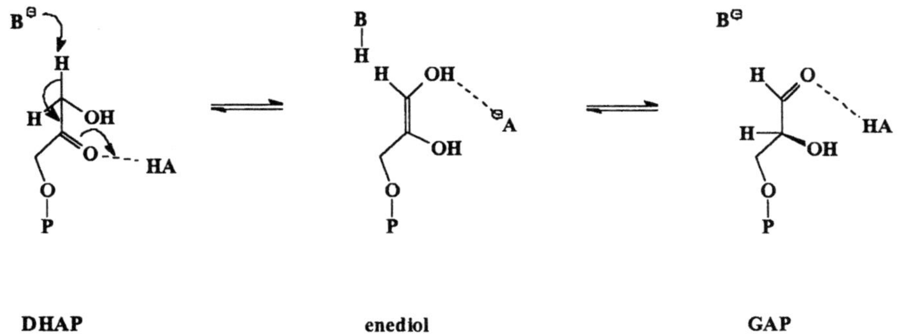

Figure 2. Schematic reaction catalysed by triosephosphtae isomerase. **A** and **B** denote acidic and basic catalytic groups.

other one on the C1 oxygen moves to the carbonyl oxygen (Knowles, 1991). The length of the path for the moving proton is reduced by the assistance of the general acid (His-95 in the enzyme) donating its proton to the C2 carbonyl oxygen, while the general base (Glu-165) accepts proton from C1. An enediol intermediate is produced that undergoes a further double proton transfer: His-95 accepts proton from the C1 oxygen, while Glu-165 donates its proton to C2 thus regenerating their original protonation states. A reliable estimate of the energetics of the process is given in Chapter 9 of this book.

Another interesting example of a general base mechanism is provided by the ring opening step of D-xylose isomerase which is displayed in Figure 3. One can try to estimate the effect of the general base on the gas-phase reaction by semiempirical calculations, and the semiempirical MNDO/PM3 model gave a barrier of 150 kJ/mol for the proton transfer from the glycosidic oxygen to the ring oxygen for the isolated substrate. This value decreases to 92 kJ/mol in the presence of the general base. It is to be stressed, however, that these numbers cannot and should not be used directly in analysing the reaction in solution and at enzyme active sites, since gas-phase calculations do not include the effect of the environment on the reaction profile.

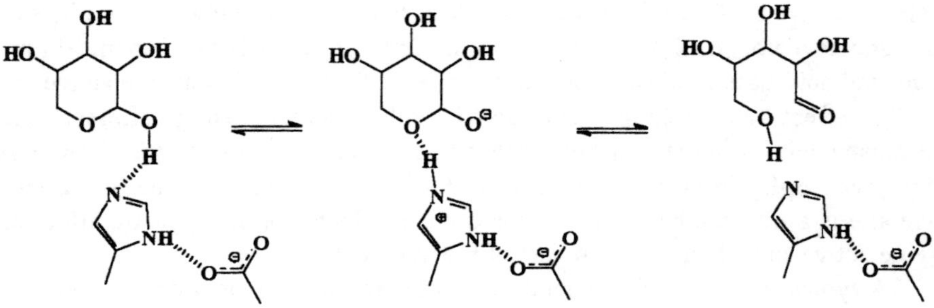

Figure 3. Schematic reaction mechanism of the ring opening step in the aldose-ketose interconversion catalysed by D-xylose isomerase (Rangarajan and Hartley, 1992)

A simple and very reliable way to estimate the "chemical" advantage of the general base mechanism in water is to compare the uncatalysed reaction (when the proton is transferred to water) to the one when the proton is transferred to the histidine base. The corresponding reduction in the free energy of the initial proton transfer step is given by the following expression: $\Delta\Delta G^{w}_{PT}$ = 2.3RT[pK^{w}_{a}(ROH) - $pK^{w}_{a}(HisH^{+})$] - 2.3RT [pK^{w}_{a}(ROH) - $pK^{w}_{a}(H_3O^{+})$] = 2.3[$pK^{w}_{a}(H_3O^{+})$ - $pK^{w}_{a}(HisH^{+})$] ≈ - 45 kJ/mol.

2.3. ELECTROSTATICS

The possibility that electrostatic effects play a major role in catalysis has not been recognised or established in the early stages of the research for the origin of enzyme catalysis. All available early experiments indicated quite clearly that electrostatic effects are *not* important. That is, as much as the electrostatic potential around the protein is concerned it was easy to see (from the interaction between surface groups influencing their pK_a's) that such an effect cannot contribute more than 2-6 kJ/mol to catalysis, while we are looking for much larger effects of up to 40 kJ/mol. A careful study of the interaction between Asp-52 and Glu-35 in the active site of lysozyme (Thoma, 1974) has indicated that even in the interior of proteins electrostatic effects are too weak to provide a major catalytic advantage. Experiments with model compounds in water (Dunn and Bruice, 1973) have indicated that electrostatic effects cannot explain the catalytic power of enzymes.

Early theoretical attempts to examine the field inside and around the protein using X-ray structures (Johannin and Kellersohn, 1972; Hayes and Kollman, 1976) suggested that the field inside proteins can be significant but these studies did not consider the induced dipoles of the protein and, more importantly, the solvent around it, that in most cases reduce electrostatic effects by a factor of 40! The fundamental problem that prevented a proper estimate of electrostatic effects in proteins was the fact that these interactions cannot be evaluated without the knowledge of the "dielectric constant" at the protein active site. Estimating this value from studies in water is not justified and assuming a low dielectric constant for the polar environment provided by proteins is also unjustified (Warshel and Russell, 1984; King et al., 1991). The first correct treatment of electrostatic energies in proteins in general and in enzymes in particular was provided by Warshel and Levitt (1976) who avoided the dielectric problems altogether by including explicitly all the electrostatic components of the system, i.e. the protein permanent and induced dipoles, the protein relaxation (by energy minimisation) and, most importantly, a microscopic representation of the surrounding solvent. This study has indicated that electrostatic effects, associated with the protein active site, play a major role in catalysis. The actual reason for this catalytic effect was further analysed by Warshel (1978) who has shown that it mainly reflects the pre-organisation of the protein polar environment.

The importance of protein electrostatics was pointed out and demonstrated by others quite early (e.g. Perutz, 1978; Hol et al., 1978; Náray-Szabó, 1979) using more qualitative considerations, and it starts to be widely accepted (cf. e.g. Sharp and Honig, 1990). Here we will try to provide several examples for the importance of such effects in enzyme catalysis and in other aspects of protein action. We will try to distinguish between very qualitative treatments that do not consider the complete environment and quantitative ones treating it correctly. This is quite crucial, since the basic issue in enzyme catalysis is the actual magnitude of the given effect and not the assumption that it can be operative in some cases.

There are several phenomena where protein electrostatics plays a major role (for reviews see Warshel and Russell, 1984; Sharp and Honig, 1990; Warshel and Åqvist, 1991; Náray-Szabó and Ferenczy, 1995). It is an important determinant of substrate specificity and binding, especially for charged substrates (Gráf et al., 1987, 1988; Wells et al., 1987; Coghlan and Vickery, 1992). Specificity is a phenomenon closely related to molecular recognition (cf. e.g. Náray-Szabó, 1993). Recognition, i.e. the complementarity (fit) between the substrate molecule and the enzyme active site crevice, is characterised by three main determinants: steric, electrostatic and hydrophobic. *Steric fit* means that interacting atoms may not approach each other beyond their van der Waals radii and, simultaneously, the free space between interacting atoms should be minimised. This determinant of recognition has no link to electrostatics, but the other two do. *Electrostatic fit* requires the maximum ionic and polar (hydrogen bonding and other dipole-dipole-type attraction) interaction between the enzyme and its substrate, while *hydrophobic fit* requires an optimal overlap between the non-polar parts of the substrate and the enzyme. Calculating the molecular electrostatic potential (MEP) and field (MEF) maps of an enzyme can provide useful information about preferred sites of electrostatic and hydrophobic interactions. While positive and negative regions of the MEP indicate favourable interactions with negative and positive ligand sites, respectively, low and high values of the MEF determine hydrophobic and hydrophilic regions inclining to associate with ligand sites of the same character (cf. the *similis simili gaudet* principle formulated by Gráf et al., 1988).

Electrostatic complementarity between the enzyme and its ligand is illustrated on the example of the binding of the Lys-15 side chain of bovine pancreatic trypsin inhibitor (BPTI) to the specificity pocket of trypsin (Figure 4.). The MEP of BPTI which is displayed on the van der Waals envelope of the Lys side chain is complementary to that displayed on the same surface but emerging from the enzyme

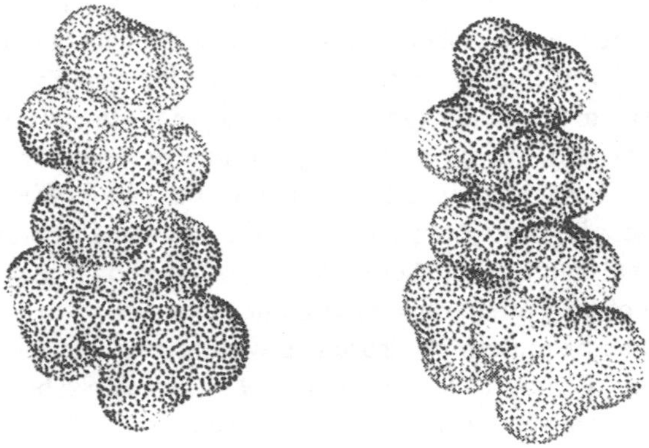

Figure 4. MEP on the van der Waals envelope of Lys-15 of BPTI provided by the inhibitor (left) and the enzyme (right). Dark grey dots represent positive, light grey dots negative potential values, respectively.

(Náray-Szabó et al., 1994). Positive and negative regions get in close contact thus maximising attraction. Several authors considered the free energy of protein-ligand complementarity at a general theoretical level and thus provided an important tool for computer-aided molecular design (Warshel and Russell, 1984; Warshel et al., 1986; Wong and McCammon, 1986; Rao et al., 1987; Lee et al., 1992; Kollman, 1993). In Section 3. we will provide further examples for the role of electrostatics in this phenomenon.

Since proton transfer reactions play a crucial role in a variety of enzymatic processes, acid-base properties of protein side chains are especially important in understanding the catalytic power of enzymes. As indicated by experimental (Sternberg et al., 1987; Alvaro and Russell, 1991) and theoretical studies (Warshel and Russell, 1984; Mehler and Eichele, 1984; Gilson and Honig, 1987; Dao-Pin et al., 1989, Åqvist and Warshel, 1992), protein electrostatics can change the pK_a of both acidic and basic amino-acid side chains. It has been shown that protonation properties can be modulated by engineering surface charges of proteins (Sternberg et al., 1987) or by altering the ionic strength of the solution where the enzymatic process takes place (Voet et al., 1981; Fersht, 1985; Alvaro and Russell, 1991). However, while the effects of ionic strength are interesting (Sharp and Honig, 1990; Davis and McCammon, 1990), these effects and those of surface charges are basically trivial and by far too small to account for the catalytic power of enzymes. This is one of the reasons why electrostatic effects on catalysis were not considered seriously until the emergency of genetic engineering of active-site groups. Apparently, it has been demonstrated by early electrostatic considerations (Warshel, 1981; Warshel, 1991) and indirectly by genetic engineering experiments (cf. e.g. Fersht et al., 1985; Craik et al., 1987) that the enzyme dipoles can significantly change the energetics of proton-transfer processes and the corresponding effective pK_a's (with up to 10 units). Thus, when we refer to a major acceleration by an enzyme, we will concentrate on the local effect of the active site rather than the effect of distant charges. It is also important to mention in this respect that theoretical studies that do not include the solvent around the protein drastically overestimate the effect of surface charges.

Another electrostatic effect which is sometimes important for enzyme action is the control of the redox potential of critical cofactors. This is done per definition by electrostatic effects (since the redox reaction involves a change in the charge of the given cofactor) and the protein can modulate the corresponding energetics by surface charges (Rees, 1980; Rogers et al., 1985) and, to a greater extent, by its local polarity (Churg and Warshel, 1986; Langen et al., 1992).

In light of the special importance of electrostatics in enzyme catalysis it is not surprising that the MEP patterns along enzyme active sites are conservative from the evolutionary point of view (Ángyán and Náray-Szabó, 1983; Desideri et al., 1992; Knighton et al., 1994). Different species of serine proteases,superoxide dismutase and dihydrofolate reductase-thymidylate synthase from *Leishmania* show a striking similarity of the MEP pattern along their active site. Thus the goal of convergent evolution in these enzyme families was not only to bring the catalytic machinery together but also to provide similar MEP distribution around them maintaining, maybe even enhancing, the accelerating effect of electrostatics in catalysis. Especially important is the stabilisation of the (- + -) charge distribution in a wide variety of enzymes (serine proteases, lipases, acetylcholinesterase, lysozyme, ribonucleases, etc.). Náray-Szabó and Gérczei (1995) used three-dimensional structures from the Protein Data Bank (Abola et al., 1987), the SYBYL software package (1993) with Pullman's net charges on protein atoms (Berthod and Pullman, 1965) and a distance-dependent dielectric constant, to calculate the protein MEP in certain points along the catalytic triad of various enzymes (see Figure 5.). It is seen that the (- + -) charge distribution is stabilised by a minimum-type MEP pattern near the active site, irrespective of the primary sequence of the enzyme differing very much for the proteins studied.

It seems likely that enzymes can stabilise polar species, like ion pairs or (- + -) charge distributions more effectively than water can, because they have dipoles that are kept in the appropriate orientation, whereas water dipoles are randomised by outer solvation shells interacting with bulk solvent (Warshel, 1978). This effect is manifested by the fact that the water dipoles have to reorient themselves in order to stabilise the transition-state charges (this involves investment of the so-called "reorganisation energy") while the enzyme dipoles are already pre-oriented toward the transition state charge distribution (Warshel, 1978; Rao et al., 1987; Warshel, 1991). This means that enzymes can act as "supersolvents" that are designed to stabilise transition states (relative to the corresponding states in water). Since it is hard to see how can van der Waals or entropy or other non-electrostatic effects help in solvating transition states, one is led to the conclusion that only electrostatic effects can do this effectively. With this in mind we conclude that enzymes work mainly by providing electrostatic stabilisation to their transition states. In the following section we will try to prove this point by giving a large number of examples that support our hypothesis.

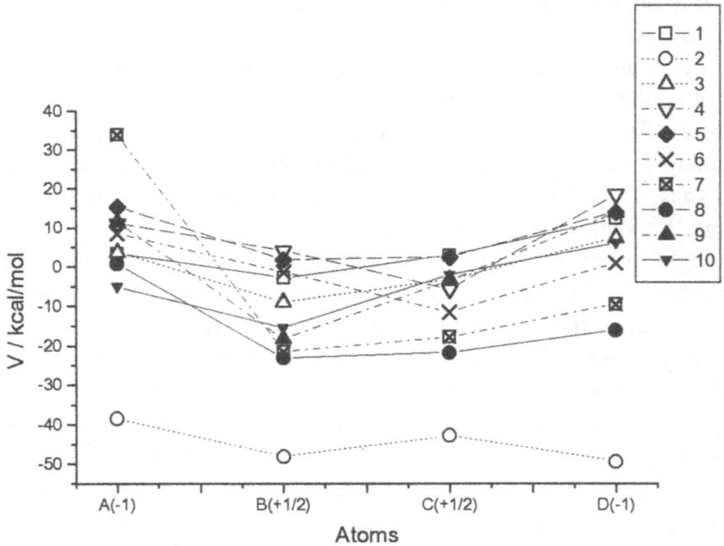

Figure 5. Variation of the protein MEP along the active sites of some enzymes. **1:** α-chymotrypsin, **2:** β-trypsin, **3:** porcine pancreatic elastase, **4:** *Streptomyces Griseus* hydrolase, **5:** α-lytic protease, **6:** subtilisin NOVO, **7:** acetylcholinesterase, **8:** lipase A, **9:** lysozyme, **10:** D-xylose isomerase. Point A is at OG of the active serine in **1-8**, at the bisector of OD1 and OD2 of Asp-52 in **9**, at O1 of the cyclic xylose in **10**. Point B is at NE2 of the catalytic histidine in **1-8**, in the first trisector of points A and D in **9**, at NE1 of His-54 in **10**. Point C is at ND1 of the catalytic histidine in **1-8** and **10**, at the second trisector of points A and D in **9**. Point D is at the bisector of the carboxylate oxygens of the catalytic Asp or Glu side chains.

3. Examples of electrostatic effects in enzyme catalysis

Many classes of enzymatic reactions involve a large change in charge distribution during the transition from their reactant to product states. Thus, it is likely that the enzymes evolve to stabilise the charge distribution of their transition states. This idea is supported by the examples given below that cover a wide range of enzymatic reactions and involve both quantitative and qualitative considerations. These examples will be used not only for demonstrating that electrostatic effects play a major role in enzyme catalysis but also for illustrating the problem associated with studies that do not consider the entire enzyme-substrate complex. We will try to explain why only quantitative approaches can be used to draw unique conclusions about the actual role of electrostatic effects in enzyme catalysis. We will also demonstrate that qualitative MEP's can provide strong support for the key role of electrostatic effects.

3.1. PROTEASES

The protease family may be conveniently classified according to their activities and functional groups (cf. Polgár, 1989 for a review). The serine and cystein (or thiol) proteases are endopeptidases that have a reactive serine and cystein residue, respectively, and pH optima around neutrality. Aspartyl (or acid) proteases are also endopeptidases that have catalytically important carboxylate side chains and work optimally at low pH values. At last, zinc proteases are metalloenzymes that function at neutral pH. Many proteases are small monomer enzymes with molecular weights between 15 and 35 kD, readily amenable to kinetic and structural study, this is why they are among the best studied enzymes. The role of electrostatic effects in the catalytic action of these enzymes has been also studied by several authors, and will be considered below.

3.1.1. *Serine Proteases*

Serine proteases (e.g. trypsin, chymotrypsin, subtilisin) catalyse the hydrolysis of ester or amide substrates through an acyl-enzyme intermediate in which the hydroxyl group of the active serine (Ser-195 in trypsin, Ser-221 in subtilisin) side chain is acylated by the substrate (cf. Figure 6., Polgár, 1989). The reaction involves a proton transfer from the serine to the catalytic histidine residue, an attack of the amide carbon of the substrate and the formation of a high-energy tetrahedral intermediate. Subsequently, the tetrahedral intermediate breaks down, the leaving group accepts the proton from the histidine residue and an acyl enzyme is formed. This acyl enzyme is then hydrolysed via the reverse route.

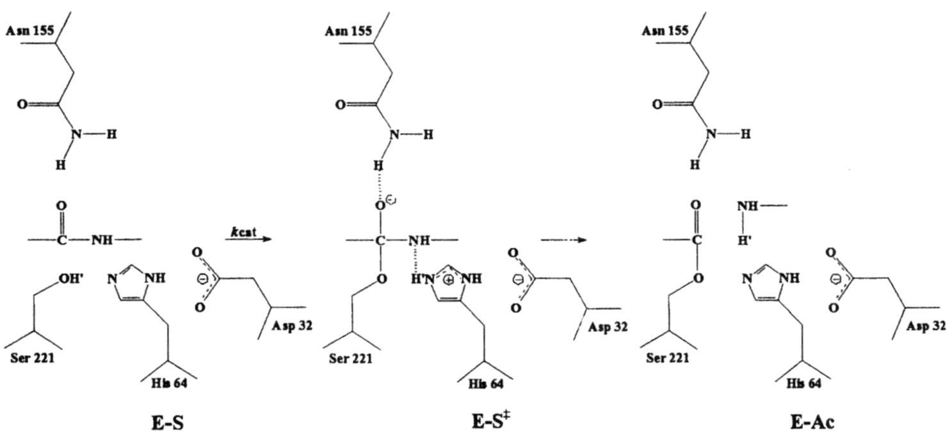

Figure 6. Schematic diagram showing the rate-limiting acylation step in the hydrolysis of peptide bonds by subtilisin. **E.S:** Michaelis complex, **$E.S^{\dagger}$:** transition-state complex, **E-Ac:** acyl enzyme.

The pioneering work of Blow and co-workers (Matthews et al., 1967) suggested that the proton from the catalytic histidine transfers to the buried aspartate during catalysis and this "charge-relay" mechanism vastly increases the nucleophilicity of the active serine through its ionisation. Subsequent workers (e.g. Hunkapiller et al., 1973) interpreted their experiments as a result of a mechanism involving a concerted transfer of two protons, from Ser to His and from His to Asp. This mechanism, depicted in many textbooks, is called the "double proton-transfer" or "charge-relay" mechanism. Later on, the hypothesis was questioned on the basis of nuclear magnetic resonance (Bachovchin and Roberts, 1978) and neutron diffraction (Kossiakoff and Spencer, 1981) studies, as well as theoretical calculations (Umeyama et al., 1981; Warshel et al., 1982; Náray-Szabó, 1982, 1983; Longo et al., 1985; Warshel and Russell, 1986, Warshel et al., 1989b). All the above studies supported the early proposal by Polgár and Bender (1969) who stated that the buried aspartate remains ionised throughout the catalytic process and its role is mainly to increase the stability of the ion pair formed by the tetrahedral intermediate and protonated histidine. Site-directed mutagenesis of trypsin (Sprang et al., 1987; Craik et al., 1987) and subtilisin (Carter and Wells, 1988) has indicated that the aspartate is important since its removal leads to a reduction of 4 orders of magnitude in the catalytic rate.

In order to shed light on the role of electrostatics, especially the ionisation state of the buried aspartate, in serine protease catalysis we performed microscopic calculations on its electrostatic contribution to the rate enhancement (Warshel et al., 1989b). Applying both the Protein Dipoles Langevin Dipoles (PDLD) method (Russell and Warshel, 1985) and the Empirical Valence Bond/Free Energy Perturbation (EVB/FEP) method (Warshel et al., 1988), these calculations reproduced the observed reduction of activation energies upon the D102N mutation in trypsin (Craik et al., 1987) and the D32A mutation in subtilisin (Carter and Wells, 1988). Table 1. displays

TABLE 1. Calculated and experimental changes in the activation energy of the reactions catalysed by mutant trypsin and subtilisin, as compared to the wild type enzymes (kJ/mol).

Term	Trypsin	Subtilisin
V_{QQ}	88	91 (109)*
$V_{Q\mu}$	0	0
$V_{Q\alpha}$	-37	-33
G_{Qw}	-29	-39
Sum	23	18
EVB-FEP (electrostatic)	23 ± 7	15 ± 5
EVB-FEP (total)	26 ± 7	17 ± 5
Exp.	26	25

*The V_{QQ} obtained by semiempirical molecular orbital calculations (Náray-Szabó, 1982) is given in brackets

results of the calculations and the breakdown of the energy change into terms emerging from the (Asp-His-Tetr) catalytic triad (V_{QQ}), from protein permanent ($V_{Q\mu}$), and polarisable dipoles ($V_{Q\alpha}$) and the solvent surrounding the enzyme (G_{Qw}). It is clear from Table 1. that the electrostatic terms govern the change in energetics of the catalytic reaction upon mutation. While most important is the direct effect of the Asp/Asn or Asp/Ala mutations, the effect of protein dipoles and surrounding water is non-negligible, if a quantitative agreement with experiment is desired. According to the EVB-FEP method the non-electrostatic part of the calculated activation energy change is negligible in both cases. Interestingly, our studies also reproduced the absolute activation barrier quite reliably (cf. Warshel and Russell, 1986). Both PDLD (Warshel et al., 1989b) and the FEP calculations (Warshel et al., 1988) have indicated that the difference between the enzyme and reference solution reaction, as well as the effects of key mutations, are almost entirely due to electrostatic effects.

Since we could reproduce the catalytic rate decrease upon point mutation of the buried aspartate from an ionisable side chain to a neutral one, it is also probable that our calculations should yield accurate results for the energetics of the hypothetical double proton-transfer mechanism. In fact, using the EVB conceptual framework we were able to provide more than just a theoretical estimate of the energetics of the alternative mechanisms but a reliable interpolation of the relevant experimental results. That is, we expressed the difference between the double and single proton-transfer mechanisms in water by the corresponding pK_a differences (plus a small Coulombic contribution) and established that the first one is more favourable than the second by about 25 kJ/mol in water (Warshel et al., 1989b). Using this fact we could express the energy differences in the protein in terms of the corresponding difference in water plus the change in solvation of the reacting fragments upon moving them from solution to the protein active site. Since the protein appears to stabilise these fragments more than water does (as is also evidenced from the observed reduction of the pK_a of the catalytic aspartate) it makes the double proton transfer mechanism even more unfavourable in the enzyme than in water (i.e. its activation energy is 50 kJ/mol higher than the corresponding energy for the single proton transfer mechanism). This approach allows one to reduce possible errors in the quantum mechanical self energy of the reacting fragments, since the same ones exist in the reference reaction and in the protein. Nevertheless, quantum mechanical calculations are also very useful and they support the crucial role of aspartate (e.g. Daggett et al., 1991). Note, however, that many early quantum mechanical calculations did not include the protein environment and drastically overestimated the effect of charged groups. Such studies could have easily supported incorrect mechanisms.

The major driving force for the catalytic effect of serine proteases appeared to be associated with the electrostatic stabilisation of the (- + -) charge distribution, representing the Ser...His...Asp triad at the active site. This hypothesis is further supported by the experiments of Corey et al. (1992) who displaced the ionised aspartate in trypsin from the site 102 to the more distant site 214 by designing the D102S/S214E(D) double mutant. The rate reduction for two different substrates, as compared to the wild-type enzyme, was found to be 3 orders of magnitude less than for the D102S mutant, where the negative charge stabilising the $HisH^{+}$...$Tetr^{-}$ ion pair, is absent. Even distant ionisable groups (Asp-36 and Asp-99) have some electrostatic effect on the catalytic rate of subtilisin BPN' (Jackson and Fersht, 1993). Depending on the ionic strength (degree of shielding of the charged side chain by counterions) the change in activation energy varies between 0 and 2.4 kJ/mol. Note, however, that these effects are much smaller than the overall effect of the enzyme or the effect of internal polar groups.

Electrostatics plays an important role also in substrate specificity of serine proteases (Náray-Szabó, 1989, 1993). Gráf et al. (1987) tested the role of Asp-189, which is located at the base of the substrate binding pocket, in determining the specificity of trypsin toward basic substrates. They replaced this residue by lysine by site-directed mutagenesis in the hope that charge specificity will change to the opposite. Though they found that D189K trypsin displays no catalytic activity toward arginyl and lysyl substrates, there is no compensatory change in specificity toward acidic substrates, no cleavage of aspartyl or glutamyl bonds was detected. Computer modelling studies indicated that this is most probably due to the drastic change in the conformation of the Lys side chain directed outside the substrate binding pocket in the mutant. Reversal of substrate charge specificity was successfully carried out for aspartate aminotransferase by site-directed mutagenesis (Cronin et al., 1987). In the R292D mutant the substrate charge specificity was reversed from the anionic (Asp ad Glu) to the cationic (Lys and Arg) amino acids. Wells et al. (1987) designed substrate specificity of subtilisin BPN' by protein engineering of electrostatic interactions. They showed that Gly/Arg and Gly/Lys mutations at position 156 and 166 may increase k_{cat}/K_m toward Glu P1 substrate by more than 3 orders of magnitude.

3.1.2. *Cystein Proteases*

Cystein proteases (papain, cathepsin B, actinidin) owe their catalytic action for peptide hydrolysis to the presence an essential Cys residue (Cys-25 in papain) in their active sites. The mechanism of action may be summarised as follows (Figure 7.). In the active site of the enzyme the side chain of the essential Cys residue forms an ion pair with the nearby His-159 with the sulphur atom lying approximately coplanar with the imidazolium ring. The formation of the ion pair is favoured by a hydrogen bond between the imidazole ring and the Asn-175 side chain. When the peptide substrate binds non-covalently to the enzyme to give **E.S**, the carbonyl group of the scissile peptide link forms a hydrogen bond with the amino group of Gln-19. The polarisation of the carbonyl bond favours the nucleophilic attack of the sulphur anion on the carbon, leading to the formation of a tetrahedral intermediate similarly as for serine proteases (Figure 6.). The imidazole side chain donates its proton to the NH of the leaving group and the peptide bond breaks giving an amine and an acyl enzyme. The second part of the reaction (deacylation) proceeds in an inverse manner: a water molecule replaces the leaving group, donates a proton to the imidazole ring and forms a second tetrahedral intermediate. Breaking of the C-S bond yields the second product (acid) and restores the enzyme in the active site.

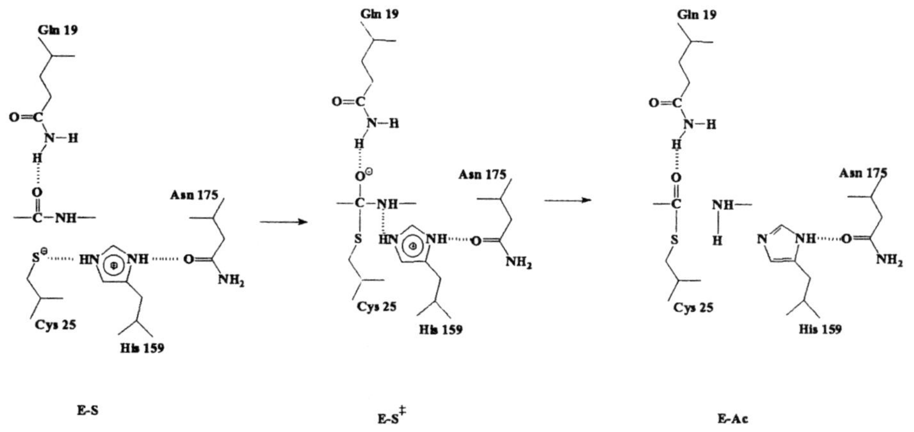

Figure 7. Schematic diagram showing the acylation step in the hydrolysis of peptide bonds by papain. **E.S:** Michaelis complex, **E.S†:** transition-state complex, **E-Ac:** acyl enzyme.

There was a debate on the ionisation state of the active cystein side chain. Some authors stated that the thiol group reacts in its non-dissociated form (Brocklehurst and Little, 1972; Campbell and Kaiser, 1973), others argued that the catalytic Cys and His form an ion pair (Polgár, 1973; Lowe and Whitworth, 1974). It is now well established that the ground state of the enzyme involves the Cys^{-}...$HisH^{+}$ ion pair. Ab initio molecular orbital calculations with a double-zeta type basis set suggested that there are two energy minima for a proton moving along the S...N line (Broer et al., 1976; Bolis et al., 1979; van Duijnen et al., 1979). One corresponds to the ion pair and one to the neutral form, the latter lying about 80 kJ/mol lower in the gas phase (cf. Table 2.)

TABLE 2. Proton transfer energies in the Im...HSMe dyad as calculated by ab initio molecular orbital theory for various models (kJ/mol). Im = imidazole, Me = methyl, ΔE is the difference between energies of the ion-pair and neutral forms (after Bolis et al., 1979)

Model	ΔE
Im...HSMe dyad in the gas phase	75
Im...HSMe dyad + Asn-175	56
Im...HSMe dyad + Asn-175 + Ala-160	22
Im...HSMe dyad + Asn-175 + Ala-160 + α-helix*	-41

*with a point-charge representation of the α-helix (van Duijnen et al., 1979)

Including the effect of Asn-175, Ala-160 and an α-helix made the ion pair more stable than the corresponding neutral pair. The importance of dipole fields provided by α-helices near the active site of enzymes has been stressed by several authors (Johannin and Kellersohn, 1972; Hol et al., 1978; Sheridan and Allen, 1980) and it has been proposed that these motifs are essential parts of the electrostatic stabilisation network present in many proteins. However, all the above calculations suffer from a major problem since they basically reflect gas-phase results. In the gas phase the helix dipole contributes ten times more than in condensed phases or in proteins (Åqvist et al., 1991). In fact, it has been shown that the effect of the helix is not associated with its macro dipole but with the last few localised hydrogen bonds (Åqvist et al., 1991) which play the same role as other preoriented hydrogen bonds, e.g. the oxyanion hole in serine proteases. Thus, although electrostatic effects are clearly crucial in papain catalysis (see Lavery et al., 1983; Arad et al. 1990), their actual magnitude cannot be assessed without including the proper protein and solvent environments and reflecting correctly their dielectric effect. A consistent study that took into account the complete protein-solvent environment (Warshel, 1991) reproduced the observed fact that the Cys^-...$HisH^+$ ion pair is more stable than the CysH...His neutral pair in the enzyme active site. This study also reproduced the fact that the neutral pair is more stable than the ion pair in water.

van Duijnen (1981) has argued that in papain the Cys^-...$HisH^+$ ion pair is stable in the active site owing to the intraproteic electrostatic field and that a similar situation is found in thiol-subtilisin (the S221C mutant of subtilisin) but not in subtilisin. Unfortunately, this suggestion was based on gas-phase calculations and, as stated above, it is impossible to draw any unequivocal conclusion without including the complete protein and solvent environment and reproducing the observed energetics of the ion pair. The inactivity of thiol-subtilisin towards proteins may be explained by the much greater basicity of the
His side chain in the complex His...Asp^- than in His..Asn. This is in agreement with the result of Arad et al. (1990) suggesting that the key role of His in serine proteases is proton abstraction, but in the cystein proteases, it is proton delivery. However, also in this case we deal with a non-conclusive gas-phase calculation.

Though the geometric structures of actinidin and papain are very similar, their active sites display different reactivity characteristics, especially in the pH-dependence. Pickersgill et al. (1988) explained this with the difference in MEP values in the active site clefts. According to their calculations papain provides a lower MEP difference (lower electrostatic field) as going from SG of Cys-25 to ND1 of His-159, than actinidin. Thus papain encourages protonation of the histidine side chain less and this could be the reason for its lower activity at low pH. Since actinidin has two negatively charged side chains (Asp-138 and Glu-121) that provide most of the stronger electrostatic field along the ion pair, changing the corresponding residues in papain (Gln-138 and Gln-118) into glutamic acid is expected to increase the activity of papain at low pH.

3.1.3. *Aspartyl Proteases*

Aspartyl proteases are an important class of digestive enzymes including pepsin, chymosin and renin, as well as a number of microbial and plant enzymes with tertiary structures showing a striking similarity to each other (Fruton, 1987). They are optimally active at acidic pH and have two aspartate side chains at the active site (Hsu et al., 1977; Subramanian et al., 1977). At present no consensus exists on the mechanism of action of aspartic proteases (Fruton, 1987; Bott et al., 1982; James and Sielecki, 1985; Polgár, 1987). All aspartic proteases appear to be highly conserved in that they all have two active aspartyl residues buried in a cleft that is large enough to accommodate polypeptides of about seven amino acid residues. As indicated by pH-dependence studies in the liquid phase, the two carboxyl groups, hydrogen-bonded in the catalytic cleft, must react in different forms, one in the ionised and the other in the unionised form.

A possible mechanism of the enzymatic reaction is depicted in Figure 8. (Polgár, 1987). The two catalytic Asp side chains are linked through a hydrogen bond. According to Polgár's mechanism, the reaction of aspartyl proteases involves a nucleophilic attack by two simultaneous proton transfers, one from the nucleophile to the dyad of the two carboxyl groups and one from the dyad to the carbonyl oxygen of the substrate (cf. Figure 8., a and b). This proposed "push-pull" general acid-base catalysis leads to the formation of a neutral rather than a negatively charged tetrahedral intermediate that may not be stabilised by the intraproteic electrostatic field.

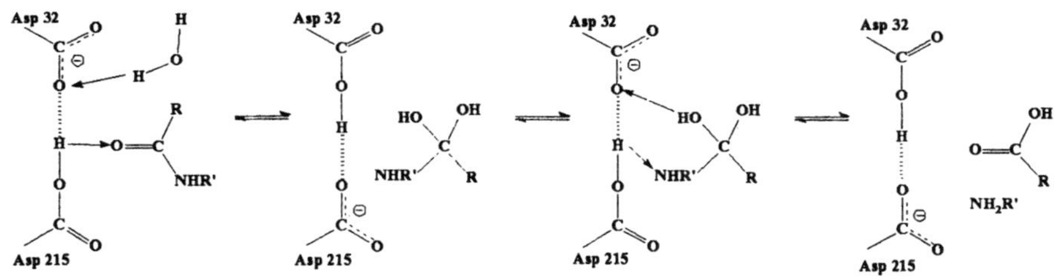

Figure 8. Mechanism of action of aspartyl proteases as proposed by Polgár (1987).

Some aspects of this mechanism has been studied by Turi and Náray-Szabó (1992) who argued that the formation of a carboxylate-hydroxonium ion pair in a non-concerted mechanism is not stabilised sufficiently by the enzyme. It is still possible, however, that the reaction is not fully concerted and its transition state involves a negatively charged oxyanion and two protonated acids. This point should be studied while considering the electrostatic effect of the entire protein-solvent system. It is also essential to use the EVB-type approaches in calibrating the energetics of the reacting fragment using pK_a's in solution, in order to obtain more unique mechanistic conclusions.

3.1.4. *Zinc Proteases*

The most studied member of zinc proteases is the digestive enzyme bovine pancreatic carboxypeptidase A (CPA) which is a metalloenzyme containing one atom of zinc bound to its single polypeptide side chain of 307 amino acids with a molecular weight of 34 kD. It is an exopeptidase, which catalyses the hydrolysis of C-terminal amino acids from polypeptide substrates, and is specific for the large hydrophobic amino acids such as phenylalanine.

Although much is known about CPA, the mechanism of hydrolysis is still not well understood. Hydrolysis is believed to proceed by attack of zinc-bound water on the peptide bond to form a tetrahedral intermediate (Christiansen and Lipscomb, 1989) but the exact nature and function of the amino acids involved is still unclear. In Figure 9. we display the mechanism proposed by Christiansen and Lipscomb (1989). According to this mechanism the Michaelis complex with the substrate carbonyl hydrogen bonded to Arg-127 allows for nucleophilic attack by a water molecule

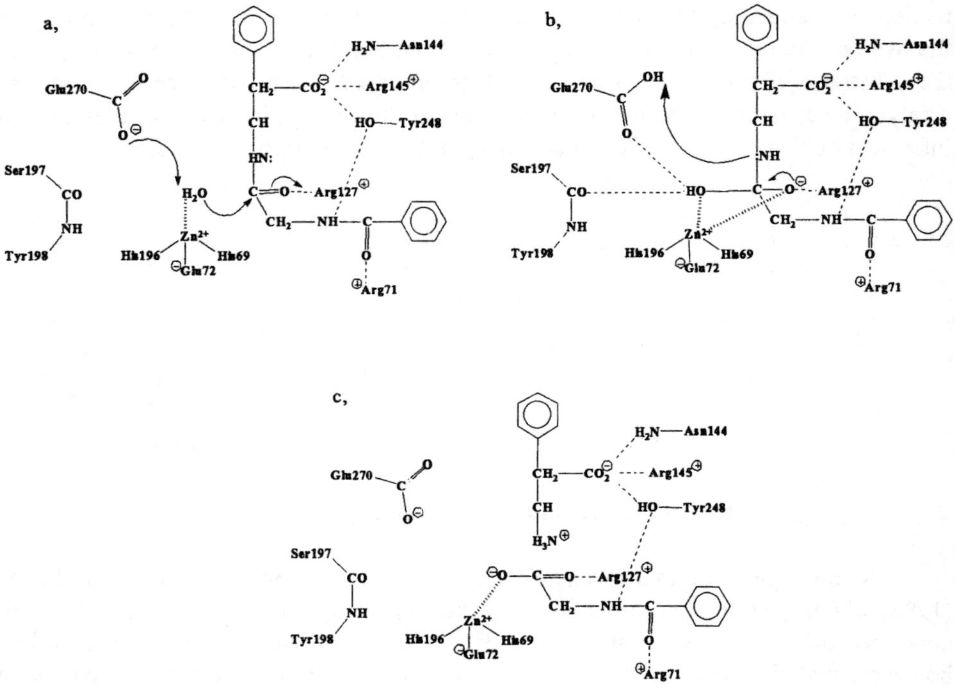

Figure 9. Catalytic mechanism for carboxypeptidase A.

promoted by zinc and assisted by Glu-270 (a). The stabilised tetrahedral intermediate collapses with proton donation by Glu-270 (b), and the final product is formed after a second proton transfer mediated by Glu-270 (c). Unfavourable electrostatic interactions between the carboxylate side chain of Glu-270 and the ionised product carboxylate may facilitate product release. Site-directed mutagenesis studies by Phillips et al. (1990) indicate that the role of Arg-127 is crucial in catalysis. It appears that the enzyme functions primarily by electrostatically stabilising the tetrahedral intermediate that forms the transition state in the rate limiting step. For small peptide and ester substrates the loss in binding energy of the rate-limiting transition state upon mutation ranges from 17 to 25 kJ/mol while it is negligible for the ground state (Phillips et al., 1990). Electrostatic calculations also suggest that the main role of Arg-127 should be transition-state stabilisation, its theoretically predicted contribution is between 25 and 33 kJ/mol in reasonable agreement with the experimental result. This large electrostatic interaction energy may be due to the fact that Arg-127 is buried in the protein, in a similar way to the catalytic aspartate in serine proteases (cf. Section 3.1.1.). In contrast to earlier speculations it was shown by site-directed mutagenesis that neither Tyr-198 nor Tyr-248 are obligatory for substrate hydrolysis though they both stabilise the transition-state complex (Gardell et al., 1985; 1987). Since stabilisation is possible through hydrogen bonding, that is of electrostatic origin (cf. e.g. Náray-Szabó et al., 1986), this effect is also explained in terms of electrostatics.

One of the early works on enzyme electrostatics involved a study of CPA (Hayes and Kollman, 1976). These authors stressed the importance of the MEP in understanding enzyme action. They suggested that the electrostatic environmental effects do contribute to the lowering the transition-state energy of the reaction. However, as stressed above, the actual calculations were basically unable to address this issue since they did not consider the protein dielectrics nor the fact that the reaction in solution involves major electrostatic stabilisation (what counts is the difference between the stabilisation in the enzyme and in water). Nevertheless, site-directed mutagenesis experiments have provided strong evidence that electrostatic effects play a major role in the reactions catalysed by carboxipeptidase A.

Umeyama and co-workers also considered protein and metal electrostatics in their calculations (Nakagawa and Umeyama, 1978; Nakagawa et al., 1981; Nakagawa and Umeyama, 1982) and they found that the proton transfer from the zinc-co-ordinated water molecule to Glu-270 is affected by electrostatic interaction with the ionic amino-acid residues. While Arg-71, Arg-127, Arg-145 and Asp-65, lying within a sphere of 1 nm radius around the active site facilitate proton transfer, the C-terminus of the substrate inhibits this process. Amino acids, lying beyond 1 nm from zinc also inhibit proton transfer. Here again the predicted value of the calculation has been quite limited since the effect of the solvent and protein dielectrics was basically neglected and the calculated effect of the ionised groups is probably an artefact since unscreened charges were used. The actual effect of the ionised residues and the metal ion is usually 20-40 times smaller than the values calculated without proper consideration of the overall dielectric effect.

3.2. LIPASES

Lipases are ester hydrolases acting on triacylglycerols. They achieve their highest catalytic rate at oil-water interfaces. Though they widely differ in size, substrate and catalytic rate and show little sequence similarity, the Ser...His...Asp(Glu) triad in the active site is a structural feature common to serine proteases (cf. Section 3.1.1.) and lipases. Their mechanism of action is not yet fully revealed. X-ray diffraction studies suggest that the putative hydrolytic site is covered by a surface loop and is therefore inaccessible to solvent (Brady et al., 1990; Winkler et al., 1990). Therefore the enzyme could be working in two stages: the hydrophobic lid is removed or displaced, possibly with interfacial activation, then the ester bond is subsequently hydrolysed by a mechanism very similar to that used by serine proteases. Since serine proteases make use of protein electrostatics in catalysis, it is most probable that lipases do the same. An indication for this is that the (- + -) charge pattern, present in the active site of serine proteases, also exists in lipases and is also stabilised by the electrostatic potential of the protein environment (cf. Figure 5.).

3.3. PHOSPHOLIPASE A_2

Phospholipase A_2 (PLA2) hydrolyses phospholipids of cell membranes to release fatty acids such as arachidonic acid and thereby initiates the production of a diverse range of potent cellular mediators such as the prostaglandins (Slotboom et al., 1982). Verheij et al. (1980) proposed a mechanism which resembles the mechanism of other hydrolytic enzymes, in particular the serine proteases. According to this proposed mechanism (Figure 10., Verheij et al., 1980) a histidine (His-48 of bovine pancreatic PLA2) and a

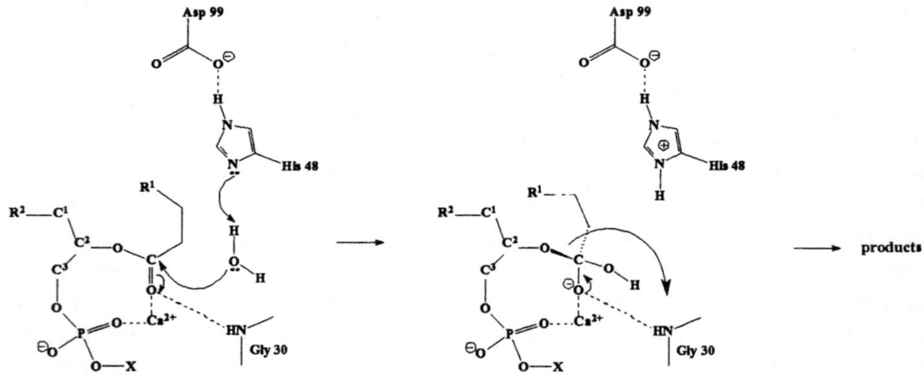

Figure 10. Proposed mechanism of action of PLA2 (after Verheij et al., 1980)

calcium ion are essential for activity. The hydrogen-bonded His-Asp couple, found within the hydrophobic wall of the active-site cleft, functions similarly to the Asp-His-Ser catalytic triad in serine proteases (cf. Section 3.1.1.). In the absence of serine, in PLA2 a water molecule is assumed to be the active nucleophile while the phospholipid substrate binds to the calcium ion via the carbonyl and phosphoryl oxygen atoms with the acyl chains interacting with the hydrophobic wall. The polarised carbonyl group is attacked by the water molecule, present in the crystal structure of bovine PLA2, following a proton transfer to His-48 and a tetrahedral intermediate is formed just like in case of serine proteases (Section 3.1.1.), lipases (Section 3.2.) or acetylcholin-esterase (Section 3.9.). Upon proton transfer from His-48 to the C^2 ether oxygen the intermediate breaks down to yield the C^2 fatty acid.

Molecular modelling has demonstrated that the phospholipid substrate can be fitted into the active site with minimal distortion of the extended conformation which the substrate adopts in the lipid membrane (Waszkowycz et al., 1990). The binding of the C^2 carbonyl oxygen and a phosphoryl oxygen to the calcium ion is presumably responsible for the alignment of the carbonyl with the water-His-Asp system. Ab initio molecular orbital calculations by Waszkowycz et al. (1990, 1991) suggested that the intermediate having the (- + -) charge distribution can be effectively stabilised by electrostatic interactions both within the calcium-oxyanion-His-Asp system and also between this system and the protein environment. This work has been based on a hybrid quantum mechanical/molecular mechanics approach resembling in many aspects to the original work of Warshel and Levitt (1976). However, the effect of the bulk solvent around the protein, as well as the repolarisation of the solvent during reaction, was not considered. This appeared to lead to a significant overestimate of the effect of the protein electrostatic field and to a negative activation barrier. That is, while the protein effect was represented in a much more reliable way than many of the works considered in this chapter, the neglect of the effect of the solvent repolarisation lead to an unrealistically low effective dielectric constant for charge-charge interactions (i.e. 2.4 rather than >20). Thus, the effect of the Ca^{2+}-ion, which is clearly important, was probably overestimated. Nevertheless, although the calculations by Waszkowycz et al. (1990, 1991) overestimated the electrostatic effect of the protein, they provided clear evidence for its importance.

3.4. LYSOZYME

This enzyme hydrolyses cell wall polysaccharides containing alternating copolymers of *N*-acetylglucosamine and *N*-acetylmuramic acid (Imoto et al., 1972). Maximal rates of hydrolysis and the highest affinities are achieved when six saccharide units are bound to the enzyme. Hydrolysis takes place between sites D and E in a cleft lined by the two

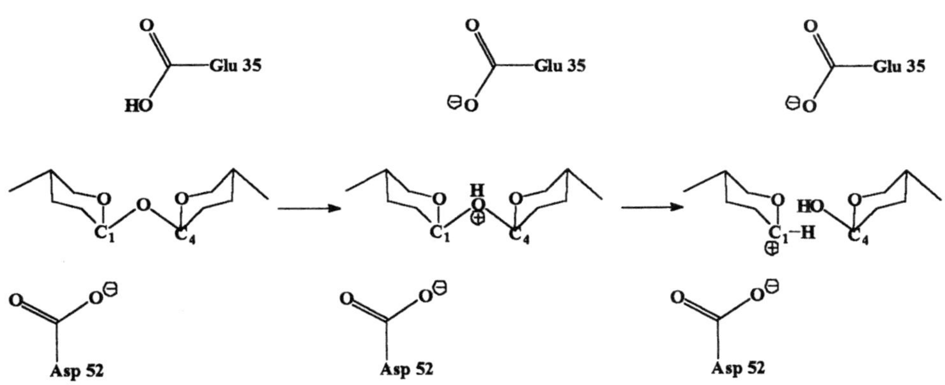

Figure 11. Schematic mechanism of the reaction catalysed by lysozyme.

catalytic residues Glu-35 and Asp-52. The mechanism involves a carbonium ion at C_1 of the D-site residue, stabilised by the carboxylate of Asp-52 with Glu-35 serving as a general acid that protonates the oxygen of the departing saccharide (Figure 11.). The reaction of a water molecule with C_1 leads to the formation of the product. Upon binding of the substrate to the active site the sugar ring in subsite D distorts from a chair towards a half-chair or sofa conformation weakening the bonds between atoms C_1 of D and $O(C_4)$ of E. This bond is further weakened as Glu-35 donates a proton to $O(C_4)$. In the transition state the same (- + -) charge distribution is formed that plays a crucial role in the catalysis by serine and cystein proteases and lipases (cf. Section 2.3.).

The elucidation of the X-ray structure of lysozyme provided the first direct information about the structure of an enzyme-substrate complex (Blake et al., 1967). This and other biochemical studies led to the proposal of three major catalytic factors: general acid catalysis, steric strain and electrostatic stabilisation. It was found that the effect of steric strain is unlikely to be important because it can be relaxed by small shifts in the substrate and enzyme geometries (Warshel and Levitt, 1976; Pincus et al., 1977). This point was further established by quantitative FEP calculations (Warshel, 1991). On the other hand, it was found that electrostatic interactions play a very important role in the enzymatic reaction. The total electrostatic stabilisation of the carbonium transition state, relative to the corresponding stabilisation in solution, was found to be ~29 kJ/mol (cf. Figure 12., Warshel, 1978). This value appears to account

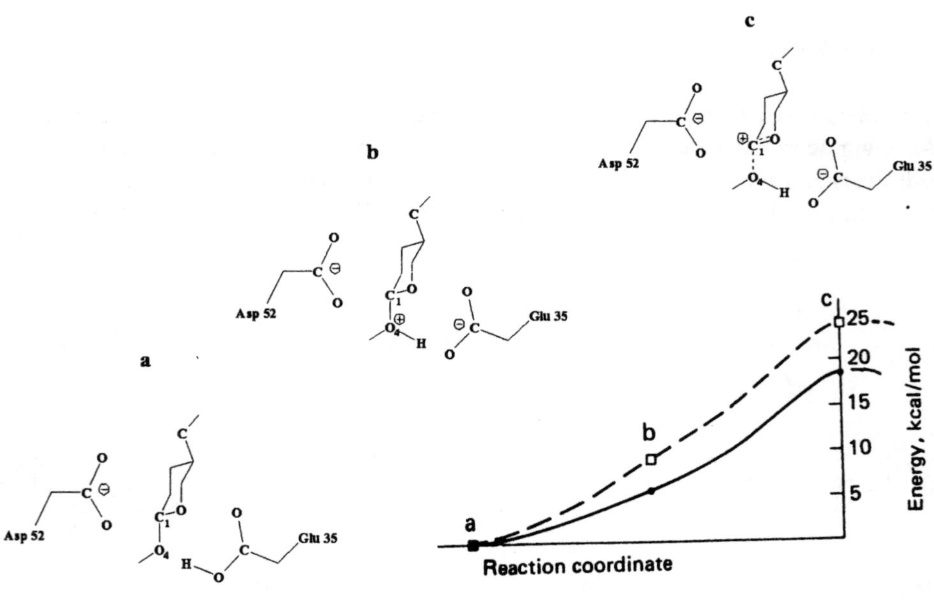

Figure 12. Schematic description of the catalytic bond cleavage of the sugar residue in the D subsite of lysozyme (Warshel, 1978). (a) Ground state of the enzyme-substrate complex, (b) $O(C_4)$ of the substrate is protonated by Glu-35, (c) the C_1-$O(C_4)$ bond is broken forming a planar carbonium ion intermediate. Full line: reaction in the enzyme active site, dashed line: reaction in aqueous solution.

for the entire observed catalytic effect of the enzyme. It was found that the ionised form of Asp-52 plays a key role in the catalytic effect (Warshel and Levitt, 1976) and that the only reason this residue is kept ionised is its highly polar local environment (Warshel and Levitt, 1976; Warshel, 1978; Warshel, 1981). This indicates that the catalytic effect of the enzyme *cannot* be described as the interaction between the ionised Asp-52 and the positively charged carbonium ion in a low dielectric environment. Careful examination of this issue led to the conclusion that the enzyme works by *solvating* the transition state charges better than water does and that this involves the use of prearranged dipoles (Warshel, 1978). It was also concluded that this type of stabilisation is particularly effective when it involves (- + -) charge configurations (e.g. $Asp_{52}^{-}...C_4^{+}...Glu_{35}^{-}$ in the case of lysozyme). This study, that was the earliest to use quantitative electrostatic considerations in the analyses of a biological process, led to the conclusion that enzyme active sites are very polar and that folding energy must be invested in forming prearranged polar active sites (Warshel, 1978).

An instructive MEP map for lysozyme has been calculated quite early by Clementi et al. (1979). A much more quantitative study was reported by Dao-Pin et al. (1989) who found that lysozymes from different origin with totally dissimilar amino-acid sequences have similar MEP patterns in their active sites (cf. Section 2.3.). The magnitudes of the MEF near the proton-donating glutamate (Glu-35 in hen egg white and human lysozymes, Glu-11 in bacteriophage T4 lysozyme) vary between 0.4 and 1.4 V/nm if no solvent or peptide partial charges are considered. The 0.5 V/nm MEF value for hen egg white lysozyme increases to 0.7 and 0.8 V/nm if bound solvent and peptide partial charges are also included in the model. These authors argued that the field from the entire enzyme rather than the active site region is crucial for the catalytic effect of the enzyme. Unfortunately, this proposal is in a *clear contrast to the fact that mutations of surface groups do not change the rate of the enzyme in a significant way.* Furthermore, while the authors argued that Asp-52 is not the most important ionised group, their own calculation of the actual effect of neutralising this acid demonstrated that it contributes for about 70% of the calculated electrostatic effect. It is also important to note that mutation experiments (Malcolm et al., 1989), that were originally interpreted as showing a small effect of Asp-52, have established that this group is crucial for catalysis with well defined substrates.

While the calculations of Dao-Pin et al (1989) were instructive in estimating the effect of mutating groups, they could not be used to establish whether or not the electrostatic effect of the enzyme lead to a more stable transition state than for the corresponding reference reaction in water. As stated above performing calculations with a proper reference state (Warshel, 1978; 1981) demonstrated that the preoriented dipoles around the (- + -) configuration of the transition state are the reason for the catalytic effect of the enzyme. The calculations suggested that most of the catalytic effect is due to the first few shells of residues around the active site. This finding is also supported by available mutation experiments, but at present we cannot exclude the possibility that long-range effects contribute to enzyme catalysis.

3.5. CARBONIC ANHYDRASE

The enzyme from mammalian source has a single polypeptide chain with a molecular weight of about 30 kD and contains one zinc cation per molecule (Lindskog, 1986). The enzyme catalyses the reversible hydration of carbon dioxide and dehydration of bicarbonate. This so-called zinc-hydroxide mechanism for action (cf. Figure 13., Silverman and Lindskog, 1988) involves two processes: Zn-water activation and an exchange between carbon dioxide and bicarbonate (cf. **a** and **b** in Figure 13.a). The so-called zinc-water activation corresponds to a proton release to the medium. The generally accepted mechanism for this process is intramolecular proton transfer between zinc-bound water and His-64 (step I in Figure 13.a) and an intermolecular proton transfer between His-64 and the buffer (step II in Figure 13.a). At low and high buffer concentrations step II and step I are rate-limiting, respectively. The exchange process corresponds to steps III, IV, and V in Figure 13.a. Step IV is rate-limiting in this process and is assumed to occur by direct nucleophilic attack of zinc-bound hydroxide on carbon dioxide. The subsequent dissociation of bicarbonate from zinc is a fast process.

The catalytic mechanism of carbonic anhydrase has been investigated by Pullman (1981), Cook et al. (1984), Liang and Lipscomb (1988), Merz et al. (1989) and Jacob et al. (1990), who studied gas-phase models using molecular orbital methods. The special importance of electrostatics in the action of this enzyme has been stressed by several authors in the last two decades (Sawaryn and Sokalski, 1979; Sheridan and Allen, 1981; Ressler, 1982; Krauss and Garmer, 1991; Åqvist and Warshel, 1992, 1993).

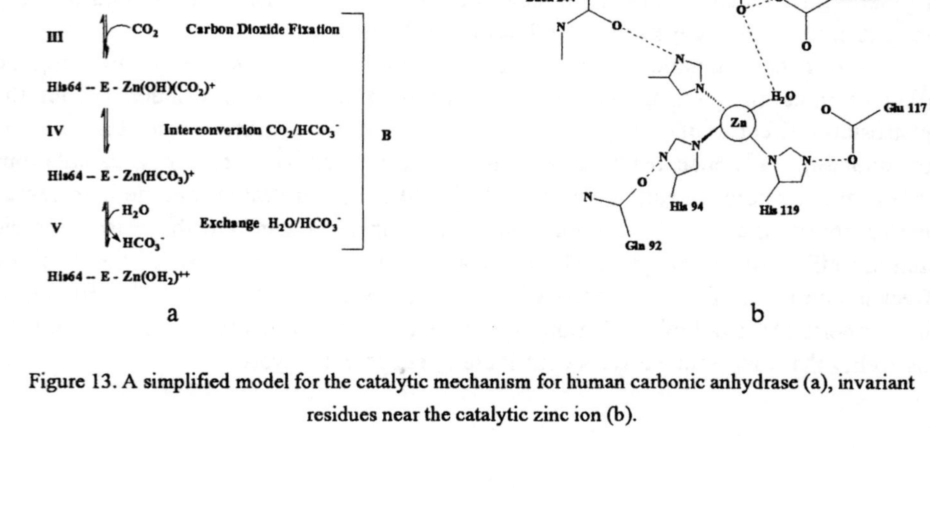

Figure 13. A simplified model for the catalytic mechanism for human carbonic anhydrase (a), invariant residues near the catalytic zinc ion (b).

Sheridan and Allen (1981) examined the active-site MEP of two isozymes of human carbonic anhydrase. They found that the MEP maps are considerably different and suggested that the charges of the surface side chain dominate the protein potential. However, this result is probably an artefact due to the neglect of the solvent around the protein which results in an overestimate of the effect of ionised groups by more than a factor of 20. Krauss and Garmer (1991) used a model that strongly suggests that when water is bound to zinc the proton transfers from the imidazole side chain of His-119 to the carboxylate of Glu-117. However, this result reflects probably the neglect of the protein and solvent dielectric effects.

Sawaryn and Sokalski (1979) considered the MEP of a few amino-acid residues around the active site and suggested that the Zn^{2+} and OH^- ions stabilise the transition state of the reaction. Unfortunately, the mechanism studied involves an attack of water rather than hydroxyl on CO_2 and the calculation did not treat correctly the protein dielectrics. The crucial role of the Zn^{2+} cation and its local environment was determined quantitatively by the Empirical Valence Bond/Free Energy Perturbation study of Åqvist and Warshel (1992) and Åqvist et al. (1993). This study demonstrated

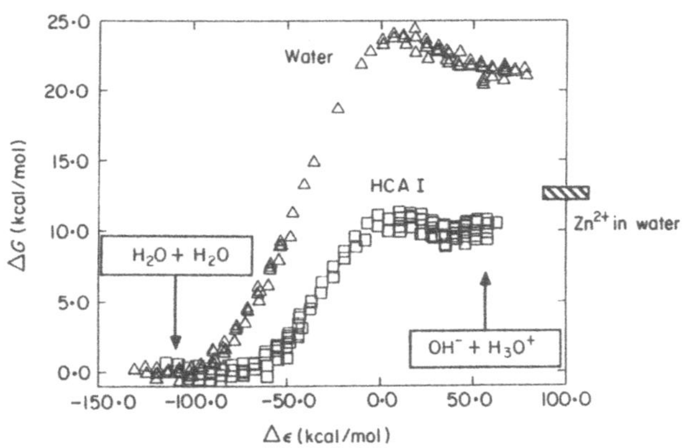

Figure 14. Calculated free energy profiles for the reference reaction, $2H_2O \leftrightarrow OH^- + H_3O^+$ (open triangles) in water and for the proton transfer step in human carbonic anhydrase I (open squares). The calculated energetics of proton transfer from a zinc-bound water molecule in aqueous solution is also shown for comparison (Åqvist and Warshel, 1992).

that the Zn^{2+} cation reduces the pK_a of the bound water by providing major electrostatic stabilisation to the hydroxyl ion (see Figure 14.). The most important point about the results in Figure 14. is that in this case the observed values of the activation barrier for the proton transfer step and of the pK_a of the bound water were reproduced without the adjustment of any parameter. All parameters were fixed and calibrated in studying the solvation of the reacting fragments in water and kept unchanged in the protein calculations. This is important since the evaluation of the effect of the Zn^{2+} ion is meaningless without taking into account the corresponding dielectric constant of the protein active site. It appears that the protein active site and, in particular, the zinc cation contribute to catalysis by both stabilising the hydroxyl ion in Step I and the HCO_3^- ion in Step IV (Åqvist et al., 1993). It is instructive to point out in this respect that the electrostatic effect of zinc should be described in terms of its interaction with the hydroxide anion rather than in terms of the ground-state polarisation of the water molecule (Åqvist and Warshel, 1992).

3.6. STAPHYLOCOCCAL NUCLEASE

Staphylococcal nuclease (SNase) is a phosphodiesterase that cleaves DNA and RNA at the 5' position of the phosphodiester bond to form 3'-phosphate monoester and a free 5'-hydroxyl group through the reaction displayed above (Tucker et al., 1978). The

$$H_2O + 5'\text{-}OP(O_2)^{-}O\text{-}3' \leftrightarrow 5'\text{-}OH + (OH)P(O_2)^{-}O\text{-}3' \quad (2)$$

enzyme consists of a single polypeptide chain with a molecular weight of 16.9 kD. It requires one Ca^{2+}-ion for its action and shows no activity when calcium is replaced by other divalent cations. The catalytic mechanism was postulated by Cotton et al. (1971; 1979) (cf. Figure 15.). First a proton is transferred from an active site water molecule to Glu-43 yielding a free hydroxide anion. This anion attacks on the phosphorus atom and forms a penta-co-ordinated transition state which breaks down by cleavage of the 5'-O-P bond.

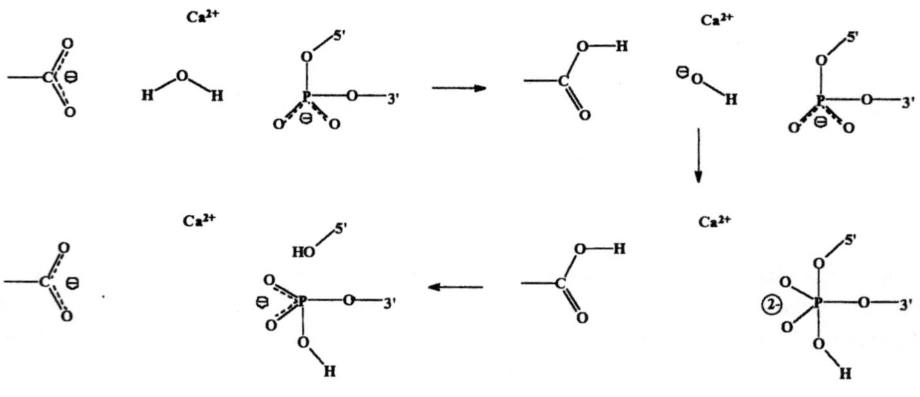

Figure 15. Reaction mechanism for staphylococcal nuclease. (a) general-base catalysis, (b) nucleophilic attack by a hydroxyl anion, (c) cleavage of the 5'-O-P bond.

Extensive calculations of the reaction free energy profile for the postulated mechanism of SNase (Åqvist and Warshel, 1989; 1990) achieved quantitative agreement between calculated and observed free energy barriers. It was concluded that the general-base mechanism itself is not a sufficiently effective way of generating the hydroxyl ion and the Ca^{2+} ion is essential for stabilisation.

The importance for an enzyme providing an electrostatically complementary environment to an ionic transition state structure is well illustrated in the case of SNase, too (Åqvist and Warshel, 1989). The stabilising effect of the positively charged Arg-35 and Arg-87 side chains and the Ca^{2+} ion on the double negative charge of the 5'-phosphate group is the major factor responsible for the reduction of the activation energy by 85 kJ/mol as compared to the corresponding reference reaction in water (the hypothetical reference reaction involves the same mechanism as the enzymatic reaction). Molecular dynamics simulations showed that hydrogen-bonding interactions between the active-site residues and 5'-phosphate are considerably strengthened as the double negative charge is formed in the transition state. Arg-87 donates two hydrogen bonds in the transition state while it is not interacting as closely with the reactants. It is also interesting to note that the calcium ion appears to be an important factor during the entire reaction. It reduces the energy of the first step by stabilising the hydroxyl ion and then helps in stabilising the phosphate charge at the transition state of the second step.

3.7. TRIOSEPHOSPHATE ISOMERASE

Triosephosphate isomerase (TIM) catalyses the interconversion of dihydroxyacetone phosphate and *R*-glyceraldehyde 3-phosphate through the formation of a *cis*-enediol or enediolate (Maister et al., 1976; Knowles, 1991). The reaction mechanism is outlined in Figure 16. A basic catalytic group (His-95) abstracts the proton on C^1 of dihydroxyacetone phosphate (DHAP in **E.S**) assisted by an acidic catalytic group (Glu-165) to produce the intermediate enediol (**E.I**). The intermediate undergoes protonation and deprotonation by the adjacent basic and acidic catalytic groups and the product, D-glyceraldehyde 3-phosphate (**E.P**) is formed. This reaction may serve as a classical example for a general acid-base catalysis (cf. Section 2.2.), where the nearby imidazole and carboxylate side chains play as mediators in proton transfer reactions.

In their early paper Alagona et al. (1984) called the attention to the role of ionised residues (particularly the Lys-13...Glu-97 ion pair) in catalysis. They evaluated the protein MEP at the substrate atoms and found that all of the atoms of DHAP, except for the *pro-R*-hydrogen at C_1, where the MEP is negative (due mainly to Glu-165), are in a positive region. This is a nice qualitative picture of the facilitation of the proton abstraction by the enzyme although the effect of ionised residues was largely over-estimated. Bash et al. (1991) stressed the importance of "solvation" of the reacting charges by the enzyme suggesting that this effect reduces the energy barrier of the reaction. This view confirmed the original proposal of Warshel (1978) that described the enzyme as a "super solvent". It should be noted that although the actual calculations used a combined quantum mechanical molecular mechanics method (that can, in principle, give a reliable description of the complete reacting system) they significantly underestimate the energies of the proton transfer process in the second step of the reaction. This problem might reflect the use of energy minimisation for evaluating the polarisation of the solvent molecule and problems with the method used

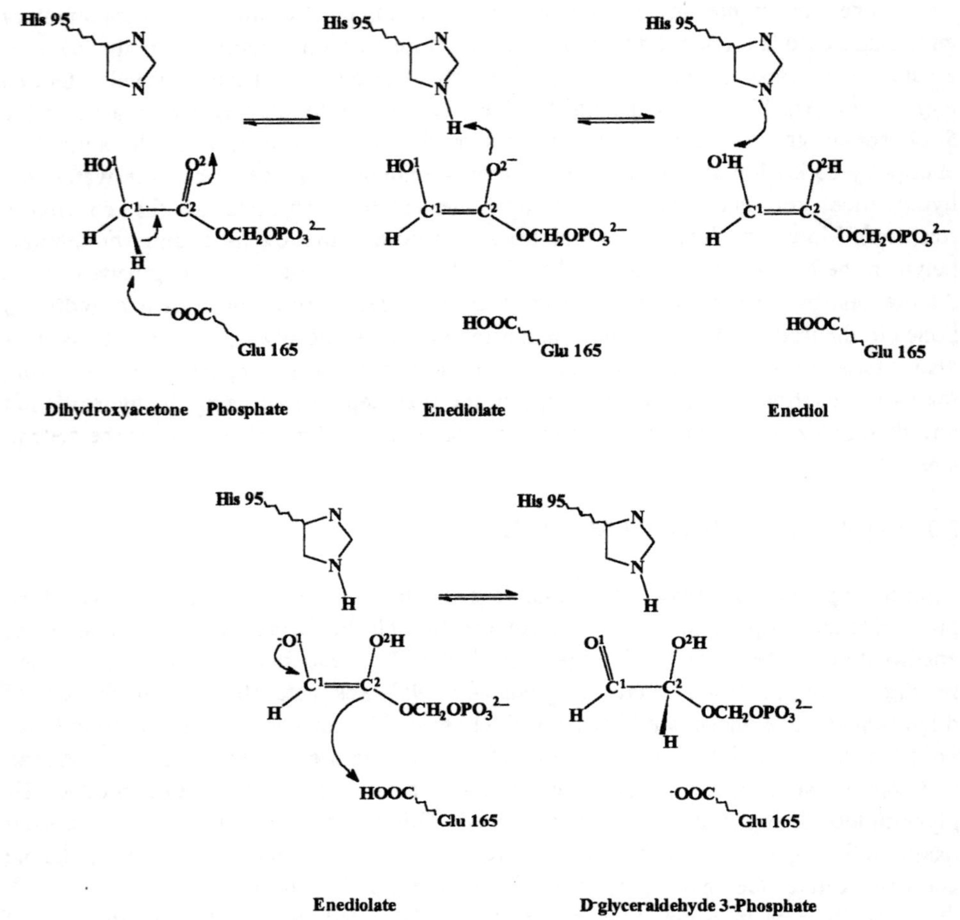

Figure 16. Putative mechanism for the isomerisation of dihydroxyacetone phosphate to D-glyceraldehyde 3-phosphate by triosephosphate isomerase.

to evaluate the energetics of the reacting fragments. Such problems do not exist in the Empirical Valence Bond approach used by Åqvist in Chapter 8 of this book, where the calculations are calibrated for the reference reaction in solution. This study has demonstrated the crucial role of the enzyme in reducing the electrostatic reorganisation energy of the reaction.

Knowles (1991) proposed that triosephosphate isomerase catalyses its reaction by using the α-helix macro dipole supporting the hypothesis of Hol et al. (1978). However, subsequent experimental studies of his group (Lodi and Knowles, 1993) indicated that the helix effect on His-95 is due to the hydrogen bonding provided by its end and not to the macro dipole. This work also considered a much smaller effect on His-103 that might be due to local amide dipoles near this residue. Apparently, the effect of the amide dipoles at the end of the helix is equivalent to that of any local dipole of the type considered by Warshel (1978) (see the discussion in Åqvist et al., 1991). This point is supported by the fact that mutations at the other end of the helix, that should neutralise the macro dipole field, have very little effect on the catalytic rate of the enzyme.

3.8. DEHYDROGENASES

3.8.1. *Lactate dehydrogenase*

Lactate dehydrogenase is a pyridine nucleotide oxidoreductase, a tetramer of 140 kD molecular weight, which has been extensively investigated (Bloxham et al., 1975; Eventoff et al., 1977). It catalyses the reversible oxidation of L-lactate to pyruvate using NAD^+ as a coenzyme. The reaction scheme with a view of the active site with bound substrate and essential amino-acid side chains are depicted in Equation (3) and in Figure 17. The probable reaction mechanism, involving proton and hydride transfers,

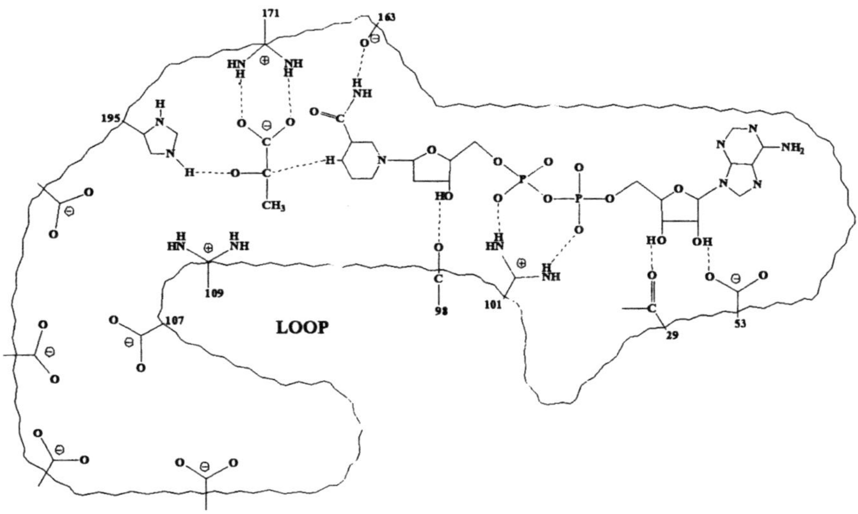

Figure 17. Schematic view of the active site of lactate dehydrogenase (after Eventoff et al., 1977). Dotted lines indicate hydrogen bonds.

$$\text{Me-CH(OH)-COO}^- + \text{NAD}^+ \leftrightarrow \text{Me-CO-COO}^- + \text{NADH} + \text{H}^+ \qquad (3)$$

can be seen in Figure 18. The carboxylate group of the substrate forms a salt bridge with the side chain of Arg-171 and the hydroxyl group of lactate forms a hydrogen bond with the unprotonated imidazole ring of His-195, acting as a general acid-base catalyst removing proton from lactate during oxidation. Though early theoretical calculations on lactate dehydrogenase mechanism (Krechl and Kuthan, 1982, 1983) did not take the protein environment into consideration, Ressler (1982) called the attention

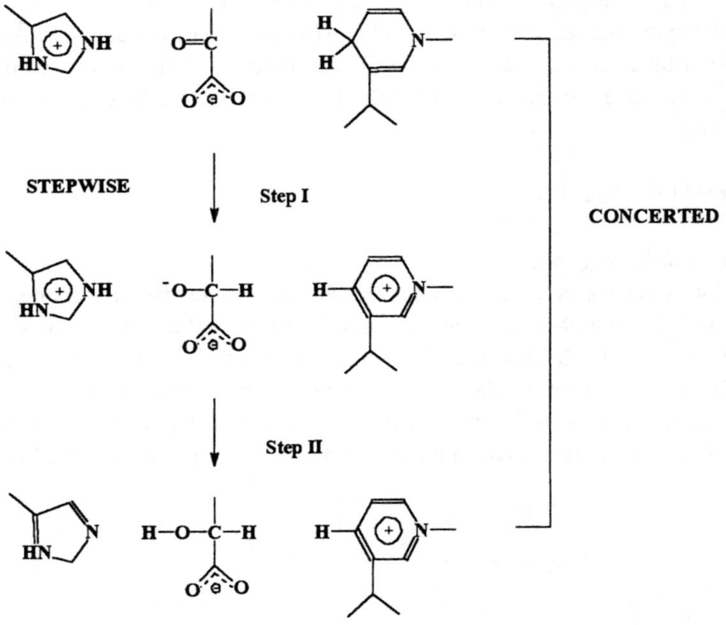

Figure 18. The schematic description of a stepwise and a concerted mechanism for the catalytic reaction of lactate dehydrogenase.

to the importance of electrostatic effects in catalytic rate acceleration. Hol et al. (1978) have shown that the enzyme binds negatively charged molecules at the N-terminus of an α-helix.

Quantitative study of this system was reported by Yadav et al. (1991). This study performed a careful Empirical Valence Bond/Free Energy Perturbation calcula-

Figure 19. Calculated contributions to the free energy functions, Δg_i, for the hydride-transfer step (I in Fig. 18.) of the lactate dehydrogenase reaction in a reference solvent cage (w, left) and in the protein active site (p, right). The Δg_i's are given as functions of the corresponding energy gaps (the $\Delta\varepsilon$'s) which are taken as the generalised reaction co-ordinates.

tion for the first step of this reaction and provided for the first time a quantitative demonstration of the role of reorganisation energy, λ, in enzyme catalysis. That is, when a given environment is involved in electrostatic stabilisation it usually pays a half of the solute-solvent interaction in solvent-solvent interaction. This penalty, which is basically the Marcus reorganisation energy (see discussion in Warshel, 1991), is minimised in enzymes where the environment is preoriented towards the transition state (see Warshel, 1978). Indeed, this effect was found to be the major catalytic effect in case of lactate dehydrogenase (see Figure 19., after Yadav et al., 1991).

3.8.2. *Alcohol dehydrogenases*

The alcohol dehydrogenases are zinc metalloenzymes of broad specificity. They oxidise a wide range of aliphatic and aromatic alcohols to their corresponding aldehydes using nicotinamide dinucleotide (NAD^+) as a coenzyme. The two most studied enzymes are those from yeast and horse liver, we deal with the latter. Horse liver alcohol dehydrogenase (LADH) is a symmetrical dimer composed of two identical chains of molecular weight of 40 kD. Each chain contains one binding site for NAD^+ and two sites for Zn^{2+}. Only one of the zinc ions is directly concerned with catalysis.

The mechanism for oxidation of alcohols, at a pH optimum of 8, has been suggested to contain at least seven different steps (cf. Figure 20.). The reaction involves the net release of one proton and the pH-dependence of the catalytic activity can be traced back to a protein-coenzyme modulated self-dissociation of water. The pK_a of the catalytic water is shifted from the unperturbed value of 16 to 9.2 upon binding to the catalytic zinc. NAD^+-binding shifts this value further to 7.6. Tapia and Johannin (1981) presented an attempt to incorporate the effect of the protein in the quantum mechanical calculation of a proton transfer from the zinc bound water to serine and then to a histidine. They argued that the protein and oxidised coenzyme exert an electric field that is sufficient to drive this process. While the qualitative value of this early work is important, it did not treat correctly the dielectric effect of the system using a reaction field model with a dielectric constant of around 2 which corresponds to a non-polar environment. This treatment neglected the large dielectric effect associated with the polarisation of the surrounding solvent and with the reorientation of the protein groups (see Warshel and Russell, 1984). This is probably the reason for the unrealistic height of the calculated barrier for the formation of this ion pair. In step 3 a ternary E...NAD^+...alcoholate complex is formed (cf. Figure 21.). The ionised alcohol displaces the zinc-bound water molecule while a strongly polar (+ - 2+) charge distribution is formed. Here again protein electrostatics should play an important role, however, we are not aware of any calculations related to this feature. For the formation of aldehyde from the tetrahedral intermediate (step 4 on Figure 20.) the most probable mechanism is hydride ion transfer. Tapia et al. (1987; 1991) presented interesting gas-phase calculations for the hydride transfer step and also considered the possible effect of a phenomenological reaction field. These calculations, however, are far too qualitative to allow any concrete conclusion. The calculated reaction profile corresponds to 125 kJ/mol exothermicity even in the presence of the

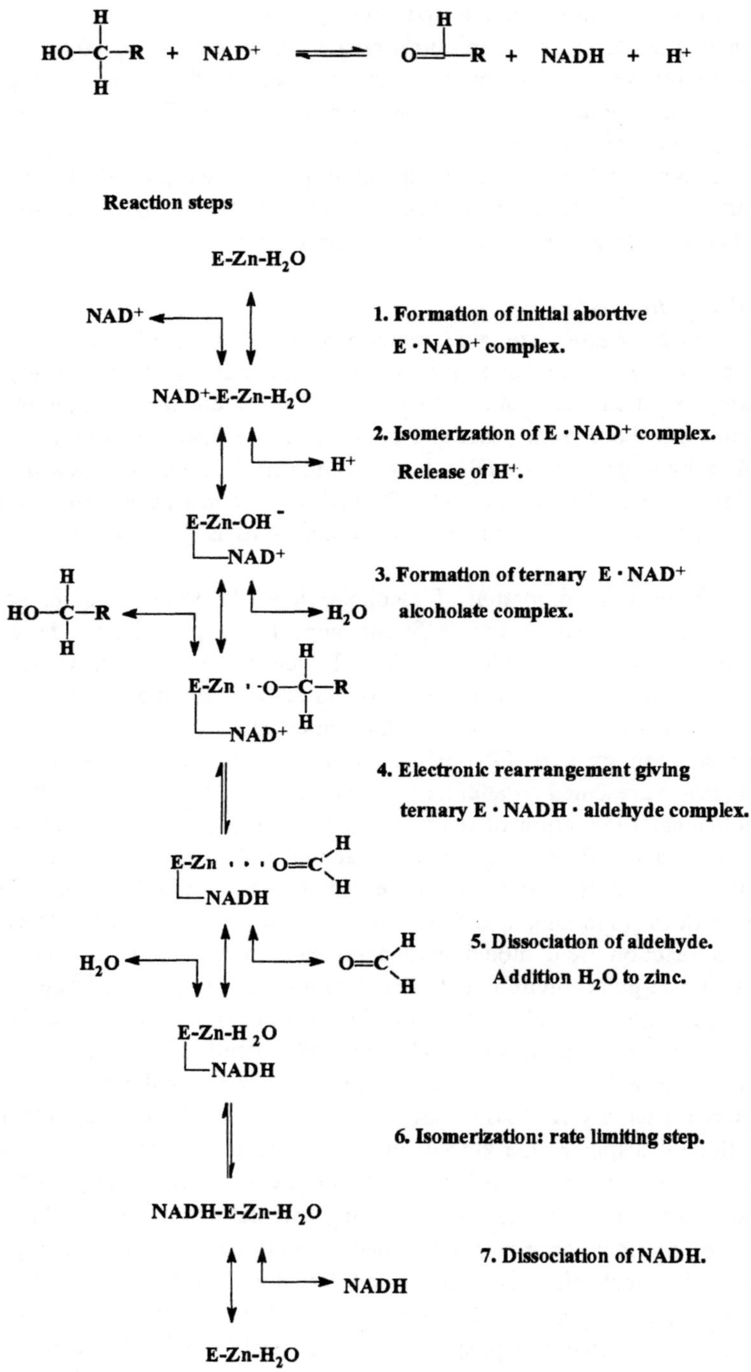

Figure 20. Schematic reaction mechanism for the oxidation of alcohols by LADH (Tapia et al., 1987).

assumed reaction field. This again emphasises the difficulties in treating charge-transfer processes without the correct description of the effect of the actual environment.

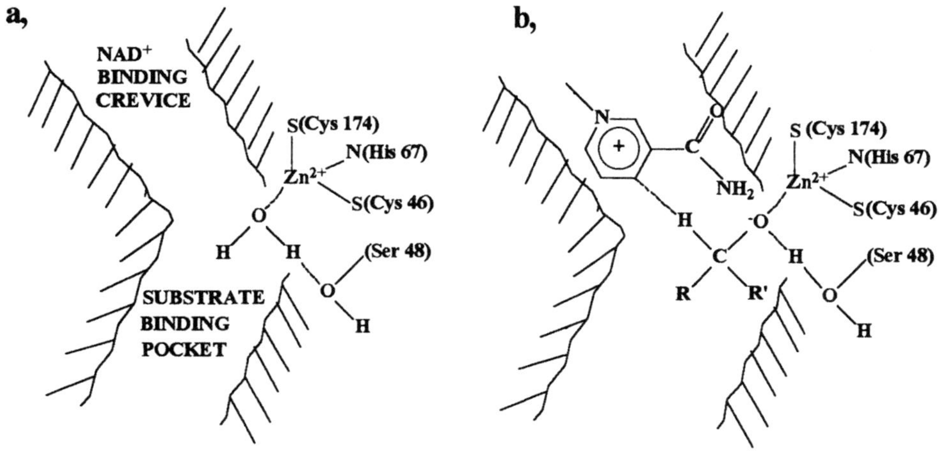

Figure 21. Models of the LADH active site in the initial state (a) and with the bound ternary complex (b).

3.8.3. *Isocitrate dehydrogenase*

Isocitrate dehydrogenase (IDH) is a key enzyme of the Krebs cycle and appears in species from man to bacteria. It catalyses dehydrogenation of isocitrate at the α-carbon atom to form a carbonyl group from a hydroxyl group and subsequent decarboxylation of the β-carbon atom (Hurley et al., 1991). The likely mechanism of action is depicted in Figure 22. The reaction is believed to occur in two steps. In the first, isocitrate is oxidised to oxalosuccinate by the removal of a proton from the hydroxyl oxygen to a base (Asp-283') and the transfer of a hydride to $NADP^+$. In the second step of the reaction, the β-carboxylate of oxalosuccinate is lost as carbon dioxide which is followed by stereospecific protonation of the β-carbon to form α-keto-glutarate. Tyr-160 and Lys-230' are hydrogen-bonded to the β-carboxylate of isocitrate where they are favourably positioned to serve as the acid catalyst protonating C^3 after decarboxylation.

The first indication that electrostatics plays an important role in IDH catalysis was that it can be inactivated by phosphorylation, an effect mediated by the negative charge of the phosphate (Thorsness and Koshland, 1987). Later on Hurley and Remington (1992) investigated the origin of the catalytic effect of the enzyme considering the protein dipoles explicitly, while treating the solvent and non-explicit contributions of the protein by a discretised continuum model. They concluded that the electrostatic field in the active site of IDH favours the transition state by 64 kJ/mol over the ground state. The increase in electrostatic transition-state stabilisation, as compared to the aqueous solution, is 54 kJ/mol. The nearby metal ion, Mg^{2+}, contributes to the stabilisation by 111 kJ/mol and this value is reduced by -47 kJ/mol due to the contribution of the enzyme charges.

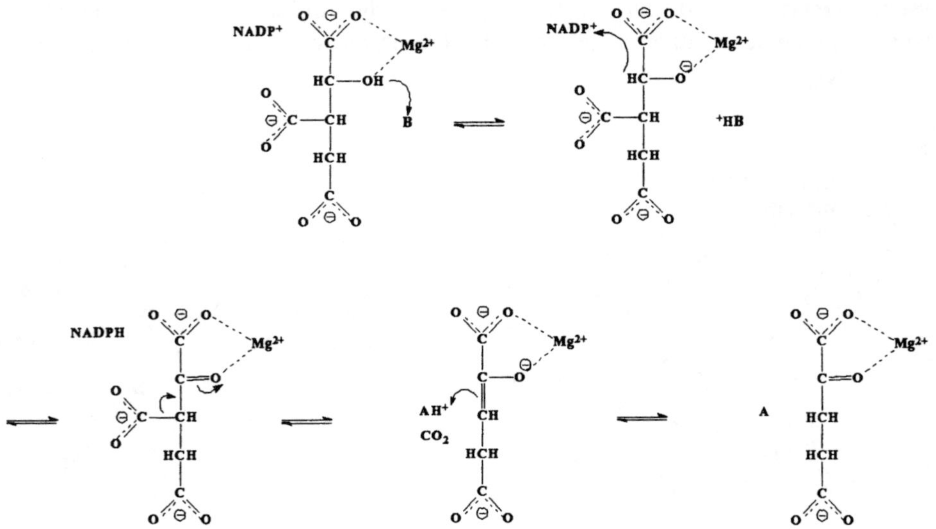

Figure 22. Likely mechanism for the reaction catalysed by IDH (Hurley et al., 1991). B is probably Asp-283', A is probably either Tyr-160 or Lys-230'.

The above authors treated the protein and solvent regions by a reasonable approximation and did consider correctly a reference reaction in solution. This led to the conclusion that the enzyme reaction is entirely different than the corresponding gas-phase reaction, in agreement with concepts introduced by Warshel and co-workers (Warshel et al., 1989a). However, other interpretations of what was basically a correct calculation are questionable. That is, as pointed out above, Hurley and Remington (1992) found that the Mg^{2+}-ion strongly stabilises the transition state and that the ionised groups of the protein reduce this stabilisation. Thus, they argued that the enzyme *destabilises* the transition state. This conclusion is apparently based on the artificial separation between the effect of the Mg^{2+}-ion and the other charges of the protein. As argued and demonstrated by Åqvist and Warshel (1989, 1990), what counts is the combined effect of the overall dipoles and charges of the enzyme that includes of course the Mg^{2+} cation. The enzyme uses this ion as the primary electrostatic factor but since ions are not stable in a non-polar environment it must stabilise the cation by its surrounding which involves in the present case negatively charged residues. Of course, these residues screen some of the effect of the metal cation but the combined effect of the metal ion and its surrounding strongly *stabilise* rather than destabilise the transition state. More importantly, even if one defines the reacting system as the substrate plus the Mg^{2+}-ion (as was done by Hurley and Remington, 1992), one will find that the enzyme (without the Mg^{2+}-ion) *stabilises* the transition state more than water does. We mention this point in order to illustrate that the elucidation of the reason why enzymes are so effective does not depend on semantics and can be established uniquely, once one uses correctly a well-defended thermodynamic cycle. When this is done, one finds that regardless of the definition used *the enzyme stabilises the transition state more than water does* and that the change in "solvation" going from water to the enzyme active site is larger in the ground state than in the transition state. Perhaps Hurley and Remington (1992) concluded that the enzyme destabilises the transition state since they overlooked the fact that water *destabilises* the transition state of the enzyme-Mg^{2+} system much more than the enzyme does. To reiterate the above points, *any* definition of the reacting system even when some enzyme charges are included in this system (as was done by Warshel, 1978, 1981) will place the transition state of the reacting system at a lower energy in the enzyme than in the corresponding solvent cage. Similarly, the activation free energy in the enzyme will always be lower than the corresponding energy in solution. This is due to the fact that the surrounding of the reacting system must stabilise the transition state more in the enzyme than in solution in order to catalyse the reaction.

It is also important to use this opportunity to comment about the suggestion (Hurley and Remington, 1992) that isocitrate dehydrogenase catalyses its reaction due to the low polarisability and strong field of its active site. The authors used in their calculation a "low dielectric" parameter but as demonstrated and explained repeatedly by Warshel and co-workers (e.g. Åqvist and Warshel, 1992; King et al., 1991), this parameter has very little to do with the dielectric constant of the system when the effects of the protein polar groups and the surrounding water are treated explicitly. Models that use a truly macroscopic approach (without treating dipoles explicitly) and assume a low dielectric constant as was indeed done by all continuum treatments until recently (cf. e.g. the discussion by Warshel, 1987) will lead to the incorrect conclusion that charged groups would never stay in the interior of proteins. Obviously, the Mg^{2+}-ion is bound strongly to the active site since it is vary polar rather than apolar.

Apparently, the overall catalytic effect in isocitrate dehydrogenase (as in many other enzymes) is due to the electrostatic stabilisation by groups with small reorganisation energy rather than to charges at low-polarisability sites. As to the effect of strong fields, in many enzymes the strong catalytic fields are due to the dipoles of the polar environment and when such fields are excerpted by charges, it is essential that these charges will be solvated by a polar environment, otherwise the ionisable group will not be ionised and metal ions will not bind.

3.9. ACETYLCHOLINESTERASE

Acetylcholinesterase (AChE) catalyses the hydrolysis of the ester bond of acetylcholine to yield choline and acetate (Sussman et al., 1991). This is a critical reaction for the termination of impulses transmitted through cholinergic synapses. It is a highly efficient catalyst, with reaction rates approaching the diffusion limit. Its overall structure resembles the lipases with an active site gorge. Above the base of the gorge is the reactive serine to be activated by the classical (Ser-200...His-440...Glu-327) catalytic triad.

The reaction between AChE and its substrate is diffusion controlled, i.e. the rate-limiting step is the encounter by diffusion between the enzyme and its substrate or inhibitor (Rosenberry, 1975). There is evidence that electrostatic forces may influence the rate of this encounter (Nolte et al., 1980). Tan et al. (1993) calculated the MEP around the AChE dimer and used it in Brownian dynamics simulations of the diffusion of the substrate, *N*-methylacridinium, towards the enzyme. Their results indicate that electrostatic steering of ligands makes a substantial contribution to the efficiency of AChE. Results for differently charged models indicate that if all enzyme charges are considered the rate increases by two orders of magnitude which suggests that ionic strength effects are important in the enzymatic process. It has to be mentioned, however, that the enzymatic effects of changing the diffusion rate are very small relative to the enormous effect of the protein active sites (Warshel and Åqvist, 1991). After the protein changes the rate by more than 10 orders of magnitude and the reaction becomes diffusion controlled the small effect of surface charges can modulate fine details of the process.

While the actual electrostatic contribution to AChE catalysis has not yet been studied, their effects on ligand binding are well established. Electrostatic calculations based on the crystal structure of AChE indicate that the enzyme has a strong dipole that is aligned with the gorge leading to its active site (Ripoll et al., 1993). Thus, a positively charged substrate will be drawn to the active site by the enzyme electrostatic field, like in *Geotrichum candidum* lipase (Section 3.2.) and in superoxide dismutase (Section 3.10.). Within the gorge, aromatic side chains appear to shield the substrate from direct interaction with most of the negatively charged residues that give rise to the dipole. The affinity of quaternary ammonium molecules for aromatic rings, coupled with this electrostatic force, may work in concert to create a selective and efficient substrate-binding site in AChE and explain why the active site is situated in the bottom of a deep gorge lined with aromatic residues.

3.10. SUPEROXIDE DISMUTASE

Aerobic metabolism generates toxic superoxide anions. The enzyme superoxide dismutase (SOD) catalyses the dismutation of two O_2^- molecules to oxygen and hydrogen peroxide removing the toxic anion and preventing the oxidative damage to the organism. SOD enzymes can be divided into two categories. One type uses a single metal ion (either Mn or Fe), the other two (Cu and Zn) and there is no sequence or structural analogy between these two enzyme classes. In the following we deal exclusively with (Cu,Zn) SOD, we are not aware of any study on the electrostatics of the first category.

SOD is a dimeric protein of 32 kD molecular weight, each monomer containing one copper and one zinc ion bridged by a histidine side chain. The copper ion is the catalytic centre. The active site is depicted in Figure 23. on the basis of X-ray structure determination (Tainer et al., 1982). Besides the metal ligands, a particularly important residue present at the entrance of the active-site cavity is an arginine (Arg-143) which is present in all known natural isoenzymes. The distinctive structural property of the active site is the bridge between the copper and the zinc ions through the imidazolate of His-63 (cf. Figure 23.).

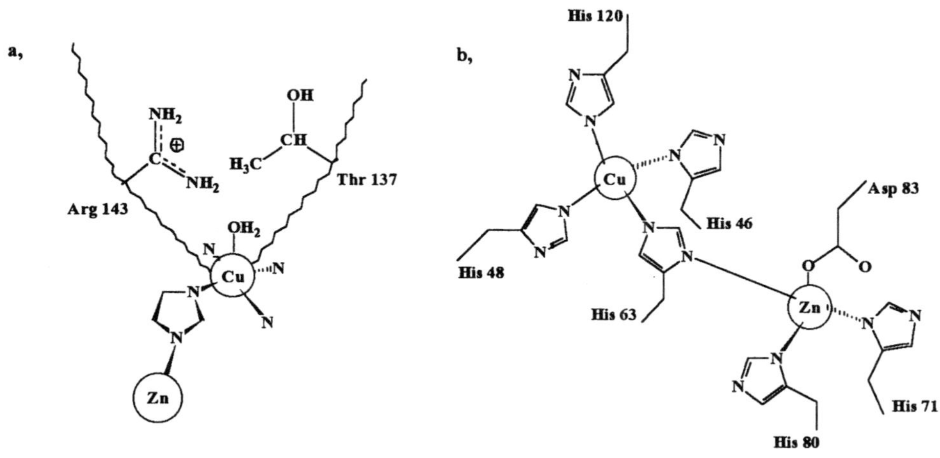

Figure 23. Schematic drawing of the SOD active site (a) and of the copper-zinc site (b). Large empty circle: Cu, large cross-hatched circle: Zn.

The details of the active-site structure and several further experiments led to a proposition for the mechanism of actions of SOD (for a review cf. Strothkamp and Lippard, 1982). It was suggested that reduction of the copper is accompanied by displacement and protonation of the bridging imidazolate side chain. Reducing the copper results in the uptake of one proton per subunit which is consistent with the proposed mechanism. The proposed role of zinc is to attenuate the pK of the His-63 side chain so that it will bind Cu(II) instead of a proton but will bind a proton in preference to Cu(I). The bridging ligand may also serve as a proton donor to the developing peroxide dianion in the second step, leading to the production of HO_2^- and reforming the imidazolate bridge.

In contrast to the above mechanism, Osman and Basch (1984) suggested that Arg-143 is crucial in catalysis by acting as general base (see Figure 24.). On the basis of molecular orbital calculations they suggested that the enzyme does not undergo a sequential reduction and oxidation by superoxide. rather it forms a relatively stable complex with superoxide (structure II in Figure 24.). Complexation with SOD is made possible primarily due to the presence of Arg-143, and it causes an activation of the superoxide anion. Superoxide is usually a weak oxidant because its reduction requires the formation of a doubly charged anion; it is also an inefficient hydrogen atom abstractor. It might be activated by the presence of the arginine side chain, thus in an indirect way it can act as an oxidant. The crucial role of Arg-143 in catalysis was also emphasised on the basis of mutagenesis studies, where its replacement by Ile, Lys or Glu resulted in a sharp decrease of activity (Beyer et al., 1987). Note, however, that the calculations of Osman and Basch (1984) have not considered the dielectric effect of the protein and therefore could not be used in any conclusive examination of the role of Arg-143 or in an attempt to determine whether or not this group really serves as a general base. Nevertheless, drawing attention to this residue has been quite useful and the large effect of mutating this group clearly demonstrated that the enzyme does much more than just attracting the superoxide to its site.

Experimental evidence in favour of an enzyme-substrate recognition process based on electrostatic interactions has been provided by selective chemical modifications (Cudd and Fridovich, 1982; Cocco et al., 1982) and by investigation of the dependence of the activity of SOD on pH and ionic strength (Argese et al., 1987). The electrostatic contribution to enzyme-substrate recognition in SOD was analysed by computer graphics (Getzoff et al., 1983). The MEP map reveals a striking positive region extending over the long, deep gorge above the catalytic copper ion forming a complementary binding site for the negatively charged superoxide substrate. Electrostatic guidance is important in attracting the negatively charged superoxide ion into the active site of SOD (Getzoff et al., 1983; Sines et al., 1990).

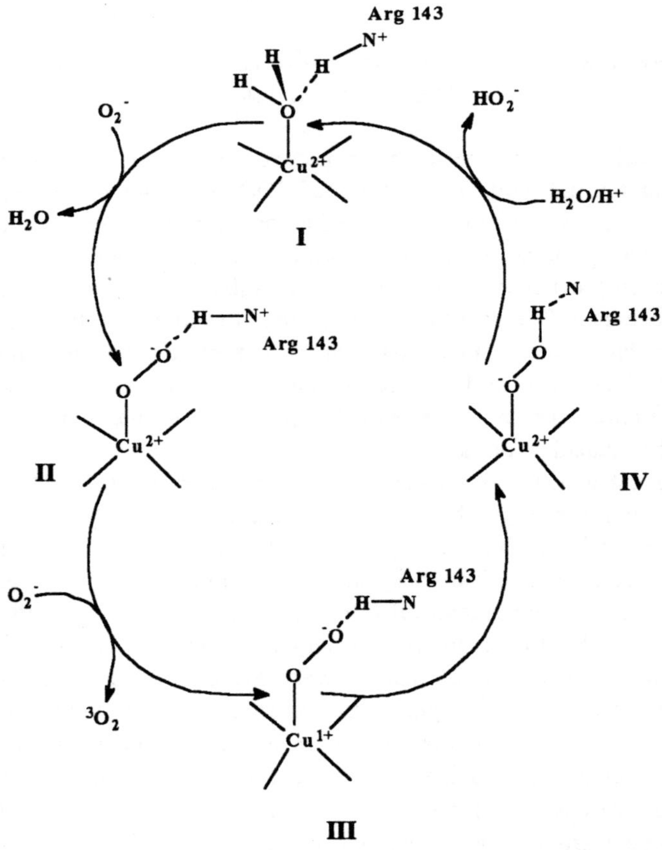

Figure 24. Reaction mechanism for SOD action as proposed by Osman and Basch (1984).

The reaction rate of the SOD...O_2^- system is quite high and decreases with increasing ionic strength of the solvent (Cudd and Fridovich, 1982; Argese et al., 1987). Because the active site represents a small fraction of the surface area of the dimeric enzyme and the electrostatic charges of the enzyme and substrate are of the same sign, it has been recognised that the approach of the substrate to the active site is likely facilitated by the enzyme electrostatic field. These results were examined by Brownian dynamics simulations (that was developed to calculate the rate at which reactant molecules diffusing in solution would collide with the appropriate orientations for reaction) of a model that represented the enzyme as a simple sphere (Allison and McCammon, 1985). Later studies of SOD reactivity involve fairly realistic models of the molecular shapes and interactions (Sharp et al., 1987; Allison et al., 1988). The calculated rate and its variation with salt concentration appeared to be in reasonable agreement with the corresponding experimental results.

Desideri et al. (1989; 1992) compared the MEP of six (Cu,Zn) SOD species displaying large differences on the net protein charge and distribution of the ionisable surface side chains. They found that the MEP pattern in the proximity of the active sites is constant in all cases. Note the analogy with the same trend for serine proteases, discussed in Section 2.3. Thus, for both classes of enzymes a general hypothesis for their evolution is suggested. It appears that the evolutional development of different species has maintained a functional spatial network of charges on the protein surface (and pre-oriented dipoles inside the protein) that results in very similar electric fields around the active sites. Apparently, the evolutional process optimises the ability of SOD to attract negatively charged molecules of small size at a diffusion-controlled rate into its active site. Beside metal-liganding side chains and the Arg-141 side chain adjacent to the metal, also a certain MEP pattern around the active site is a constant feature in the molecular evolution of SOD species.

It should be noted, however, that the effect of the enzyme in changing the diffusion rate is by far smaller than the catalytic effect of the active site on the reacting fragments. The evolution of the enzyme led probably to an active site that is able to reduce the activation barrier by so much that the reaction became diffusion controlled. It seems to us that the origin of the reduction of the activation barrier is the real question, rather than the modulation of the diffusion rate. As much as the reduction of the activation barrier is concerned it is very likely that the enzyme accomplished this important task by its electrostatic field.

3.11. RIBULOSE-1,5-BISPHOSPHATE CARBOXYLASE

This enzyme (Rubisco in a shorthand notation) catalyses the initial step in the photosynthetic fixation of atmospheric carbon dioxide, the carboxylation of ribulose-1,5-bisphosphate (RuBP), yielding two molecules of phosphoglycerate (Andrews and Lorimer, 1987). It can also catalyse the oxygenation of RuBP by utilising atmospheric oxygen instead of carbon dioxide as the gaseous substrate. This reaction initiates the photorespiratory pathway leading to the formation of carbon dioxide and the release of energy. In higher plants Rubisco contains of eight large (L with M = 52-55 kD) and eight small (S with M = 14 kD) subunits, which form an L_8S_8 complex. The catalytic activities reside on the large subunit. Both the oxygenation and the carboxylation reaction require activation of the enzyme, i.e. carbamylation of the Lys-191 side chain at the active site by a carbon dioxide molecule differing from the substrate. The carbamate is then stabilised through the binding of a Mg^{2+} ion (Lorimer, 1981).

The overall catalytic mechanism of Rubisco is depicted in Figure 25. After the enzyme is activated by formation of carbamate at Lys-191 the first, and sometimes the rate-limiting, step is enolisation which is followed by carboxylation or oxygenation. The reaction mechanism has been studied by Tapia and co-workers by molecular orbital methods (Andrés et al., 1992; 1993; Tapia and Andrés, 1992; Tapia et al., 1994a). However, their calculations considered only the reacting fragments in the gas phase and therefore could not be used to deduce the catalytic effect of the enzyme.

MEP calculations on non-activated Rubisco from two different sources, as well as some of its complexes with activator and substrate indicated the special importance of electrostatics in enzymatic action (Lu et al., 1992). The overall MEP of the enzyme

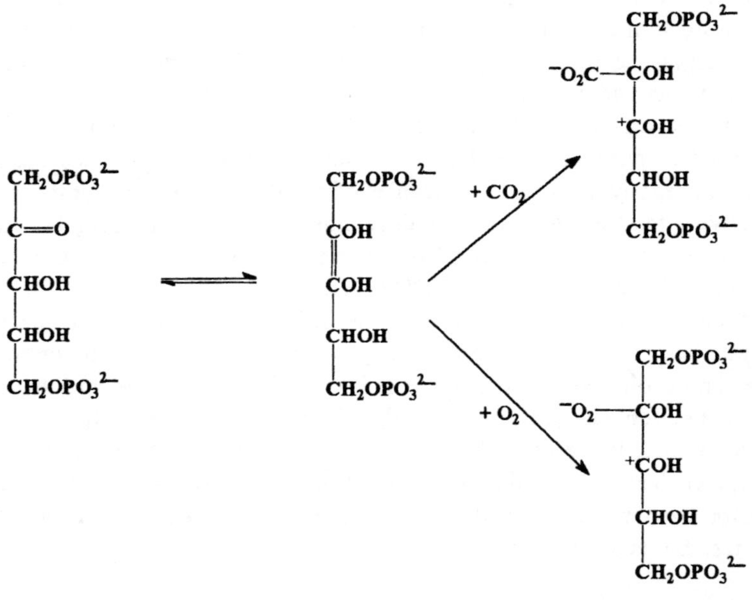

Figure 25. Schematic reaction mechanism for the carboxylation of RuBP by Rubisco (Andrews and Lorimer, 1987).

from the photosynthetic bacterium *Rhodospirillium rubrum* shows that the core of the dimer, consisting of the two C-terminal domains, has a predominantly positive potential. These domains provide the binding sites for the negatively chargedphosphate groups of the substrate. The two N-terminal domains have mainly negative potential. At the active site, situated between the C-terminal domain of one subunit and the N-terminal domain of the second subunit, a large potential gradient (MEF) at the substrate-binding site is found. This might be important for polarisation of chemical bonds of the substrate and enhancement of transfer of protons during catalysis. The close environment of the activator Lys-191 provide a positive MEP region which might lower the pK value of this residue. This observation suggests that the electrostatic field at the active site is responsible for the specific carbamylation of the ε-amino group of the activated enzyme. The overall shape of the MEP in the L_2 building block of the Rubisco from spinach is, despite only 30 per cent amino-acid homology for the L-chains, similar to that of the same type from *Rhodospirillium rubrum*, suggesting that a similar electrostatic evolution took place for Rubisco as for various species of serine proteases (Sec. 3.1.1.), superoxide dismutase (Sec. 3.10.) and lysozyme (Sec. 3.4.).

3.12. XYLOSE ISOMERASE

D-xylose isomerase catalyses the reversible conversion of D-xylose to D-xylulose, and it is capable of converting other sugars from aldose to ketose (Yamanaka, 1968). The enzyme requires bivalent metal cations (Mg^{2+}, Co^{2+} or Mn^{2+}) for activation while Zn^{2+}, Ca^{2+}, Ba^{2+} and Cu^{2+} inhibit catalysis (Callens et al., 1988). On the basis of crystallographic data (Collyer et al., 1990; Rangarajan and Hartley, 1992) and molecular orbital calculations (Zheng et al., 1993, Fuxreiter et al., 1995) the probable mechanism is outlined in Figure 26. The xylose-xylulose reaction takes place in two steps, the opening of the α-pyranose ring followed by isomerisation of the acyclic intermediate. On the basis of kinetic studies Rangarajan and Hartley (1992) concluded that a significant energy barrier appears for the not for the ring opening but for the hydride transfer. This statement has been confirmed by molecular orbital calculations on large models partly including environmental effects (Fuxreiter et al., 1995).

As it is seen from Figure 26.a (- + -) charge distribution, a common pattern in several enzymes (cf. Section 2.3.) emerges during the ring opening reaction step with the participation of Asp-57, His-54 and the ionised substrate (Fábián et al., 1994). While His-54 acts as a general base mediating proton from O^1 to O^5 of the substrate, Asp-57 does not participate directly in the reaction. Its role is to stabilise the protonated imidazole formed during the process. By performing calculations on models with larger radius around the transition-state complex lower activation energies were obtained for the rate-limiting step. This indicates the importance of electrostatic stabilisation by the protein environment (Fuxreiter et al., 1995). Another indication for the importance of electrostatic effects is that charge transfer to the catalytic metal ion, that is small for activating and large for inactivating metal ions, was found to be crucial for the reaction. Large charge transfer diminishes the catalytic stabilisation of the transition state by reducing the net charge on the metal and thus its electrostatic effect.

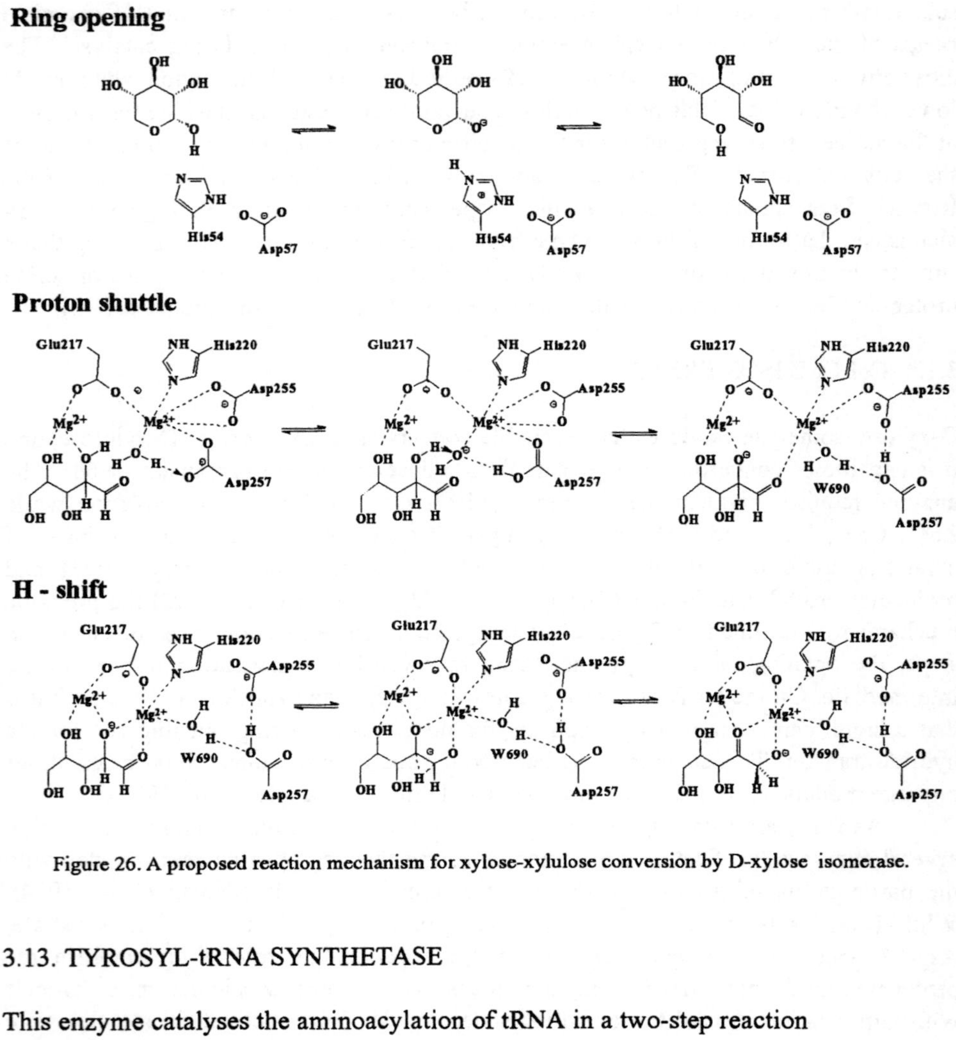

Figure 26. A proposed reaction mechanism for xylose-xylulose conversion by D-xylose isomerase.

3.13. TYROSYL-tRNA SYNTHETASE

This enzyme catalyses the aminoacylation of tRNA in a two-step reaction

$$E + Tyr + ATP \leftrightarrow E.Tyr\text{-}AMP + \tag{4}$$

$$E.Tyr\text{-}AMP + tRNA^{Tyr} \rightarrow Tyr\text{-}tRNA^{Tyr} + AMP + PP_i \tag{5}$$

Activation in step 1, Eq. (4) is followed by transfer according to Eq. (5). The enzyme is a symmetrical dimer of molecular weight of 2×47.5 kD, the schematic view of the active site for binding of tyrosyl adenylate is displayed in Figure 27. (Rubin and Blow, 1981). As Fersht et al. (1985) pointed out hydrogen bonding (known to be electrostatic in nature, cf. e.g. Náray-Szabó et al., 1986) is crucial in the enzymatic process. Replacement of one or another side chain hydrogen-bonded to the substrate by non-hydrogen bonding residues (by site-directed mutagenesis) causes a rate decrease between a factor of 2 and 1,000. Most important determinants of binding are Gln-195 and Tyr-169. As it can be seen from Figure 27. hydrogen bonds considerable contribute

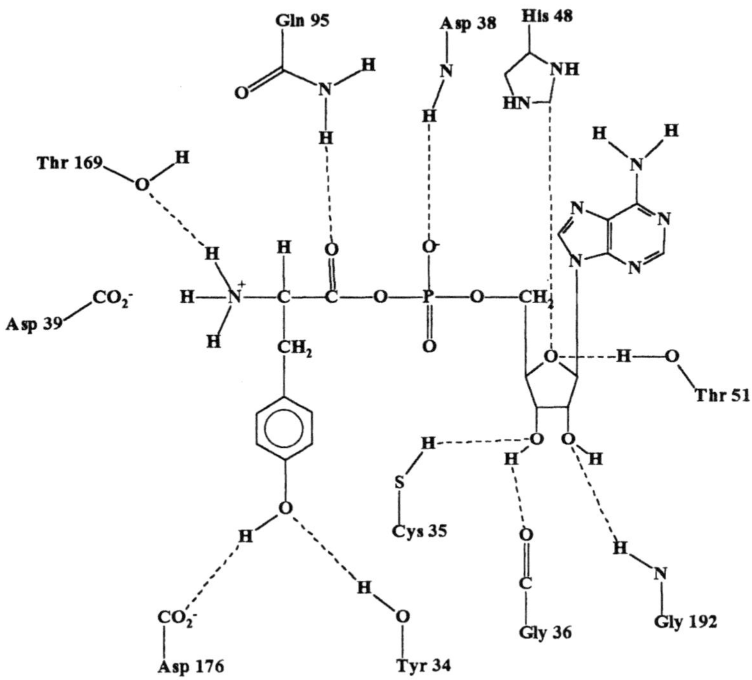

Figure 27. Schematic view of the active site of tyrosyl tRNA synthethase with bound substrate. Hydrogen bonds are indicated by dashed lines.

to complementarity by providing an "electrostatic lock" for the substrate. These studies were the first where an enzyme with known three-dimensional structure could be studied by protein engineering. As we have seen in preceding sections, site-directed mutagenesis proved to be a powerful tool in detecting and characterising electrostatic effects on enzyme catalysis.

The study of Fersht et al. (1985) provided an interesting Linear Free Energy Relationship (LFER) between the catalytic effect of the enzyme with its effect on the free energy of the reaction. The theoretical basis of this effect has been provided by a recent work (Warshel et al., 1994).

3.14. RIBONUCLEASE

Bovine pancreatic ribonuclease (RNase) catalyses the hydrolysis of RNA by a two-step process in which a phosphate intermediate is formed (Richards and Wyckoff, 1971). The enzyme consists of a single polypeptide chain of molecular weight of 13.7 kD. The bond between Ala-20 and Ser-21 may be cleaved by subtilisin, but the resulting peptide remains attached to the rest of the protein by non-covalent bonds. The modified protein, called RNase S, and the native protein, termed RNase A, have identical catalytic activities. The probable mechanism of these enzymes (Roberts et al., 1969) is depicted in Figure 28. The first step is a transphosphorylation process that involves addition of the 2'-OH group on the 3'-ribose to the phosphate group and a cleavage of the ribonucleic acid chain at the 5'-end, yielding a 2',3' cyclic phosphate and a free 5'-OH group. In the second step, hydrolysis, involves addition of water to the cyclic intermediate, yielding a terminal 3'-phosphate monoester. In the transphosphorylation step a charge shift occurs as proceeding from the ground state towards the transition state (cf. Figure 29.). In the first stage the substrate is bound to the enzyme and in the

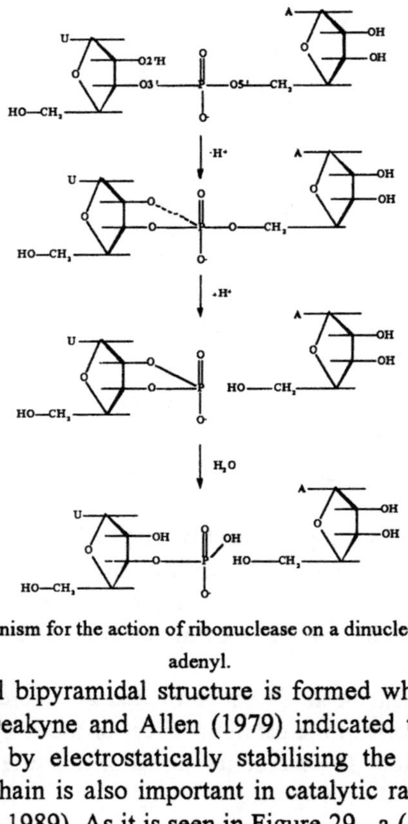

Figure 28 . The probable mechanism for the action of ribonuclease on a dinucleotide substrate. U = uridyl, A = adenyl.

transition state a trigonal bipyramidal structure is formed which breaks down to the product. The study of Deakyne and Allen (1979) indicated that Lys-41 may play a crucial role in catalysis by electrostatically stabilising the transition state and the proximal Asp-121 side chain is also important in catalytic rate acceleration (Ressler, 1982; Náray-Szabó et al., 1989). As it is seen in Figure 29., a (- + -) charge distribution is formed in the transition state within the Asp-121. .His-119...substrate triad that is similar to the electrostatic pattern found in many other enzymes (cf. Figure 5.). Since both Lys-41 and Asp-121 play important roles in catalysis, it is interesting to examine the effect of the K41A and D121A mutations on the catalytic rate using a simple electrostatic approach. This was done by Náray-Szabó et al. (1989) who found using simple electrostatic calculations that while the rate should increase in a hypothetical D121A mutant ($\Delta \log k_{cat} = 0.9$), the K41A mutation probably decreases the rate by

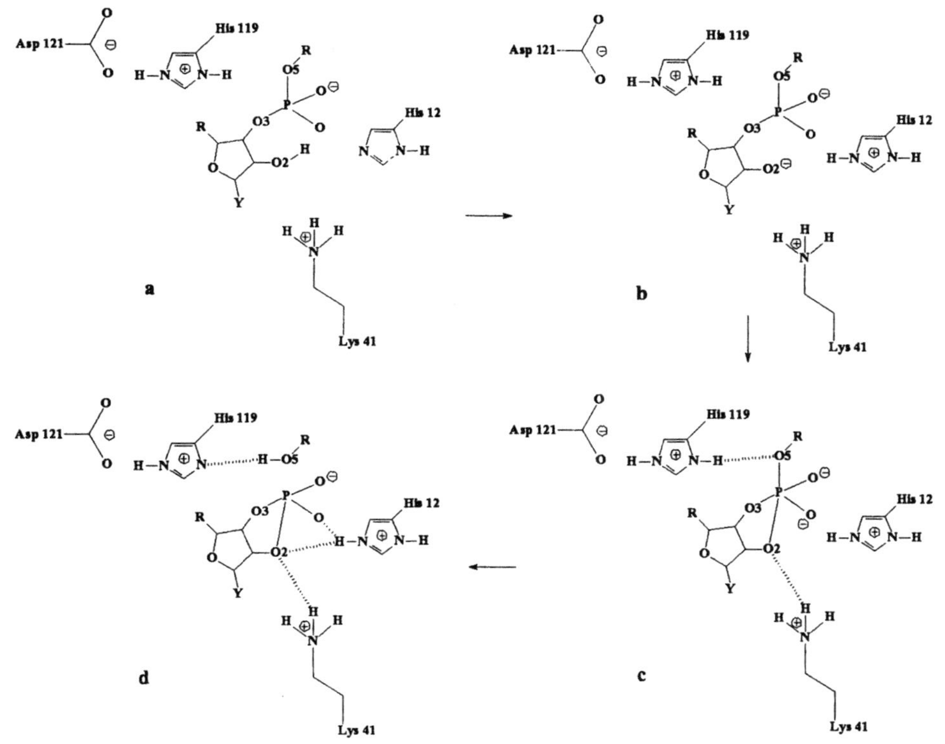

Figure 29. Geometric representation of the ground (a), transition (c) and final (d) states in the trans-phosphorylation step of the RNase catalysed reaction. N denotes nucleoside.

about two orders of magnitude ($\Delta \log k_{cat}$ = -1.7). A simple explanation for this finding is that the (+ -) pair representing the His...substrate couple in the ground state changes to a (+ 2- +) pattern in the transition state. In other words, a (- +) dipole is switched on in the transition state and is destabilised by the negatively charged Asp-121 side chain lying closer to the negative end. In the D121A mutant the monopole-dipole repulsion vanishes thus the transition state becomes stabilised, i.e. k_{cat} increases. On the other hand, in the K41A mutant the Lys positive charge is switched to zero, so the stabilisation of the (- +) dipole and the catalytic rate decreases.

3.15. CYTOCHROME C PEROXIDASE

Cytochrome C peroxidase (CCP) is a 294-residue haem-containing enzyme which catalyses the reduction of peroxides by ferrocytochrome *c*. The exact function of CCP is somewhat obscure but it probably serves to protect yeast mitochondria from the toxic build-up of peroxides. The overall scheme for the enzymatic reaction is given below (Pelletier and Kraut, 1992). In step 1, Eq. (6) CCP reacts with hydrogen peroxide to

$$CCP(Fe^{III}) + H_2O_2 \rightarrow \text{compound-I}(Fe^{IV}O^{2-}, \text{Trp-191}^{\bullet}) + H_2O \qquad (6)$$

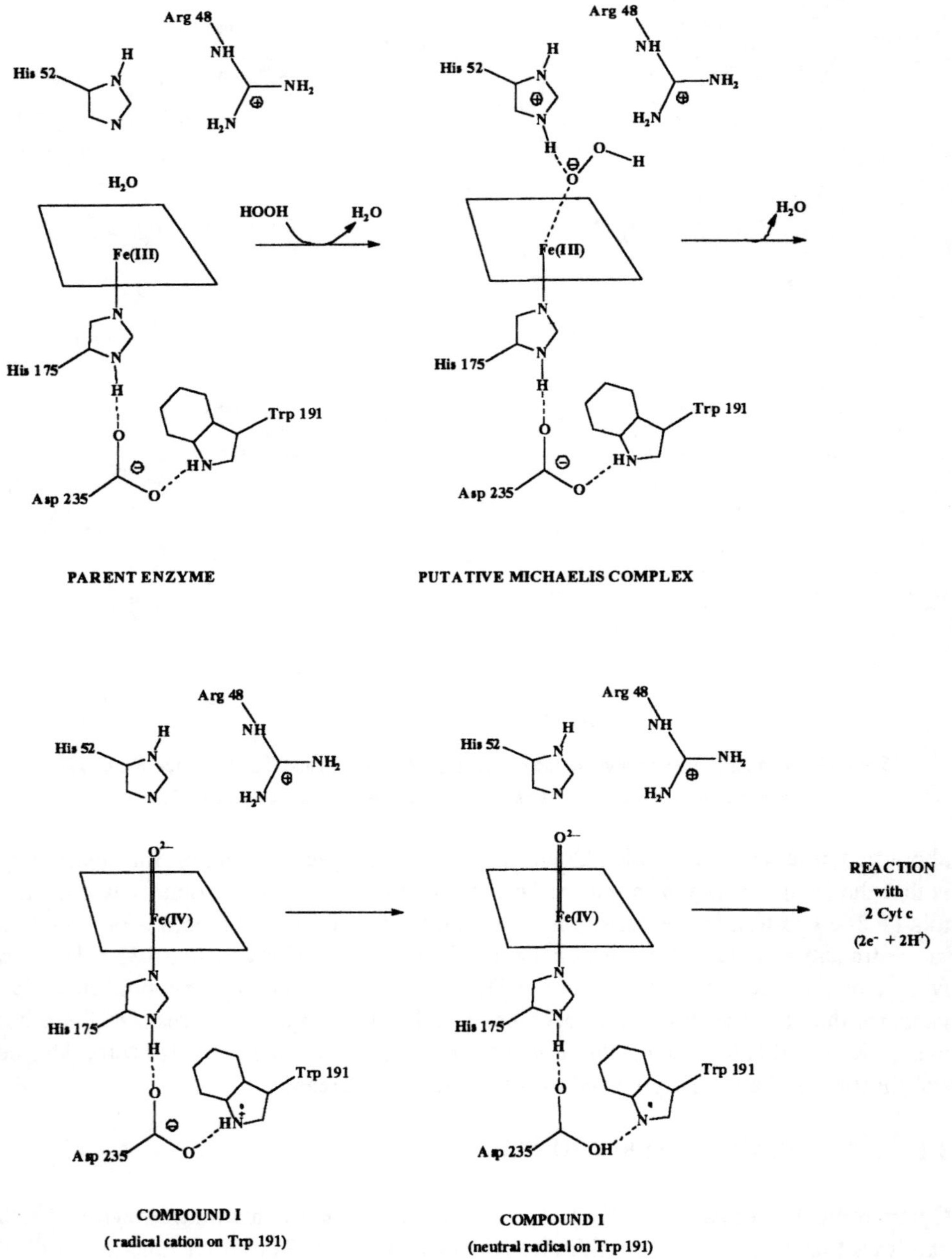

Figure 30. Schematic reaction of formation of compound I in CCP with parent enzyme (a), putative Michaelis complex (b), and compound I (c).

generate a doubly oxidised enzyme intermediate (called compound I) containing an oxyferryl group (Chance et al., 1986) and a free radical associated with Trp-191 (Sivaraja et al., 1989). In the second step this intermediate reacts with a molecule of ferrocytochrome *c* to form compound II (Equation 7). At last, in step 3 (Equation 8) compound II obtains another electron from a second molecule of ferrocytochrome *c*

$$\text{compound-I}(\text{Fe}^{IV}\text{O}^{2-},\text{Trp-191}^{\bullet}) + \text{cyt-}c(\text{Fe}^{II}) + \text{H}^{+} \rightarrow \\ \rightarrow \text{compound-II}(\text{Fe}^{IV}\text{O}^{2-}) + \text{cyt-}c(\text{Fe}^{III}) \qquad (7)$$

$$\text{compound-II}(\text{Fe}^{IV}\text{O}^{2-}) + \text{cyt-}c(\text{Fe}^{II}) + \text{H}^{+} \rightarrow \text{CCP}(\text{Fe}^{III}) + \text{cyt-}c(\text{Fe}^{III}) + \text{H}_2\text{O} \qquad (8)$$

and it returns to the ground state. The scheme for the formation of compound I is depicted in Figure 30 (Poulos and Kraut, 1980; Fülöp et al., 1994).

There was a long debate about the location of the free radical. While earlier it was thought that it is the haem ring which looses an electron during reaction (Chance et al., 1986), more recent studies indicate that it is the indole ring of Trp-191 (Mauro et al., 1988; Sivaraja et al., 1992). Miller et al. (1994) used discretised continuum calculations (see e.g. Sharp and Honig, 1990) for this enzyme and concluded that the protonated radical cation state of Trp-191 is stabilised by the strong negative MEP in its vicinity. However, it is unclear whether this stabilisation effect is enough for pushing the proton from Asp-235 to Trp-191. Clearly, it is protein electrostatics that determines the site of radical formation because in the gas phase the Asp-235...Trp-191 couple is more stable in its neutral form (Krauss and Garmer, 1993). A similar effect was found for serine proteases where theoretical calculations gave evidence that in the gas phase the neutral form of the Asp...His dyad is favoured to the ion pair and it is just the protein environment that stabilises the latter (Náray-Szabó and Polgár, 1980; Umeyama et al., 1981; Warshel et al., 1982; Warshel and Russell, 1986).

3.16. OTHER ENZYMES

The concept of electrostatic catalysis has been applied indirectly by Schramm and co-workers for the design of transition-state analogues in reactions by nucleoside hydrolase (Horenstein et al., 1991; Horenstein and Schramm, 1993) and AMP deaminase (Kline and Schramm, 1994). They stated, in accordance with our previous examples, that the MEP provides novel insight into transition-state structure and the forces associated with substrate binding and release.

Nucleoside hydrolase has been proposed to participate in purine salvage in the trypanosome *Crithidia fasciculata*. The enzyme hydrolyses the N-glycosidic linkage of the naturally occurring purine and pyrimidine nucleosides. A geometric model of the transition state for nucleoside hydrolase for the reaction (Horenstein et al., 1991)

$$\text{inosine} + \text{H}_2\text{O} \rightarrow \text{hypoxanthine} + \text{ribose} \qquad (9)$$

indicated that there is substantial oxycarbonium ion character in it. Horenstein and Schramm (1993) applied MEP calculations to transition-state structures on the basis of experimentally determined kinetic isotope effects and deduced some unique features of the transition state. The results indicate that nucleoside hydrolase could provide electrostatic stabilisation of glycosyl-cation transition states by interactions at sites other than the oxonium group. This statement is supported by the inability of the enzyme to hydrolyse 2'-, 3'-, and 5'-deoxynucleosides (Parkin et al., 1991).

AMP deaminase is a zinc-containing enzyme that catalyses the deamination reaction

$$AMP + H_2O \rightarrow IMP + NH_3 \quad (10)$$

This enzyme contains a single AMP binding site, a single zinc, and two AMP regulatory sites per monomer in the tetrameric enzyme (Merkler and Schramm, 1990). The enzyme is inhibited by purine analogues that contain a tetrahedral secondary alcohol of *R*-configuration at a position equivalent to C^6 of the adenine. The proposal that these inhibitors are transition-state analogues is supported by the X-ray crystal structure for mouse adenosine deaminase. (*R*)-Coformycin, (*R*)-2'-deocoformycin 5'-mono-phosphate and (*R*)-coformycin 5'-monophosphate are powerful transition-state inhibitors and comparison of the MEP of the N-3 derivative of the (*R*)-coformycin ring reveals that it resembles the early transition state more closely than the amino-protonated intermediate. Protonation of the exocyclic amino group introduces substantial changes in the MEP thus it no longer resembles that of the (*R*)-coformycin ring. These findings are consistent with the structure of the transition state deduced from kinetic isotope effects.

A similar analysis has been performed for the complex between *Enterobacter cloacae* β-lactamase (Lobkovsky et al., 1994). These enzymes catalyse the hydrolysis of β-lactam antibiotics causing bacterial resistance against these drugs. The clinically important β-lactamases are serine hydrolases acting by a double displacement mechanism involving an acyl-enzyme intermediate like the one in serine proteases (cf. Section 3.1.1.). It is known that phosphonate monoesters, $RCONHCH_2PO_2^-$-OAr, are appropriate transition-state analogues. The anionic phosphonyl derivative of the enzyme, E-O-PO_2^--CH_2NHCOR, should be structurally analogous to the tetrahedral intermediates and associated transition states in turnover of analogous acyclic substrates. MEP calculations were performed on three serine hydrolases: γ-chymotrypsin, the class A β-lactamase of *Bacillus licheniformis* and the class C β-lactamase of *Enterobacter cloacae*. The results indicate that the positive MEP is larger in the case of the β-lactamases than chymotrypsin. This difference may explain the susceptibility of β-lactamases to phosphonate monoester monoanions. Subtle differences between inhibition of class A and class C β-lactamases, however, cannot be that well rationalised by the calculated MEP.

In a recent paper Knighton et al. (1994) proposed an "electrostatic highway" that channels the substrate across the surface of the protein in the bifunctional thymidylate synthase-dihydrofolate reductase (TS-DHFR) complex. The folate binding sites of TS and DHFR are 4 nm apart and this is too far for direct transfer of substrate from one active site to the other without a major conformational change in the protein. Since dihydrofolate has a net charge of -2 and becomes further negatively charged by the addition of several glutamates, the negatively charged folates should be sensitive to the electrostatic field around the enzyme. The MEP on the envelope of the TS-DHFR complex clearly shows a strongly positive electrostatic "highway" (Stroud, 1994) across the surface linking both active sites. Perhaps the overall positive MEP between the two active sites surrounded by a generally repulsive negative MEP promotes a guided diffusion from one site to another, though there is another possible mechanism that involves a simple torsional rotation of the Arg-287 side chain in the TS domain that could position its guanidinium group 1 nm from the side-chain nitrogen of Lys-73 in the DHFR domain.

4. Concluding Remarks

This chapter reviewed computer modelling studies of enzymatic reactions focusing on electrostatic contributions. Evidently, there are many ways to estimate electrostatic effects ranging from simple MEP calculations to hybrid QM/MM and to quantitative EVB/PDLD and EVB/FEP methods. However, as we tried to emphasise here, it seems to us that only quantitative approaches can be used to draw definite conclusions about the actual importance of electrostatic effects. For example, one might conclude that a helix macro dipole accounts for the catalytic effects of an enzyme while using unsolvated protein as a model for the given enzyme. On the other hand, correct calculations might yield a high dielectric that will completely screen the helix effect and even lead to much less stabilisation than that provided by the solvent for the reference reaction in water. Similarly, modelling a reaction with an unscreened metal ion can lead one to believe that this ion provides enormous catalytic effect, but this does not mean that the same ion is able to account for the catalytic effect of the actual enzyme, where the field of the metal is strongly screened. In view of the problems associated with incomplete models we can only draw unique conclusions from quantitative studies that consider the entire enzyme-substrate-solvent system and evaluate energetics of the reaction in the enzyme relative to the energetics of the corresponding reference reaction in water. Such studies have demonstrated that electrostatic effects can account for the major part of the catalytic effect of the enzymes. These findings are probably quite reliable since they were obtained without assuming any adjustable dielectric constant and without adjusting any parameter in calculating the difference between the energy of the transition state in the protein and in water.

The idea that the most important factor in enzyme catalysis is associated with electrostatic effects, has received strong support from simple MEP calculations. Such calculations indicated that enzymes provide complimentary environments to the charge distribution of their transition states. In fact, the crucial importance of electrostatic effects receives overwhelming support from simple logical considerations. That is, in order to catalyse reactions, enzymes must interact with the changes of the reacting systems during the reaction. This involves changes in structure, flexibility and charges, which can lead to strain, entropy and electrostatic effects, respectively. Strain and entropy can only be effective in ground-state destabilisation effects and such effects cannot help optimising k_{cat}/K_M (see Section 2.1). Thus we are left only with transition-state stabilisation of changing charges (steric effect can only destabilise the transition state and the attractive r^{-6} van der Waals term provides a very similar and small contribution to the activation barriers in solution and proteins).

One may try to salvage the idea that the transition state is stabilised by non-electrostatic effects by adopting the recent proposal that enzymes stabilise the transition states by the so-called "low-barrier hydrogen bond (LBHB)" effects, where hydrogen-bond (HB) donors are supposed to form a partial covalent bond with negatively charged transition states (Cleland and Kreevoy, 1994; Frey et al., 1994). This proposal has become popular as a substitute for the previously accepted strain and entropy proposals (Jencks, 1969) due to the accumulation of experimental and theoretical evidence that transition-state stabilisation by HB's can play a key role in catalysis (Wells et al., 1986; Fersht et al., 1986; Warshel, 1978). Perhaps because of the reluctance to accept electrostatic factors (that include HB's) as major catalytic effects, it has been proposed (e.g. Cleland and Kreevoy, 1994) that HB's stabilise charged transition states by forming partial covalent bonds (e.g. the $-X^-$ transition state is supposed to be stabilised by the $-X^-$ H-Y $\leftrightarrow$ -X-H $-Y^-$ resonance stabilisation) rather than by the electrostatic effect of preoriented HB's. It is, however, quite simple to show that the LBHB proposal is fundamentally incorrect and that it leads to anti catalysis rather than catalysis. That is, in water the $-X^-$ charge distribution is stabilised relative to the delocalised $-X^{-1/2}...H...-Y^{-1/2}$ charge distribution (which corresponds to an LBHB) by well-known solvation effects. Since the enzyme must stabilise the transition state more than water does, it must do so by providing more solvation and making the LBHB even less stable than in water (Warshel et al., 1995). Basically, any energy-based considerations will show that the LBHB catalytic concept contradicts the first law of thermodynamics.

Apparently, one of the main problems in the elucidation of the origin of enzyme catalysis has been associated with the inability to estimate solvation effects. Without knowing the contributions of solvation energies it is impossible to apply any energy-based analysis of enzyme catalysis and any discussion and proposal becomes circular. The ability to estimate solvation effects in a quantitative way (Warshel and Levitt, 1976) allows one to finally use thermodynamic cycles in studies of enzyme catalysis. Doing so provided unique conclusions on several open questions. For example it was possible to show that enzymes work by being better solvents than water (Warshel et al., 1989b) and not by desolvation (Dewar and Storch, 1985).This conclusion would be obtained by anyone who tries to construct a thermodynamic cycle for the catalytic process. Similarly, it is possible to show that the enzyme must stabilise transition states that already include metal ions more than water does in contrast to some proposals (see discussion of the isocitrate dehydrogenase case in Section 3.8.2.).

Thermodynamic considerations establish that enzyme active sites must be polar rather than low-dielectric non-polar environment. A case in point is the proposal by

Krishtalik and Topolev (1984) that enzymes work by providing low reorganisation energy in a *low dielectric* environment. Careful examination of this proposal establishes clearly that such an enzyme would destabilise the transition state (see the discussion in the footnote 56 of Yadav et al., 1991).

Careful analysis of the energetics of enzyme catalysis leads one to conclude that it must involve electrostatic stabilisation of transition states. It is also becoming clear that this is done by active sites with preoriented polar environment that can be viewed as a "supersolvent" for the transition state (Warshel, 1981).

5. Acknowledgements

This work was supported by Grants 222/92a from the U.S.-Hungarian Science and Technology Joint Fund, GM24492 from the National Institute of Health and DE-FG03-94ER1945 from the Department of Energy.

6. References

Abola, E.E., Bernstein, F.C., Bryant, S.H., Koetzle, T.F., and Weng, J. (1987) in Allen, F.H., Bergerhoff, G., and Sievers, R., (eds.) *Crystallographic Databases - Information Content, Software Systems, Scientific Applications,* Data Commission of the Union of Crystallography, Bonn, p. 107.

Alagona, G., Desmeules, P., Ghio, C., and Kollman, P.A. (1984) *J. Am. Chem. Soc.* **106,** 3623.

Allison, S.A., Bacquet, R.J., and McCammon, J.A. (1988) *Biopolymers* **27,** 251.

Allison, S.A. and McCammon, J.A. (1985) *J. Phys. Chem.* **89,** 1072.

Alvaro, G. and Russell, A.J. (1991) *Meth. Enzymol.* **202,** 620.

Andrés, J., Safont, V.S., and Tapia O. (1992) *Chem. Phys. Lett.* **198,** 515.

Andrés, J., Safont, V.S., Queralt, J., and Tapia, O. (1993) *J. Phys. Chem.* **97,** 7888.

Andrews, T.J. and Lorimer, G.H. (1987) in Hatch, M.D. (ed.) *The Biochemistry of Plants, Vol. 10,* Academic Press, Orlando, p. 131.

Ángyán, J. and Náray-Szabó, G. (1983) *J. Theor. Biol.* **103,** 349.

Åqvist, J. and Warshel, A. (1989) *Biochemistry* **28,** 4680.

Åqvist, J. and Warshel, A. (1990) *J. Am. Chem. Soc.* **112,** 2860.

Åqvist, J. and Warshel, A. (1992) *J. Mol. Biol.* **224,** 7.

Åqvist, J. and Warshel, A. (1993) *Chem. Rev.* **93,** 2523.

Åqvist, J., Luecke, H., Quiocho, F.A., and Warshel, A. (1991) *Proc. Natl. Acad. Sci. USA* **88,** 2026.

Åqvist, J., Fothergill, M., and Warshel, A. (1993) *J. Am. Chem. Soc.* **115,** 631.

Arad, D., Langridge, R., and Kollman, P.A. (1990) *J. Am. Chem. Soc.* **112,** 491.

Argese, E., Viglino, P., Rotilio, G., Scarpa, M., and Rigo, A. (1987) *Biochemistry* **26,** 3224.

Bachovchin, W.W. and Roberts, J.D. (1978) *J. Am. Chem. Soc.* **100,** 8041.

Bash, P.A., Field, M.J., Davenport, R.C., Petsko, G.A., Ringe, D., and Karplus, M. (1991) *Biochemistry* **30,** 5826.

Berthod, H. and Pullman, A. (1965) *J. Chem. Phys.* **62,** 942.

Beyer, W.F., Fridovich, I., Müllenbach, G.T., and Halewell, R.A. (1987) *J. Biol. Chem.* **262,** 11182.

Blake, C.C.F., Johnson, L.N., Mair, G.A., North, A.C.T., Phillips, D.C., and Sarma, V.R. (1967) *Proc. Roy. Soc. Ser. B* **167,** 378.

Bloxham, D.F., Giles, I.G., Wilton, D.C., and Akhtar, M. (1975) *Biochemistry* **14**, 2235.

Bolis, G., Clementi, E., Ragazzi, M., Salvederi, D., and Ferro, D.R. (1979) *Int. J. Quant. Chem.* **14**, 815.

Brady, L., Brzozewski, A.M., Derewenda, Z.S., Dodson, E., Dodson, G., Tolley, S., Turkenburg, J.P., Christiansen, L., Huge-Jensen, B., Norskov, L., Thim, L., and Menge, U. (1990) *Nature* **343**, 767.

Bott, R., Subramanian, E., and Davies, D.R. (1982) *Biochemistry* **21**, 6956.

Brocklehurst, K. and Little, G. (1972) *Biochem. J.* **128**, 471.

Broer, R., van Duijnen, P.T., and Nieuwpoort (1976) *Chem. Phys. Lett.* **42**, 525.

Bruice, T.C. and Sturtevant, J.M. (1959) *J. Am. Chem. Soc.* **81**, 2860.

Callens, M., Tomme, P., Kerstens-Hilderson, H., Cornelis, W., Vangrysperre, W., and De Bruyne, C.K. (1988) *Biochem. J.* **250**, 285.

Campbell, P. and Kaiser, E.T. (1973) *J. Am. Chem. Soc.* **95**, 3735.

Carter, P. and Wells, J.A. (1988) *Nature* **332**, 564.

Chance, M., Powers, L., Poulos, T.L., and Chance, B. (1986) *Biochemistry* **25**, 1266.

Christiansen, D.W. and Lispcomb, W.N. (1989) *Acc. Chem. Res.* **33**, 62.

Churg, A.K. and Warshel, A. (1986) *Biochemistry* **25**, 1675.

Cleland, W.W. and Kreevoy, M.M. (1994) *Science* **264**, 1887..

Clementi, E., Ranghino, G., and Scordamaglia, R. (1979) *Chem. Phys. Lett.* **49**, 218.

Cocco, D., Rossi, L., Barra, D., Bossa, F., and Rotilio, G. (1982) *FEBS Letters* **150**, 303.

Coghlan, V.M. and Vickery, L.E. (1992) *J. Biol. Chem.* **267**, 8932.

Collyer, C.A., Henrick, K., and Blow, D.M. (1990) *J. Mol. Biol.* **212**, 211.

Cook, C.M., Haydock, K., Lee, R.H., and Allen, L.C. (1984) *J. Phys. Chem.* **88**, 4875.

Corey, D.R., McGrath, M.E., Vásquez, J.R., Fletterick, R.J., and Craik, C.S. (1992) *J. Am. Chem. Soc.* **114**, 4905.

Cotton, F.A., Bier, C.J., Day, V.W., Hazen, E.E., jr., and Larsen, S. (1971) *Cold Spring Harbor Symp. Quant. Biol.* **36**, 243.

Cotton, F.A., Hazen, E.E., jr., and Legg, M.J. (1979) *Proc. Natl. Acad. Sci. USA* **76**, 2551.

Craik, C.S., Roczniak, S., Largman, C., and Rutter, W.J. (1987) *Science* **237**, 909.

Cronin, C.N., Malcolm, B.A., and J.F. Kirsch (1987) *J. Am. Chem. Soc.* **109**, 2222.

Cudd, A. and Fridovich, I. (1982) *J. Biol. Chem.* **257**, 11443.

Daggett, V., Schröder, A., and Kollman, P.A. (1991) *J. Am. Chem. Soc.* **113**, 8926.

Dao-Pin, S., Liao, D.I., and Remington, S.J. (1989) *Proc. Natl. Acad. Sci. USA* **86**, 5361.

Davis, M.E. and McCammon, J.A. (1990) *Chem. Rev.* **90**, 509.

Deakyne, C.A. and Allen, L.C. (1979) *J. Am. Chem. Soc.* **101**, 3951.

Desideri, A., Falconi, M., Parisi, V., and Rotilio, G. (1989) *FEBS Lett.* **250**, 45.

Desideri A., Falconi, M., Polticelli, F., Bolognesi, M., Djinovic, K., and Rotilio, G. (1992) *J. Mol. Biol.* **223**, 337.

Dewar, M.J.S. and Storch, D.M. (1985) *Proc. Natl. Acad. Sci. U.S.A.* **82**, 2225.

Dunn, B.M. and Bruice, T.C. (1973) *Adv. Enz. Related Areas in Mol. Biol.* **37**, 1.

Eventoff, W., Rossman, M.G., Taylor, S., Torff, H.J., Meyer, H., Keil, W., and Klitz, H.H. (1977) *Proc. Natl. Acad. Sci. USA* **74**, 2677.

Fábián, P., Asbóth, B., and Náray-Szabó, G. (1994) *J. Mol. Struct. THEOCHEM* **307**, 171.

Fersht, A.R. (1985) *Enzyme Structure and Mechanism*, Freeman, New York.

Fersht, A.R. and Kirby, A.J. (1968) *J. Am. Chem. Soc.* **90**, 5833.

Fersht, A.R., Shi, J.P., Knill-Jones, J., Lowe, D.M., Wilkinson, A.J., Blow, D.M., Brick, P., Carter, P., Waye, M.M.Y., and Winter, G. (1985) *Nature* **314**, 235.

Fersht, A.R., Leatherbarrow, R.J., and Wells, J.N.C. (1986) *Nature* **322**, 284.

Fischer, E. (1894) *Ber. Deutsch. Chem. Ges.* **27**, 2984.

Frey, P.A., Whitt, S.A., and Tobin, J.B. (1994) *Science* **264**, 1924.

Fruton (1987) in Neuberger, A. and Brocklehurst, K. (eds.), *New Comprehensive Biochemistry,* Elsevier, Amsterdam, Chap. 1.

Fuxreiter, M., Farkas, Ö., and Náray-Szabó, G. (1995) *Protein Engng.* **8**, 925.

Fülöp, V., Phizackerley, R.P., Soltis, S.M., Clifton, I.J., Wakatsuki, S., Erman, J., Hajdu, J., and Edwards, S.L. (1994) *Structure* **2**, 201.

Gardell, S.J., Craik, C.S., Hilvert, D., Urdea, M.S., and Rutter, W.J. (1985) *Nature* **317**, 551.

Gardell, S.J., Hilvert, D., Barnett, J., Kaiser, E.T., and Rutter, W.J. (1987) *J. Biol. Chem.* **262**, 576.

Getzoff, E.D., Tainer, J.A., Weiner, P.K., Kollman, P.A., Richardson, J.S., and Richardson, D.C. (1983) *Nature* **306**, 287.

Gilson, M.K. and Honig, B.H. (1987) *Nature* **330**, 84.

Gráf, L., Craik, C.S., Patthy, A., Roczniak, S., Fletterick, R.J., and Rutter, W.J. (1987) *Biochemistry* **26**, 2616.

Gráf, L., Jancsó, Á., Szilágyi, L., Hegyi, G., Pintér, K., Náray-Szabó, G., Hepp, J., Medzihradszky, K., and Rutter, W.J. (1988) *Proc. Natl. Acad. Sci. USA* **85**, 4961.

Haldane, J.B. S. (1930) in *Enzymes,* Longmans, London, p. 182.

Hayes, D.M. and Kollman, P.A. (1976) *J. Am. Chem. Soc.* **98**, 3335, 7811.

Hol, W.G.J., van Duijnen, P.T., and Berendsen, H.J.C. (1978) *Nature* **273**, 443.

Horenstein, B.A. and Schramm, V.L. (1993) *Biochemistry* **32**, 7089.

Horenstein, B.A., Parkin, D.W., Estupinan, B., and Schramm, V.L. (1991) *Biochemistry* **30**, 10788.

Hsu, I.N., Delbaere, L.T.J., and James, M.N.G. (1977) *Nature* **266**, 140.

Hunkapiller, M.W., Smallcombe, S.H., Whitaker, D.R., and Richards, J.H. (1973) *Biochemistry* **12**, 4732.

Hurley, J.H. and Remington, S.J. (1992) *J. Am. Chem. Soc.* **114**, 4769.

Hurley, J.H., Dean, A.M., Koshland, D.E., Jr., and Stroud, R.M. (1991) *Biochemistry* **30**, 8671.

Imoto, T., Johnson, L.N., North, A.C.T., Phillips, D.C., and Rupley, J.A. (1972) in Boyer, P.D. (ed.) *Enzymes,* Academic Press, New York, p. 665.

Jackson, D.Y., Jacobs, J.W., Sugasawara, R., Reich, S.H., Bartlett, P.A., and Schultz, P.G. (1988) *J. Am. Chem. Soc.* **110**, 4841.

Jackson, S.E. and Fersht, A.R. (1993) *Biochemistry* **32**, 13909.

Jacob, O., Cardenas, R., and Tapia, O. (1990) *J. Am. Chem. Soc.* **112**, 8692.

James, M.N.G. and Sielecki, A.R. (1985) *Biochemistry* **24**, 3701.

Jencks, W.P. (1969) *Catalysis in Chemistry and Enzymology,* McGarww Hill, New York.

Jencks, W.P. (1975) *Adv. Enzymol.* **43**, 219.

Johannin, G. and Kellersohn, N. (1972) *Biochim. Biophys. Res. Commun.* **49**, 321.

King, G., Lee, F.S., and Warshel, A. (1991) *J. Chem. Phys.* **95**, 4366.

Kline, P.C. and Schramm, V.L. (1994) *J. Biol. Chem.* **35**, 22385.

Knighton, D.R., Kan, C.C., Howland, E., Janson, C.A., Hostomska, Z., Welsh, K.M., and Matthews, D.M. (1994) *Nature Struct. Biol.* **1**, 186.

Knowles, J.R. (1991) *Nature* **350**, 121.

Kollman, P.A. (1993) *Chem. Rev.* **93**, 2395.

Kossiakoff, A.A. and Spencer, S.A. (1981) *Biochemistry* **20**, 6462.

Krauss, M. and Garmer, D.R. (1991) *J. Am. Chem. Soc.* **113**, 6426.

Krauss, M. and Garmer, D.R. (1993) *J. Phys. Chem.* **97**, 831.

Kraut, J.(1977) *Annu. Rev. Biochem.* **46**, 331.

Krechl, J. and Kuthan, J. (1982) *Int. J. Quant. Chem.* **21**, **1029**.

Krechl, J. and Kuthan, J. (1983) *Int. J. Quant. Chem.* **24**, 479.

Krishtalik, L.I. and Topolev, V.V. (1984) *Mol. Biol. (Moscow)* **18**, 721.

Langen, R., Brayer, G.D., Berghuis, A.M., McLendon, G., Sherman, F., and Warshel, A (1992) *J. Mol. Biol.* **224**, 589.

Larsen, J.W. (1973) *Biochem. Biophys. Res. Commun.* **50**, 839.

Lavery, R., Pullman, B., and Wen, Y.K. (1983) *Int. J. Quant. Chem.* **24**, 353.

Lee, F.S.; Chu, Z.T.; Bolger, M.B., and Warshel, A. (1992) *Prot. Eng.* **5**, 215.

Lerner, R.A., Benkovic, S.J., and Schultz, P.G. (1991) *Science* **253**, 659.

Liang, J.Y. and Lipscomb, W.N. (1988) *Biochemistry* **27**, 8676.

Lindskog, S. (1986) in *Zinc Enzymes,* Bertini, I., Luchinat, C., Maret, W., and Zeppezauer, M. (eds.), Birkhauser, Boston, p. 307.

Lobkovsky, E., Billings, E.M., Moews, P.C., Rahil, J., Pratt, R.F., and Knox, J.R. (1994) *Biochemistry* **33**, 6762.

Lodi, P.J. and Knowles, J.R. (1993) *Biochemistry* **32**, 4338.

Longo, E., Stamato, F.M.L.G., Ferreira, R., and Tapia, O. (1985) *J Theor. Biol.* **112**, 783.

Lorimer, G. (1981) *Biochemistry* **20**, 1236.

Lowe, G. and Whitworth, A.S. (1974) *Biochem. J.* **141**, 503.

Lu, G., Lindquist, Y., and Schneider, G. (1992) *Proteins* **12**, 117.

Maister, S.G., Pett, C.P., Albery, W.J., and Knowles, J.R. (1976) *Biochemistry* **15**, 5607.

Malcolm, B.A., Rosenberg, S., Corey, M.J., Allen, J.S., de Baetselier, A., and Kirsch, J.F. (1989) *Proc. Natl. Acad. Sci. U.S.A.* **86**, 133.

Matthews, B.W., Sigler, P.B., Henderson, R., and Blow, D.M. (1967) *Nature* **214**, 652.

Mauro, J.M., Fishel, L.A., Hazzard, J.T., Meyer, T.E., Tollin, G., Cusanovich, M.A. and Kraut, J. (1988) *Biochemistry* **27**, 6243.

Mehler, E.L. and Eichele, G. (1984) *Biochemistry* **23**, 3887.

Menger, F.M. (1992) *Biochemistry* **31**, 5368.

Merkler, D.J. and Schramm, V.L. (1990) *J. Biol. Chem.* **265**, 4420.

Merz, K.M., jr., Hoffmann, R., and Dewar, M.J.S. (1989) *J. Am. Chem. Soc.* **111**, 5636.

Miller, M.A., Han, G.W. and Kraut, J. (1994) *Proc. Natl. Acad. Sci. USA* **91**, 11118.

Nakagawa, S. and Umeyama, H. (1978) *J. Am. Chem. Soc.* **100**, 7716.

Nakagawa, S. and Umeyama, H. (1982) *J.Theor. Biol.* **96**, 473.

Nakagawa, S., Umeyama, H., Kitaura, and Morokuma, K. (1981) *Chem. Pharm. Bull.* **29**, 1.

Náray-Szabó, G. (1979) *Int. J. Quant. Chem.* **16**, 265.

Náray-Szabó, G. (1982) *Int. J. Quant. Chem.* **22**, 575.

Náray-Szabó, G. (1983) *Int. J. Quant. Chem.* **23**, 723.

Náray-Szabó, G. (1989) *J. Mol. Graph.* **7**, 76.

Náray-Szabó, G. (1993) *J. Mol. Recogn.* **6**, 205.

Náray-Szabó, G. and Ferenczy, G.G. (1995) *Chem. Rev.* **95**, 829.

Náray-Szabó, G. and Gérczei, T. (1995) *Croat. Chem. Acta*, in press.

Náray-Szabó, G. and Polgár, L. (1980) *Int. J. Quant. Chem. Quantum Biol. Symp.* **7**, 397.

Náray-Szabó, G., Surján, P.R., and Ángyán, J.G. (1986) *Applied Quantum Chemistry*, Akadémiai Kiadó-Reidel, Budapest-Dordrecht.

Náray-Szabó, G., Nagy, J., and Bérces, A. (1989) *J. Mol. Struct. THEOCHEM* **200**, 401.

Náray-Szabó, G., Tóth, G., Ferenczy, G.G., and Csonka, G. (1994) *Int. J. Quant. Chem. Quantum Biol. Symp.* **21**, 227.

Nolte, H.J., Rosenberry, T.L., and Neumann, E. (1980) *Biochemistry* **19**, 3705.

Osman, R. and Basch, H. (1984) *J. Am. Chem. Soc.* **106**, 5710.

Page, M.I. and Jencks, W.P. (1971) *Proc. Natl. Acad. Sci. USA* **68**, 1678.

Parkin, D.W., Horenstein, B.A., Abdulah, D.R., Estupinan, B., and Schramm, V.L. (1991) *J. Biol. Chem.* **266**, 20658.

Pauling, L. (1946) *Chem. Eng. News* **24**, 1375.

Pelletier, H. and Kraut, J. (1992) *Science* **258**, 1748.

Perutz, M.F. (1978) *Science* **201**, 1187.

Phillips, M.A., Fletterick, R., and Rutter, W.J. (1990) *J. Biol. Chem.* **265**, 20692.

Pickersgill, R.W., Goodenough, P.W., Sumner, I.G., and Collins, M.E. (1988) *Biochem. J.* **254**, 235.

Pincus, M.R., Zimmerman, S.S., and Scheraga, H.A. (1977) *Proc. Natl. Acad. Sci. USA* **74**, 2629.

Polgár, L. (1973) *Eur. J. Biochem.* **33**, 104.

Polgár, L. (1987) *FEBS Lett.* **219**, 1.

Polgár, L. (1989) *Mechanisms of Protease Action*, CRC Press, Boca Raton, Florida.

Polgár, L. and Bender, M.L. (1969) *Proc. Natl. Acad. Sci. USA* **64**, 1335.

Poulos, T.L. and Kraut, J. (1980) *J. Biol. Chem.* **255**, 8199.

Pullman, A. (1981) *Ann. N.Y. Acad. Sci.* **367**, 340.

Rangarajan, M. and Hartley, B.S. (1992) *Biochem. J.* **283**, 223.

Rao, S.N.; Singh, U.C.; Bash, P.A., and Kollman, P.A. (1987) *Nature* **328**, 551.

Rees, D.C. (1980) *J. Mol. Biol.* **141**, 323.

Ressler, N. (1982) *J. Theor. Biol.* **97**, 195.

Richards, F.M. and Wyckoff, H.W. (1971) in Boyer, P.D. (ed.) *Enzymes*, Academic Press, New York, Vol. 4, p. 647.

Ripoll, D.R., Faerman, C.H., Axelsen, P.H., Silman, I., and Sussman, J.L. (1993) *Proc. Natl. Acad. Sci. USA* **90**, 5128.

Roberts, G.C.K., Dennis, E.A., Meadows, D.H., Cohen, J.S., and Jardetzky, O. (1969) *Proc. Natl. Acad. Sci. USA* **62**, 1151.

Rogers, N.K., Moore, G.R. and Sternberg, M.J.E. (1985) *J. Mol. Biol.* **182**, 613.

Rosenberry, T.L. (1975) *Adv. Enzymol. Relat. Areas Mol. Biol.* **43**, 103.

Rubin, J. and Blow, D.M. (1981) *J. Mol. Biol.* **145**, 489.

Russell, S.T. and Warshel, A. (1985) *J. Mol. Biol.* **185**, 389.

Sawaryn, A. and Sokalski, W.A. (1979) *Int. J. Quant. Chem.* **16**, 293.

Scheiner, S. and Kar, T. (1995) *J. Am. Chem. Soc.* **117**, 6970.

Schowen, R.L. (1978) in R.D. Gandour, and R.L. Schowen (eds.), *Transition States of Biochemical Processes*, Plenum, New York, Ch. 2.

Schultz, P.G. (1989) *Angew. Chem. Intl. Ed. Engl.* **28**, 1283.

Sharp, K.A. and Honig, B. (1990) *Annu. Rev. Biophys. Chem.* **19**, 301.

Sharp, K.A., Fine, R., and Honig, B. (1987) *Science* **236**, 1460.

Sheridan, R.P. and Allen, L.C. (1980) *Biophys. Chem.* **11,** 133.

Sheridan, R.P. and Allen, L.C. (1981) *J. Am. Chem. Soc.* **103,** 1544.

Silverman, D.N. and Lindskog, S. (1988) *Acc. Chem. Res.* **21,** 30.

Sines, J.J., Allison, S.A., and McCammon, J.A. (1990) *Biochemistry* **29,** 9403.

Sivaraja, M., Goodin, D.B., Smith, M., and Hoffman, B.M. (1989) *Science* **245,** 738.

Slotboom, A.J., Verheij, H.M., and de Haas, G.H. (1982) *New Comprehensive Biochemistry* **4,** 354.

Sprang, S., Standing, T., Fletterick, R.J., Finer-Moore, J., Stroud, R.M., Xuong, N.H., Hamlin, R., Rutter, W.J., and Craik, C.S. (1987) *Science* **237,** 905.

Sternberg, M.J.E., Hayes, F.R.F., Russell, A.J., Thomas, P.G., and Fersht, A.R. (1987) *Nature* **330,** 86.

Storm, D.R. and Koshland, D.E., Jr. (1970) *Proc. Natl. Acad. Sci. USA* **66,** 445.

Strothkamp, K.G. and Lippard, S.J. (1982) *Acc. Chem. Res.* **15,** 318.

Stroud, R.M. (1994) *Nature Struct. Biol.* **1,** 131.

Subramanian, E., Swan, I.D.A., Liu, M., Davies, D.R., Jenkins, J.A., Tickle, I.J., and Blundell, T.L. (1977) *Proc. Natl. Acad. Sci. USA* **74,** 556.

Sussman, J.L., Harel, M., Frolow, F., Oefner, C., Goldman, A., Toker, L., and Silman, I. (1991) *Science* **253,** 872.

SYBYL (1993) Version 6.0a, TRIPOS Associates, St. Louis, MO, February.

Tainer, J.A., Getzoff, E.D., Beem, K.M., Richardson, J.S., and Richardson, D.C. (1982) *J. Mol. Biol.* **160,** 181.

Tan, R.C., Truong, T.N., McCammon, J.A., and Sussman, J.L. (1993) *Biochemistry* **32,** 401.

Tapia, O. and Andrés, J. (1992) *Mol. Engng.* **2,** 37.

Tapia, O. and Johannin, G. (1981) *J. Chem. Phys.* **75,** 3624.

Tapia, O., Eklund, H., and Brändén, C.I. (1987) in Náray-Szabó, G. and Simon, K. (eds.) *Steric Aspects of Biomolecular Interactions,* CRC Press, Boca Raton, p. 159.

Tapia, O., Cardenas, R., Andrés, J., Krechl., Campillo, M., and Colonna-Cesari, F. (1991) *Int. J. Quant. Chem.* **39,** 767.

Tapia, O., Andrés, J., and Safont, V.S. (1994a) *J. Phys. Chem.* **98,** 4821.

Tapia, O., Paulino, M., and Stamato, F.M.L.G. (1994b) *Mol. Eng.* **3,** 377.

Thoma, J.A. (1974) *J. Theor. Biol.* **44,** 305.

Thorsness, P.E. and Koshland, D.E., Jr. (1987) *J. Biol. Chem.* **262,** 10422.

Tucker, P.W., Hazen, E.E., jr., and Cotton, F.A. (1978) *Mol. Cell. Biochem.* **22,** 67.

Turi, L. and Náray-Szabó, G. (1992) *Int. J. Quant. Chem.* **42,** 1537.

Umeyama, H., Nakagawa, S., and Kudo, T. (1981) *J. Mol. Biol.* **150,** 409.

van Duijnen, P.T. (1981) *Biophys. Chem.* **13,** 133.

van Duijnen, P.T., Thole, B.T., and Hol, W.G.J. (1979) *Biophys. Chem.* **9,** 273.

Verheij, H.M., Volwerk, J.J., Jansen, E.H., Puyk, W.C., Dijkstra, B.W., Drenth, J., and de Haas, G.H. (1980) *Biochemistry* **19,** 743.

Voet, J.G., Coe, J., Epstein, J., Matossian, V., and Shipley, T. (1981) *Biochemistry* **20,** 7182.

Warshel, A. (1978) *Proc. Natl. Acad. Sci. USA* **75,** 5250.

Warshel, A. (1981) *Biochemistry* **20,** 3167.

Warshel, A. (1987) *Nature* **333,** 15.

Warshel, A. (1991) *Computer Modeling of Chemic al Reactions in Enzymes and in Solutions,* Wiley, New York.

Warshel, A. and Åqvist, J. (1991) *Annu. Rev. Biophys. Chem.* **20,** 267.

Warshel, A. and Levitt, M. (1976) *J. Mol. Biol.* **103,** 227.

Warshel, A. and Russell, S.T. (1984) *Q. Rev. Biophys.* **17,** 283.

Warshel, A. and Russell, S.T. (1986) *J. Am. Chem. Soc.* **108,** 6569.

Warshel, A., Russell, S.T., and Weiss, R.M. (1982) in B.S. Green, Y. Ashani, and D. Chipman, (eds.), *Chemical Approaches to Understanding Enzyme Catalysis: Biomimetic Chemistry and Transition-State Analogs,* Elsevier, Amsterdam, 267.

Warshel, A., Russell, S., and Sussman, F. (1986) *Isr. J. Chem.* **27,** 217.

Warshel, A., Sussman, F., and Hwang, J.K. (1988) *J. Mol. Biol.* **201,** 139.

Warshel, A., Åqvist, J., and Creighton, S. (1989a) *Proc. Natl. Acad. Sci. USA* **86,** 5820.

Warshel, A., Náray-Szabó, G., Sussman, F., and Hwang, J.K. (1989b) *Biochemistry* **28,** 3629.

Warshel, A., Schweins, T., and Fothergill, M. (1994) *J. Am. Chem. Soc.* **116,** 8438.

Waszkowycz, B., Hillier, I.H., Gensmantel, N., and Payling, D.W. (1990) *J. Chem. Soc. Perkin Trans. 2,* 1259.

Waszkowycz, B., Hillier, I.H., Gensmantel, N., and Payling, D.W. (1991) *J. Chem. Soc. Perkin Trans. 2,* 225.

Wells, J.A., Cunningham, B.C., Graycar, T.P., and Estell, D.A. (1986) *Philos. Trans. R. Soc. London Ser. A* **317,** 415.

Wells, J.A., Powers, D.B., Bott, R.R., Graycar, T.P., Estell, D.A. (1987) *Proc. Natl. Acad. Sci. USA* **84,** 1219.

Westheimer, F.H. (1962) *Adv. Enzymol.* **24,** 456.

Winkler, F.K., D'Arcy, A., and Hunziker, W. (1990) *Nature* **343,** 771.

Wong, C.F. and McCammon, J.A. (1986) *J. Am. Chem. Soc.* **108,** 3830.

Yadav, A., Jackson, R.M., Holbrook, J.J., and Warshel, A. (1991) *J. Am. Chem. Soc.* **113,** 4800.

Yamanaka, K. (1968) *Biochim. Biophys. Acta* **151,** 670.

Zheng, Y.J., Merz, K.M., jr., and Farber, G.K. (1993) *Protein Engng.* **6,** 479.

ON THE MECHANISMS OF PROTEINASES

A. GOLDBLUM
Department of Pharmaceutical Chemistry
School of Pharmacy, Hebrew University of Jerusalem
Jerusalem, Israel 91120

1. Introduction

Proteinases (proteases) are a large group of enzymes that hydrolyze proteins and peptides, and many of them cleave amides and esters at measurable rates as well. They are classified mainly according to the principal functional group to which catalytic activity is attributed. Four major classes are known: serine, cysteine (thiol), aspartic (carboxyl) and metallo (Fersht, 1985; Polgar, 1990). The first three are polar amino acids, while the fourth is a "prosthetic group" which is not part of the primary sequence of proteins. The common catalyzed reaction of all proteinases is the cleavage of a peptide bond by water to obtain an acid (eq. 1, with R being the original N-terminal part of the peptide) and an amine (where R' is the original C-terminal of the peptide), where both are in the ionized form at neutral pH:

(1) $$\text{R-C(=O)-N(H)-R'} + \text{H2O} \text{ ------> } \text{R-C(=O)-O}^{-} + \text{H3N-R'}^{+}$$

Studies of proteinase activities comprise some of the most important current research efforts in the field of theoretical enzyme mechanisms. Results from crystallography and kinetics in the 70's and 80's paved the way for such theoretical studies, mainly of the serine proteinase family. Such studies are extending nowadays, as more structures of proteinases are solved with high resolution and more detailed kinetic studies are conducted. But, while earlier structural results were available for the native structures alone, recent crystallographic evidence is available for complexes with peptide analogs, with intermediate analogs and with mutant enzymes. When these structural studies are coupled with results of kinetic research, a large database is formed for the theoretician to consider as a basis for construction, simulation and analysis by computer models.

A major reason for the expanding interest in some of the proteinases is the growing evidence of the involvement of many of them in disease conditions. In most of these illnesses, excessive activity of a specific proteinase should be blocked. It is assumed that "mechanism based" inhibitors have the greatest potential to become drugs for the relevant disease (Sandler and Smith, 1989). Such inhibitors should best mimic the structure of the appropriate transition states. But protein crystallography reflects relatively stable and long-lived structures, while transition states that determine catalytic

G. Náray-Szabó and A. Warshel (eds.), Computational Approaches to Biochemical Reactivity, 295–340.

rates are short-lived and inaccessible to direct experiments. Theoretical studies are not limited to stable structures alone but can also follow reaction pathways to unstable intermediates and transition states, and thus have become indispensable for providing the basis for rational drug design.

In the last few years, kinetic studies with proteinase mutants taught us much more about the complexity and intricacy of structure-function relations of proteinases. They focused, in many cases, on the effects of single residues on k_{cat} and K_m. Such details of the mechanism encourage theoreticians to construct sophisticated models that are able to answer specific questions and to report energies that may be compared to experiments.

The increase in central processing power of computers and of memory size contributed much to our ability to study larger models and to extend their relevance. Methodologies continue to improve and to adapt to the strengthening of technology. In some cases, computational results reached a state of high reliability and predictive potential. Surely, this trend will gain more importance in the following years. The interplay between experimental and theoretical approaches is demonstrated both in specific references of one to the other and in collaborations that result in publications where experimentalists join forces with theoreticians or supplement their experiments with theoretical results.

This chapter summarizes recent theoretical research of proteinase mechanisms as well as novel experimental results which correspond to theoretical suggestions or should be relevant to future theoretical studies. Two major approaches are employed to study enzymatic reactions: quantum mechanics and molecular mechanics ("force field") methods. The accuracy of quantum mechanical (QM) methods is still much more diverse than that of force-field approaches, but the most accurate QM methods can be applied only to small systems. This imposes a minimalist approach for the construction of enzyme models for QM studies.

The main problem with construction of models for proteins is the structural input. It is yet impossible to "fold" a primary sequence to its correct 3-D structure from first principles. X-ray crystallography has thus become the most important source for model construction, but suffers itself of a few limitations. First, resolution may be too low to assign some positions of atoms, in particular - positions of some water molecules that may be relevant to binding or to activity. In addition, the "heavy atoms" for which coordinates are given have thermal factors that imply some vibrational movement around their given positions. The structure of a crystallized protein is certainly more "rigid" than its solution structure, and even if we may safely assume that the structures differ only slightly, the implications of such differences for reactivity may be large. Once the crystal's atomic positions are employed to construct a model of a protein or only of its active site, it should not be easily applied to any state other than that extracted from the crystal. A native structure or that of a complex with an inhibitor are not similar to complexes with substrates, so one should use great care in modifying the experimental structures. Introducing a substrate into coordinates of a native structure is based, implicitly or explicitly, on many assumptions and rationalizations, which should best be tested and verified.

Proton positions are not "seen" in the crytallographic results due to their low electronic density. Correct positioning of protons is important for three sets of atoms: those of the residues that may be involved in proton transfers along the catalytic pathway, protons of ionizable residues for which the pKa may change with respect to their measured acidity (as "free" or as "blocked" amino acids), and protons of hydroxylic residues that may be involved in alternative hydrogen bonding relations by rotation around the C-OH bond. These must be evaluated by some theoretical procedure for correct positioning. But even the best procedures for adding protons to protein coordinates may only be appropriate for the experimental structure. Any model that modifies that structure, such as by introduction of a substrate or along a reaction pathway, should reconsider the proton positions.

Nuclear magnetic resonance is gaining strength as a tool for studying native protein structure and complexes with ligands, while it also reflects the mobility of the structure compared to its crystallographic rigidity. It also has the potential of reporting proton positions, based on their mutual interactions at short distances. Neutron diffraction studies have been employed to determine all the atomic positions, including protons, in a protein structure, but only very few enzymes have been studied as yet. These two methods are employed to study structural aspects that may be more more closely related to the activity of enzymes than the crystal environment. Molecular dynamics (MD) studies are most appropriate for simulation of solution structures and to account for solution effects such as pH, temperature, etc. They are less adapted for studying chemical reactivity, such as catalysis and bond breaking and making, since the variations in electron density can not be explicitly considered.

In many of the enzymes, the surrounding solution is in direct contact with their active sites, and the rest of the protein is immersed in water. The "environment" has an effect on several aspects of the mechanisms. Surface charges of ionized residues may be neutralized by counterions and their associated water spheres. Ligands approach the protein through the solution, and large changes in solvation take place along the binding. Both the protein and the ligand loose water of solvation, which may be accompanied by conformational changes in both. Binding is related to the catalytic efficiency because it imposes spatial relations between the enzyme and the substrate.

The binding of a substrate is usually discussed in terms of a few interaction components, mostly electrostatics, Van-der-Waals and hydrophobic. Part of the binding is of the substrate's side chains to less clearly defined "enzyme pockets". The peptide residues are assigned according to Berger and Schecter (1970) so that the cleavage occurs between residues P1 and P1'. P1 is on the N-terminal side of the cleaved peptide bond, and residues that are further towards the N-terminal are assigned as P2, P3, etc., while the residues following P1' towards the C-terminal are P2', P3', etc. The side chain of residue P1 is in contact with a "pocket" of the enzyme that is designated S1, P2 with S2, etc. This chapter deals with the reaction mechanism alone, and not with the specificities of the enzymes or with the effect of binding on the mechanism. Many experimental results dealing with these issues have been published, and those require a separate review.

2. Serine Proteinases

Serine proteinases (SP) are mostly classified according to their resemblance to trypsin ("trypsin super family" - in which chymotrypsin, elastase, thrombin and others are included) or to the subtilisin super family. The SP have in common two major elements of the catalytic machinery. The catalytic triad consists of serine, histidine and aspartic acid. The "oxyanion hole" is modified between the two super families of SP. While main chain bonds (N-H) comprise this "hole" in trypsins, the subtilisins have both main and side chain contributions to it. Also, positions of the catalytic triad in the primary structure vary between the two. In chymotrypsin they are Ser-195, His-57 and Asp-102, while these residues occupy, respectively, positions 221, 64 and 32 in subtilisin-BPN'. Crystallographic studies of proteins from the two families demonstrate that the three dimensional fold brings those three residues into close proximity, and very similar spatial relations between these residues is found in the different enzymes. But, the primary structures of trypsins and subtilisins differ greatly. Thus it is believed that the two are an example of "convergent evolution". Important reviews of SP activity may be found in reviews by Kraut (1977), Huber and Bode (1978), Steitz and Shulman (1982), Polgar and Halasz (1982) Schowen (1988) and in the books of Fersht (1985) and Warshel (1991). A new family of SP, prolyl oligopeptidase and its homologues (Polgar, 1992) as well as the more distant LexA enzyme (Slilaty and Vu, 1991) are not discussed in this chapter.

The specificity of SP varies in the same family. In the trypsins, while trypsin cleaves best proteins and peptides that have Arg or Lys in position P1, chymotrypsin is specific for hydrophobic residues in this position (Trp, Tyr, Phe) and elastase prefers, at P1, small amino-acid side chains such as Gly or Ala. In addition to the catalytic triad, other residues have been shown to affect the catalytic activity. The "oxyanion hole" in the trypsin family is that region that is in close contact with the carbonyl oxygen of the cleaving peptide bond. This interaction is attributed to two N-H from the main chain, one belonging to the catalytic residue Ser-195, the other from Gly-193. Also, there are a few residues that are in close contact with Asp-102 of the catalytic triad, presumably stabilizing its negative charge by hydrogen bond donation.

From both structural and kinetic data it is known that peptide cleavage by SP proceeds in two major steps (figure 1). In the first (figure 1, steps A-C), an acyl-enzyme is formed (The acyl group is the original N-terminal part of the cleaved peptide) and in the second step (figure 1, D) this acyl-enzyme is hydrolyzed by water and the active site is ready to cleave another substrate. Each step is composed of a few bond-breaking and bond-making movements of atoms and, taken together, they comprise the mechanism of the reaction. Each step has an associated energy requirement: the pathway is in need of activation energy in some steps while it releases energy in others. The quantitative energy changes are the main focus for the theoretician, who wishes to determine the preferential path from reactants to products, based on their energies. The common assumption is that the reaction will always move along the lowest energy pathway that is available to its constituents.

In the next few paragraphs, the status of the present consensus about SP mechanism will be delineated. However, many of the details are not yet proven by experiments or by theory.

Following the binding of a substrate, it is assumed that Ser-195 attacks the carbonyl carbon of the P1 residue and its oxygen (OG) forms a covalent bond to this carbon (figure 1, B), thus breaking the double bond of the carbonyl and transferring electron density to the carbonyl's oxygen, which increases its negative charge. During the attack, the carbonyl changes hybridization from trigonal sp^2 to tetrahedral sp^3, and a "tetrahedral intermediate" (TI) is formed. Towards the formation of the TI, Ser-195 should loose its hydroxylic proton. It is most probable that the proton is not lost in the native state of the enzyme, due to the relatively high pKa of Ser. The acidity of Ser-195 may be increased along the binding of a substrate, as the peptide carbonyl carbon that carries a partial positive charge gets closer to the oxygen of serine. At some stage, the proton is transferred from oxygen, probably to NE of His-57. This phase could also occur in a concerted manner along the attack of Ser on the carbonyl.

The presence of an ionized aspartic acid next to histidine raises its basicity and facilitates proton transfer from Ser-195 in the direction of His-57. If the proton resides on histidine, this side chain is positively charged. The charge is stabilized by the neighboring Asp-102, which may also stabilize the conformation of His-57. In some earlier proposals for SP mechanisms, it was assumed that a "double PT" or "charge relay" system is operating between the catalytic triad residues, and two PTs, the second being from His to Asp, take place.

As the tetrahedral intermediate is formed, the character of the peptide C-N bond changes together with that of the carbonyl: it is no more an amide, in which the nitrogen shares its lone pair with the carbonyl, but is closer to a single C-N bond which carries a negatively charged oxygen on the carbon. Nitrogen should be more basic in the tetrahedral intermediate, compared to the amide of the substrate. This increased basicity allows the nitrogen to extract a proton, presumably from the more acidic histidine. To achieve such proton transfer, the nitrogen may have to undergo inversion or rotation, and the imidazole ring of histidine may have to rotate with respect to the conformation that received the proton from serine. This proton transfer could be accompanied or followed by C-N bond breaking, leading to the final cleavage of the peptide. Electronic reorganization about the carbonyl leads to the first stable product, an acyl enzyme (figure 1, C).

The C-terminal part of the original substrate leaves the active site and makes place for water molecules to enter. A water molecule can now be positioned with its oxygen close to the carbonyl carbon of the acyl moiety and may also act as a H-bond donor to NE of histidine (figure 1, D). The previous role of Ser-195 can now be "reproduced" by the water molecule, while the "oxyanion hole" resumes its role of the first part of the reaction sequence. As water approaches the carbonyl, it loses a proton to histidine and, becoming

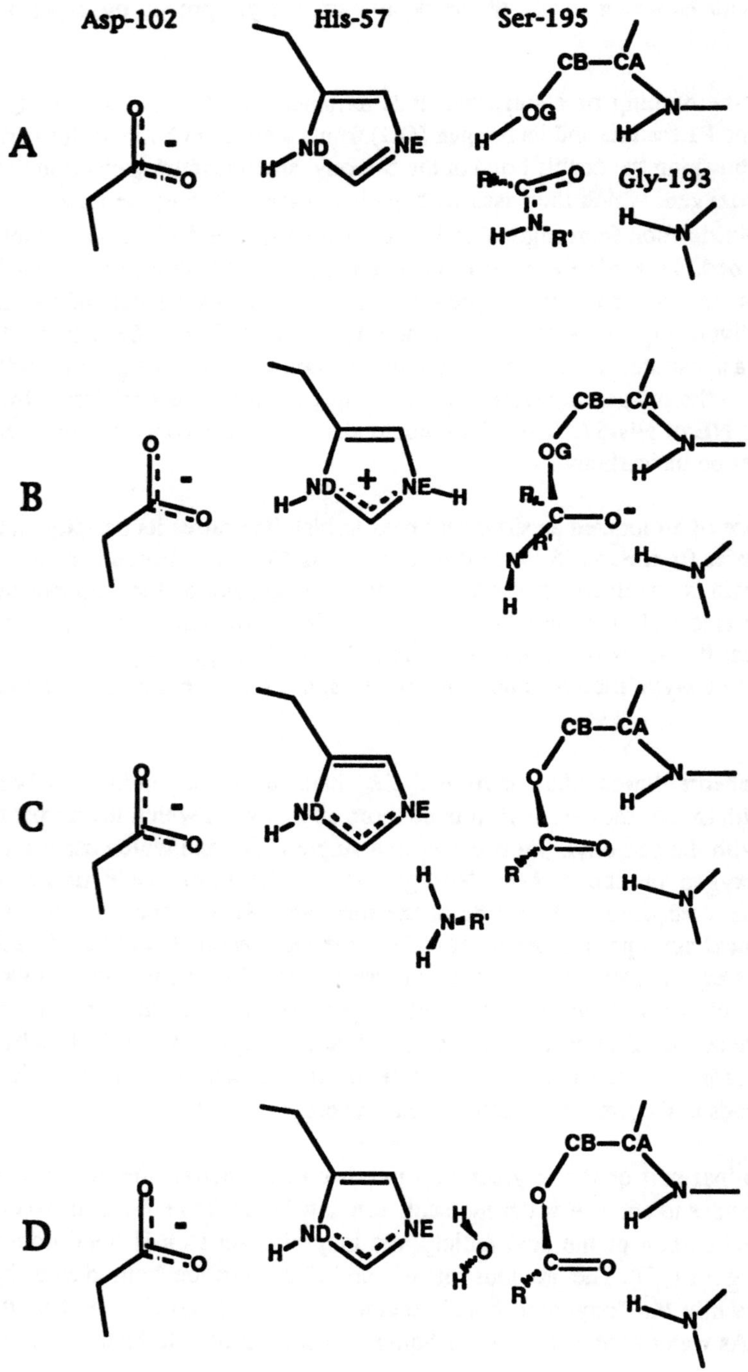

Figure 1. A few of the assumed steps in the catalytic cycle of serine proteinases

more of a hydroxyl anion, attacks the carbonyl more easily, transferring this carbon again from planar to tetrahedral. Another TI is formed, from which bond breaking of C-OG continues the pathway. As this bond breaks - the negative charge and basicity of OG increase and it attracts back the acidic proton of His NE. In the last step, the acidic carboxyl product is evicted from the active site, leaving the enzyme ready for a new cycle. This second part of the mechanism, the hydrolysis of the acyl-enzyme, is faster than the first acylation reaction with most peptide substrates. However, with ester substrates, the rate relations are reversed - and the hydrolytic deacylation step becomes rate determining.

The above detailed account of the mechanism is not decisive on many issues. First and foremost: which step is the rate-determining one, the "bottleneck" of the reaction, and by how much? Is there more than one step that requires a large activation energy? Is it possible that PT from Ser does not reach NE of histidine but is directed towards the nitrogen of the substrate? Could this happen in a concerted fashion, along with bond formation between Ser-OG to the carbonyl? or maybe with no formation of a bond between Ser and the substrate? Does the nitrogen undergo inversion or rotation in order to acquire the proton? Is only one proton being transferred during each part of the reaction? What is the effect of some residues in the vicinity of the active site on the rate of peptide cleavage? Why is deacylation faster with peptides than with esters ? What is the contribution of other residues to the activation energy (such as those of the "oxyanion hole")? How would mutations effect the structure and the energetics of the cleavage?

Many of the above questions have been discussed in recent experiments and theoretical investigations of the mechanism, and the main points in these studies are presented below.

Daggett et al. (1991) followed the reaction pathway of SP based on the 1.5Å resolution structure of bovine trypsin (Chambers and Stroud, 1977). An initial molecular mechanics (MM) refinement of the structure was conducted prior to a "gas phase" semiempirical QM study of a limited set of atoms that represent some of the catalytic relevant groups. The MM was mainly employed for removing bad contacts between hydrogen atoms that were added to the structure. It was also required for refining the enzyme's interactions with a tripeptide substrate (Ac-Phe-Val-Lys-Nme) that was aligned in the binding pocket. In addition, counterions were positioned near ionized residues on the surface (to an overall charge of -1) and TIP3P water molecules were added to a distance of 20Å from the hydroxyl of the active Ser-177. The minimized structure formed the basis for extracting the QM model positions of atoms and for MD studies. Active site residues were modeled with PM3 calculations (Stewart, 1989) by the following small molecules: methanol for Ser-177, methyl-imidazole for His-40, acetate for Asp-84, two water molecules for the N-H of Ser-177 and Gly-175 ("oxyanion hole") and N-methyl acetamide for the substrate's Lys-NMe group. An ester model, methyl acetate, was constructed to compare its reaction to the peptide cleavage. A study of proton affinities of isolated molecules and of their relative affinities was done prior to the construction of the reaction model (Schroder et al., 1991). It had been demonstrated that the quality of PM3 calculations for the modeled molecules is close to experimental values (for proton affinities) and to high level ab-initio studies with the basis set 6-31G*/MP2, for interactions between the constituents.

To begin the reaction sequence, the distance between OG from serine and the carbonyl carbon of the substrate was optimized. Proton transfer from serine to histidine was studied by the "reaction coordinate" method at the different distances of OG to the carbonyl carbon. This "grid approach" was employed to aid in the decision on the lower energy path for PT and for O---C bond formation. At each of the O---C distances (3.30, 2.70, 2.10 and 1.50Å) three positions of the proton were tested, at 1.77, 1.40 and 1.00Å from the nitrogen of histidine. Proton transfer is unfavorable at the larger O---C distances but becomes favorable between 2.1 and 1.5Å. Thus, it is a "late" proton transfer, preventing the formation of "free" methoxide. The formation of the O-C bond requires an activation enthalpy of 31.5 kcal/mol with the amide, and 22.3 kcal/mol with an ester model. This is clearly much larger than the experimental values, and the errors may be explained by the missing residues of the protein and by the lack of solvation effects. PT was found to be concerted with this bond formation. The authors beleive that this is consistent with kinetic isotope effect results (Schowen, 1988). However, an isotope effect has not been calculated. Moreover, When PT energies are overestimated, the potential surface should become more favorable to "concerted" reactions, which are not necessarily existent.

The tetrahedral intermediate (TI) formed at this stage is considered to represent the transition state structure, i.e., it is the highest energy species along the reaction path and should at least be close to the TS. This intermediate has its carbonyl oxygen hydrogen-bonded to the two water molecules that mimic the "oxyanion hole" and the nitrogen lone pair position may be judged from the geometry of the substituents on nitrogen to be nearly anti-periplanar to the newly formed O-C bond (Dutler and Bizzozero, 1989). There is a substantial contribution by both the "oxyanion hole" and the Asp anion to the reduction of the barrier: without the water molecules that mimic the N-H bonds of Ser-177 and Gly-175, the TI is raised by about 14 kcal/mol. Without Asp, it is raised by more than 25 kcal/mol. The effect of the two is additive. Most of the stabilization of TI by Asp is due to its interaction with the positively charged histidine. The residues that stabilize Asp-102 through hydrogen bonding have not been included in the calculation, probably due to computer time limits.

Histidine carries one proton on each of its nitrogens at this stage. It could transfer the proton on ND to an oxygen of Asp (this constitutes the "charge relay" system). Such a PT requires a small activation energy, of 3.9 kcal/mol, but gains nearly 16 kcal/mol due to the stability of the structure which has neutral His and Asp. If the path to TS requires about 30 kcal/mol, it should be accompanied by this second PT that requires about 1/10 of that energy. However, this calculated low barrier may be an artifact of the "gas phase" calculation that tends to oppose charge separations. Also, the residues that stabilize ionized Asp are missing in this calculation. The neutron diffraction study of a trypsin complex with a "transition state analog" (Kossiakof and Spencer, 1981) that found an ionized Asp still does not exclude the possibility that proton transfer to Asp may occur during the TS.

Subsequent steps towards acylation are characterized by energy reductions. Rotation around the single bond (1.53Å) of C-N (the amide bond is 1.34Å in the substrate model) has no barrier and brings the nitrogen to 2.84Å from the proton of histidine. This conformer of the TI is 3.9 kcal/mol more stable than the TI. It is not clear why this conformation was not achieved by minimization of TI. Also, the subsequent PT from histidine to the nitrogen of the substrate requires no activation energy and the C-N bond cleaves while the resulting methylamine product (or methanol, in an ester model) remains hydrogen bonded to NE of histidine. This structure, which is the acylated enzyme model and includes binding of the first product before it is displaced by water, is 11 kcal/mol more stable than the TI. Stabilization is much more prominent with an ester substrate, ~23 kcal/mol.

For the study of the second step, hydrolysis of the acyl-enzyme, a water molecule was positioned by superimposing it on the NH2 portion of the leaving product, methylamine, and its initial position was optimized. The water's oxygen is 2.52Å from the acyl carbon and its proton has an H-bond to NE of histidine, at a H---N distance of 1.78Å. The nucleophilic attack of water on the ester is accompanied by a concerted proton transfer to histidine and requires an energy of activation of 15.6 kcal/mol. An initial proton transfer to produce a "bare" hydroxide nucleophile is much more expensive energetically, requiring ~39 kcal/mol. The final step, PT from histidine to serine, reproduces the native state of the enzyme as the carboxylic acid product is formed by breaking the bond between Ser-OG and the carbonyl. This PT requires an activation energy of 14.2 kcal/mol, about half of the activation energy for the first step of the reaction. The product is hydrogen bonded to serine and to the oxyanion hole. The enzyme-product complex is de-stabilized by the aspartate, an effect which might be needed to regenerate the native enzyme.

Molecular mechanics (AMBER, Singh et al., 1986) was employed by the authors to study the effect of the whole system (protein + solvent) on the scissile bond in the Michaelis complex and in the TI. They find that interaction enthalpy is more favorable in the TI due to stabilization by the catalytic triad, by the oxyanion hole and by the environment, a total of ~26 kcal/mol. The breakdown to interaction components demonstrates the importance of electrostatic stabilization of the TI. More reliable estimates of these interactions require free energy perturbations or related free energy methods.

Thus, the results of Daggett et al. demonstrate the importance of those residues that were included in this model for the catalytic activity, and a TI for the reaction has been suggested. However, a TS has not been characterized in the regular manner, i.e., by the Hessian matrix at the point where the first derivative of the energy vs. reaction coordinate is zero (Flanigan et al., 1977). This is a problematic procedure for models of larger systems that have some "frozen" atoms, and should not be required for such models. Also, a difference between the rate limiting step for amides and esters has, unfortunately, not been detected.

Warshel et al. (1988) employed the EVB method (Warshel, 1981) to evaluate the effect of mutations in trypsin (G216A and G226A) and that of a few mutations of Asn-155 in subtilisin. The trypsin mutations cause deformation of the oxyanion hole, resulting in a change in catalytic energy of ~5 kcal/mol, which amounts to a rate reduction of $4X10^3$ (compared to $2X10^3$ from experiment). Results for mutations in subtilisin (Asn-155 to Thr, Ala, Leu and Asn, Wells et al. 1986; Bryan et al., 1986) were also found to be very close to experimental values for both binding and catalysis in the mutants relative to wt subtilisin. In an earlier paper (Warshel and Russel, 1986) the PT energy from Ser to His was calculated to be ~13kcal/mol from full EVB calculations. This PT was achieved along the attack of Ser-OG on the carbonyl of the substrate, which required itself an activation energy of ~21 kcal/mol to reach the TS. The experimental energy was estimated from kinetics (Fersht, 1985) to be around 18 kcal/mol. The charge-relay (double PT) mechanism was shown to be improbable due to the low pKa of Asp-102 in the protein (exprimental= ~3, calculated to be ~0 and therefore harder to protonate in the enzyme).

The role of aspartic acid in the mechanism has been questioned in a subsequent paper by Warshel et al. (1989), where a microscopic model was employed with the water solvent represented as Langevin dipoles (PDLD, Russel and Warshel, 1985). The enzyme was divided into three regions: the catalytic triad, the rest of the protein, and water molecules surrounding the active site. The catalytic triad was assumed to move along the reaction from the ground state to the tetrahedral intermediate via proton transfer and a transition state. Variable residual charges represented the catalytic triad, permanent residual charges and atomic induced dipoles represented the rest of the protein, and Langevin-type dipoles represented water molecules. Electrostatic interaction with the different charge distributions in the triad was calculated by summing the coulomb interactions in the triad, the interactions of the triad with permanent dipoles of the protein, the energy of polarization of protein-induced dipoles in the protein by the triad charges, and the free energy of polarization of the Langevin dipoles of surrounding water. For the free energy between any state and the ground state, an additional non-electrostatic contribution was added, for the formation of that state from the ground state. The effect of mutations can be easily accommodated into this scheme.

The outcome of mutating Asp to Ala (Craik et al., 1987; Carter and Wells, 1988) in the model is to raise the TS energy by more than 4 kcal/mol. This replacement reduces considerably the negative potential on the catalytic histidine. The results with PDLD were compared to more extensive calculations with empirical valence bond and free energy perturbation (EVB/FEP, Warshel et al., 1988) and were found to be quite similar. The free energy for the double proton transfer mechanism was evaluated in two steps, first by estimating the free energy for the reference reaction in water (~6 kcal/mol). The solvent was replaced by the protein environment, and this raised the energy so that the double PT mechanism was disfavored by about 12 kcal/mol with respect to the electrostatic (ionized Asp) one. The authors argue that ionized Asp is even more stable in the protein environment than in water, due to its hydrogen bonding relations with neighbor residues and as evident from its experimental pKa (~3 in chymotrypsin). To these hydrogen bonds with Asp, the term "aspartate hole" was suggested. It was also recommended to mutate Ser-214 (trypsin) to test the relevance of this idea. Modeling of SP mechanism by the empirical valence bond method and free energy perturbations has been reviewed by Warshel (1991).

The polar cavity of Asp-102 also includes two buried water molecules, as well as backbone amides of Ala-56 and His-57. In about 200 SP, all but three have this Ser-214 residue. It is solvent inaccessible, and its OG forms a hydrogen bond to Asp-102. In an extensive experimental study by McGrath et al.(1992), Ser-214 has been mutated to Ala, Glu and Lys in rat anionic trypsin and both kinetics and crystallography were employed to study the structural implications. The side chains of the Glu and Lys mutants are buried. The terminal N of Lys is solvent inaccessible and is 5Å from Asp-102. S214K has 1% the catalytic activity of the wild-type on a tripeptide substrate and S214E has 44%. S214A, however, was more active than trypsin (130%). This could be the result of an increased polarizability at the active site due to an additional water molecule replacing OG of Ser-214 in the Ala mutant. K_m values increased ~30 times in the Lys and Glu mutants, but remain unchanged with the Ala mutant. The decrease in activity of the Lys mutant could be due either to structural changes or to the electrostatics exerted by Lys-214. Delphi calculations (Gilson and Honig, 1988) suggest that between 46-79% of the loss in k_{cat} could be due to electrostatic effects.

The authors suggest that Ser-214 may be called the fourth member of the catalytic triad in SP. However, removal of its -OH group (such as in the Ala mutant) does not reduce activity. The Lys and Glu side chains were thought to replace the two water molecules near Ser-214, but instead they displace the side chain of Trp-215, which results in inefficient recognition of substrates. This is an example that any mutation may result in structural changes, even though local, as the size of side chains changes. It is important to remember that water molecules are also part of the structure in many cases. It was recently discovered (Fitzpatrick et al., 1993) that structural water molecules bind strongly to subtilisin Carlsberg, so that most of them (99 out of 119) are not displaced even in anhydrous acetonitrile, and hydrogen bonds between the active site triad residues remained intact under such extreme conditions.

In addition to the stabilizing electrostatic effect on the aspartate, Braxton and Wells (1991) demonstrated a stabilizing effect on the transition state in the oxyanion hole of subtilisin BPN' by OG from Thr-220, which is next to the catalytic Ser-221, but at a distance of 4Å away from the developing oxyanion. Another donor in the oxyanion hole of subtilisin, Asn-155, contributes 3-5 kcal/mol stabilization to the TS (Wells et al., 1986; Bryan et al., 1986), with a minor contribution to binding of the substrate. Earlier molecular dynamics and free energy perturbation studies (Rao et al., 1987) suggested that Thr-220 could hydrogen bond to the oxyanion, and this prediction was tested by mutating it to Ser, Cys, Val and Ala. T220S, retaining OG, was only 0.5 kcal/mol less stabilizing, while removal of OG in the other mutations reduced the stabilization by 1.8-2.1 kcal/mol. All the effect stems from changes in k_{cat} and not K_m. A double mutant, T220A + N155A, demonstrates the independent effects of those two (Thr-220 and Asn-155) contributors to TS stabilization. T220A is less than 10 times slower than wt, which is interpreted as a possible participation of a water molecule that can occupy the empty space created by the smaller side chain. Even if Thr-220 is not directly H-bonding to the oxyanion at the crystallographic distance, it may be regarded as an "oriented dipole" (Warshel, 1987) which is highly conserved in the subtilisin family. With the use of Warshel's relation for dipole-charge interaction (Warshel and Russel, 1986) : $\Delta G = 332Q\mu/\varepsilon r^2$ (Q is the charge= -1, μ is the OH dipole= -0.467, r is the separation= 4Å and ε is the effective dielectric = ~4) one gets 2.0 kcal/mol for the effect of Thr-220 and 4.5 for the effect of Asn-155. This charge dipole interaction is operating on the TS.

Support for an even more remote electrostatic effect on the mechanism is supplied by the studies of Jackson and Fersht (1993). They mutated charged residues on the surface of Subtilisin BPN', that are 13-15Å from the active site, to either neutral or oppositely charged residues. The effect of those mutations on the inhibition constant, Ki, of a trifluoromethyl ketone, was compared for wt and mutated subtilisin. The mutations were: Asp-36 (located on a surface loop outside the active site cleft and separated from His-64 by about 15-16Å) to Gln, and of Asp-99 (about 12-13Å from His-64) to Ser and to Lys. The active site of subtilisin includes Ser-221, His-64 and Asp-32.

The effect of the mutations on Ki was employed to characterize the charge state of the His-oxyanion pair in the tetrahedral intermediate with the inhibitor, regarded as a "TS model". All the mutations increased Ki, and the affinity to the TS analog is mildly reduced under a hundred-fold increase of ionic strength (0.01-1M). The measured effect was smaller than 1 kcal/mol [RTln(Ki,mut/Ki,wt)]. Thus, it is much smaller than the effects of the closer residues, those that have hydrogen bonds to the catalytica triad. The experimental results were compared to theoretical calculations of electrostatic interactions (with DelPhi, Gilson and Honig, 1988), and the best fit was found with both His-64 (positive charge) and the oxyanion being ionized. Both Asp-36 and Asp-99 are closer to His-64 than to the oxyanion, and their replacements with a neutral (D36N, D99S) residue loses the small stabilizing effect by the negative charges, while a positively charged residue at this position (D99K) is more destabilizing. The effect is stronger for the transition state structure, because water is excluded from the active site when a substrate enters, while it is smaller for the ground state, where the solution may enter the site and "screen" some of the electrostatic interactions.

The contribution of Asp to maintaining His in such an orientation that can accept a proton from Ser is considered to be of minor importance since the electrostatic contribution to the difference in free energy between native and mutant serine proteinases accounts for most of the calculated effects of the mutations (Asp to Asn in trypsin, Asp to Ala in subtilisin). The other stabilizing effect is of the oxyanion hole on the negatively charged TI, which is also electrostatic in nature.

Stamato and Tapia (1988) discussed the catalytic mechanism of ester hydrolysis, which is known from experiments to have a different slow step than the cleavage of peptides and amides. Ab-initio computations were employed with a minimal representation of serine by methanol, histidine by ammonia, and the substrate by methyl acetate (original coordinates were extracted from chymotrypsin). Reaction field effects were introduced by computing the coupling factor of the system with the surrounding (solute-solvent coupling) that included the dielectric constant of the medium and the radius of the spherical cavity in which the system was placed. values of the dielectric constant and of the distance r were varied in order to study the effect of the medium on the energy gap between different protonation states of the catalytic dyad (with Asp missing). Computations were carried out at a split valence shell 3-21G basis set level in the RHF (restricted hartree-fock) formalism. Three aspects of the mechanism were studied: approach of the substrate to the active site, stability of ionized vs. neutral in the His-Ser pair (methanol - ammonia), and the effects of reaction fields of increasing strength on the active site-substrate interaction. It was found that for a CH3OH-substrate distance smaller than 2.5Å, repulsive forces build up against further approach. However, with deprotonated CH3OH, no such repulsion is present. The stabilization with the methoxide is 16 kcal/mol at a separation distance of 2.4Å compared to the interaction at 5.0Å. The nucleophilic addition in the gas phase proceeds without an activation barrier. Solvation effects are necessary to create a thermodynamic barrier for the reaction in solution. Such a barrier has been shown by a previous study that included the effect of solvent water (Weiner et al., 1986).

Although neutral methanol and ammonia are more stable in vacuo than their ions, the reaction field is capable of inverting this gap. At 3.0Å as the spherical cavity radius, the di-ionic form becomes more stable. The tetrahedral substrate can approach the dyad to a shorter distance than the planar substrate. The repulsive barrier occurs at distances shorter than 2.5Å for the planar, but only at 2.0Å for the tetrahedral. The tetrahedral substrate is more stabilized by the reaction field effect than the planar substrate, due to an increase in the in-vacuo dipole moment of the tetrahedral. The reaction field is supposed to mimic the protein surrounding, and it is proposed that the protein stabilizes the diionic form even though the simulation of the reaction field is not sufficient to obtain a realistic interpretation. This study indicates a tendency to tetrahedralization of the model substrate at distances characteristic of the Michaelis-Menten complex formation. The authors believe that this must affect intermolecular interactions of large substrates.

The mobility of histidine and conformational fluctuations in the catalytic cycle were addressed by Sumi and Ulstrup (1988) who constructed a model based on electron and atom transfer theory, incorporating the dynamics of conformational nuclear modes as a new element, based on experimental evidence for the mobility variations of His-57 upon enzyme-adduct formation. The model has been applied to the acylation step in SP. Conformational modes of the residues were represented in the form of harmonic oscillators. From these, the authors estimated the contribution of the conformational modes to activation free energy and to nuclear proton tunneling. As a result of charge alignment of aspartate, protonated histidine and the tetrahedral intermediate, the curvature of the potential surface in this state is much lower than in both the initial state (free enzyme) and in the acylated state. This results in storing of the energy liberated on first proton transfer in the conformational system , being utilized towards the second proton transfer, without being dissipated to the surrounding protein atoms or to the solution. This, however, may result from the values that were employed for the partial model. In other investigations of the dynamical effects on protein reactions, it was found that the relaxation times of enzymatic reactions and the corresponding reaction in solution are not different in a fundamental way, and thus it is not probable that enzymes use dynamical effects as an important factor in catalysis (Warshel, 1991).

The approach of the substrate towards the active site of SP was studied with electrostatic field potentials (ESP) by Lamotte-Brasseur et al. (1990). They examined the force exerted by the active sites of the serine proteinases alpha-chymotrypsin and subtilisin on an approaching substrate. About 20 residues from each of these enzymes were employed to construct electrostatic potential maps, using point charges derived from CNDO calculations (Pople and Beveridge, 1970) with $\varepsilon = 1$. Net charges were obtained from a mulliken analysis of the deorthogonalized MO coefficients. ESP was calculated with a complete CNDO scheme and nuclear attraction integrals were approximated by the opposite of the repulsion between s orbitals. The most salient feature of the 3D maps is the negative volume or nucleophilic suction pump that is generated by the active site. This volume bears a narrow funnel which is located in the immediate vicinity of the active site Ser-195 in chymotrypsin and Ser-221 in subtilisin, and which orients itself towards Ser-214 in chymotrypsin or Ser-125 in subtilisin. The authors suggest that this is important for orienting the ligand peptide, the nucleophilic "suction pump" acting as a magnet towards the peptide's electrophilic part.

Chymotrypsin hydrolysis of spin-labeled ester substrates was studied by Electron Nuclear Double Resonance and molecular modeling methods (Wells et al., 1994). The spin-labeled acyl-enzyme was stabilized in low temperatures, and conformations of the substrate in the active site have been assigned from the experiments - both free in solution and in the active site. Conclusions from this study are that significant torsional alteration in the substrate's structrue occurs between its "free" form in solution and its bound form in the active site. The enzyme does not "recognize" the solution structure, but an altered one, that is stereospecifically complementary to the surface of the active site.

One study of the preferred state of the residues in the "resting state" of SP suggests a very different mechanism. In this earlier study, By Rauk et al. (1987), a partial model for the active site was employed and the method of choice was MNDOC (Thiel, 1981). Methanol, 4-methylimidazole and acetic acid represented the active site, but no substrate, oxyanion hole or other protein/solvent effects were considered, and planarity was imposed on the molecules, while allowing full optimization of their relative positions. Heats of formation were evaluated after addition of correlation energy from second order Brillouin-Wigner perturbation theory with Epstien-Nesbet denominators involving all molecular orbitals. They found that protonated (neutral) acetic acid is favored by 16.6 kcal/mol in its interaction with a negatively charged imidazole, compared to the interaction of neutral imidazole with ionized acetate. A similar preference for a negatively charged imidazole is found for its interaction with methanol. In an arrangement of the three molecules of which one carries a negative charge, the lowest energy is found with a negatively charged imidazole while the two others are neutral, 18.7 kcal/mol more stable than an alternative arrangement with ionized acetate. The authors argue that the proton should be transferred to acetate in a "charge transfer" system due to the hydrophobic nature of the space around the triad, where only few molecules can enter. Negatively charged His should be reactive and attack the substrate itself, rather than the expected attack by serine. An acylated histidine is thus formed. Acyl-imidazoles are unstable species and decompose readily, and this could be the reason for not isolating these acyl-enzyme derivatives. Rauk et al. interpret the existence of well known acylated serine derivatives of active site directed reagents to signify that acylated Ser deacylates more slowly than the major His acylated species. The value for the preferred enthalpy of the protonated acetate in its interaction with His/Ser is quite similar to the result of Dagget et al., who found 15.7 kcal/mol stabilization for the state with protonated Asp, albeit with a larger model that has a neutral histidine and also includes the tetrahedral intermediate and the oxyanion hole. Apart from that comparison, this study of Rauk et al. differs from all others in its results and analysis. It is upsetting that the most expected arrangement with two anionic species (Ser and Asp) interacting with a positive histidine between them, has not been mentioned in this study. Thus, the initial choice should have been to study the Asp(-)---His---Ser and then to test the proton transfers to Asp(-)---His(+)---Ser(-) and to Asp---His---Ser(-). Also, there is no report of the proton affinities of each of the species employed for the study, so that errors due to over- or underestimation of the relative affinities by MNDOC can not be evaluated. This method is not known to be successful for hydrogen-bonded species, on both energies and geometries. Furthermore, it is hard to determine the energies of different ionization states without including some representation of the solvent.

The second step of the mechanism is deacylation, which involves an attack by water on the carbonyl of the newly formed ester. This step is rate-determining with ester substrates but is faster than acylation with peptides.

Tonge and Carey (1992) extended a correlation between $\nu_{C=O}$ and logk_3 (the rate of deacylation) over a range of 16,300-fold reactivity ratio (k_3). The data indicate that such an increase of k_3 is associated with a shift in $\nu_{C=O}$ of -54 cm^{-1}. This is also correlated with a bond length change of 0.025Å - representing approx. 11% of the change expected from a formal double bond to a single bond (C--O). In the transition state, the bond length of the carbonyl is expected to be close to a C-O (single bond) length. Site directed mutagenesis in subtilisin BPN' (of Asn-155) demonstrates that the increase in k_3 and the decrease in $\nu_{C=O}$ may be attributed to stronger H-bonding of the acyl carbonyl in the oxyanion hole. Sufficient energy is available in the range of H-bonding strengths to account for the change in activation energies in this series. Enthalpies of H-bonds are also correlated with the ΔG of activation: the equivalence of the enthalpy and entropy supports the intuitive notion that entropy is not an important factor in deacylation. Also, the 16300-fold modulation is caused by changes in ground states of the acyl enzymes and the TS throughout the series is invariant.

Bemis et al. (1992) investigated the difference of ~60-fold between the hydrolysis of two acyl-chymotrypsin derivatives, following the reaction with enantiomeric lactones that leads to β-substituted β-phenylpropionic esters. The S-ester (k_d = 0.17 min^{-1}) is cleaved faster than the R-ester (0.0029 min^{-1}). By modeling the reactions of the enantiomers, the authors hope to avoid the errors that might be encountered when different molecules are modeled.

Molecular mechanics minimization and molecular dynamics were chosen to examine the possible conformations for the two acyl enzymes and conclusions were drawn from the time evolution of the two systems. The starting point was a crystal structure of phenylethaneboronic acid bound to alpha-chymotrypsin. QUANTA/CHARMM (Brooks et al., 1983) was employed for the calculations. Ninety-five water molecules from the X-ray structure were included. Distance monitoring and the creation of H-bonds were the main criteria for differentiating between the two molecules. Both acyl enzymes have their ketone carbonyls H-bonded to Gly-216 NH. Both start with their ester carbonyl in the oxyanion hole (H-bonded to Ser-195 and to Gly-193). The R-acyl enzyme looses both of these hydrogen bonds during the simulation. Attack of water on the R-species should, thus, be less frequently successful. Values for differences in energy were not used because of a small difference, of 6.0 °K, between the average simulation temperatures for the two reactions. If this temperature difference is ignored, it may be stated that the R structure (that is less readily hydrolyzed) is less stabilized in the TI structure with respect to the acyl-enzyme "ground state", compared to the S structure. The TI structure was used in this study as a TS analog. However, the energy differences are much too large than needed to account for the experimental differences. And, since the calculations did not include FEP or a related evaluation of energy, they should be regarded as mostly qualitative.

Nakagawa et al. (1993) employed stochastic boundary molecular dynamics simulations to study the motion of water molecules, of the acyl moiety, the catalytic triad and the oxyanion binding site in acyl-chymotrypsin. They employed the refined structure of γ-chymotrypsin that was inactivated by 3-benzyl-6-chloro-2-pyrone, replacing the acyl moiety by N-acetyl phenylalanyl, the acyl part of this highly reactive ethyl ester. γ-chymotrypsin was previously studied by crystallizing it from a range of pH values and solving at refinements of 1.6-1.8Å resolution and has been shown to be fairly invariable at large pH differences (Dixon et al., 1991). The available X-ray and neutron diffraction structures do not provide adequate data concerning the solvent in the active site due to the disorder in most of the water surrounding the protein.

Following the addition of water molecules by several consecutive procedures and after deletion of more remote parts of the protein, the protein was divided into a reaction region, a buffer region and a reservoir region, and separate equations of motion were applied to each. In the final model, about 40% of the enzyme was retained. Two trajectories have been investigated, which differ in the Langevin random force applied to water molecules that cross the boundary between the buffer and reservoir region.

The time series for distance fluctuations shows a water molecule hydrogen bonding to NE of His-57 with an Ow-----N average distance of 3.05Å and a Hw----N average distance of 2.4Å. These distances correlate well with the crystallographic position reported for water. The average distance of the water's oxygen to the carbonyl of the acyl moiety is 3.95Å, a somewhat large value. A water molecule was also found in an acyl-enzyme complex of trypsin (Singer et al., 1993) but was missing in cryoenzymology studies of an acyl-enzyme complex of elastase (Ding et al., 1994). Thus, the authors find in the equilibrium simulations that active site residues, substrate and water are stabilized in the proximity of each other, which may compensate for the entropic cost required for such reactions in bulk solvent. An exchange of this water molecule with solvent takes place in the other trajectory. This may support the expectancy that water molecules should be mobile in the active site and enable substrate binding. On the basis of the dynamic production results the authors conclude that configurations that are necessary for a concerted mechanism (water attack on the carbonyl and PT from water) were sampled. The authors state, however, that such a method, that is based on empirical potential without being designed for reactions, can not probe covalent bond forming or breaking. In the simulations, His-57 remains in the gauche form during most of the simulation and does not have to rotate much in order to align the water molecule correctly for attacking the carbonyl. The average energy of interaction between His and water is -3.9 kcal/mol. In addition to His-57, the other active site residues are fairly rigid due to hydrogen bonding with neighbor residues. The active site side-chains are 40% less fluctuating then the rest of the side-chains. This reduced flexibility is consistent with the idea of "substrate freezing at reaction center" in enzymes (Mildvan, 1974). Three hydrogen bonds anchor the buried carboxylate of Asp-102 (from His-57, Ser-214 and Ala-56).

While in the trypsins the oxyanion hole is comprised of main chain N-H groups, in subtilisin-BPN' one N-H is donated by the side chain of Asn-155. Whiting and Peticolas (1994) investigated the nature of the binding of an acyl moiety to the oxyanion hole in subtilisin by Raman difference spectroscopy. On the basis of the results, they postulate that a tetrahedrally distorted carbonyl exists in the intermediate acyl-enzyme and that this distortion is essential for strong oxyanion hole hydrogen bonding.

A peptide (73 residues) which has chymotrypsin-like esterase activity has been designed and synthesized by Hahn et al. (1990). The peptide consists of a bundle of four short helical sequences, and the catalytic residues were synthesized at the amino end of the bundle in spatial relationships similar to chymotrypsin. Also, the oxyanion hole and substrate binding pocket for acetyltyrosine ethyl ester, a chymotrypsin substrate, have been included in the design. the peptide cleaves this ester at a rate that is 1/100 of the chymotrypsin rate. It is also inhibited by chymotrypsin inhibitors.

Tong and D'Souza (1995) calculated two smaller molecular models of acyl-chymotrypsin that were designed earlier by experimentalists to demonstrate the effect of a carboxylate on the hydrolysis of the ester. Both models include an imidazole, an acidic moiety that can hydrogen bond to the imidazole, and an ester group to be hydrolyzed (figure 2). The oxyanion hole is missing from both models. In the model of Bruce (figure 2, left; Rogers et al., 1974) the effect of the acid was found to be negligible, while in the model of Bender (figure 2, right; Mallick et al., 1984) the effect was very large, more than five orders of magnitude acceleration.

Calculations were performed with MACROMODEL and BATCHMIN (Homamadi et al., 1990). TS study was done with restraining interatomic distances and setting huge force constants to those distances. Initial, intermediate and transition states were minimized. A combination of aqueous accessible surface area based treatment and the generalized born equation were used to simulate the aqueous experimental conditions of reported results. Solvent-solvent, VDW solvent -solute and solvent-solute electrostatic polarization were fully considered. Similar TS structures were assumed for all 4 species studied. Results from computations with these small models suggest that the functional groups present in an enzyme should decrease the distance between the reacting atoms, and the energy of this conformation should be low enough for these two atoms to spend sufficient time in that state for effective rate acceleration. A serine proteinase without a carboxylate group should increase the flexibility of the histidine and reduce its effectivity as a general base catalyst. The carboxylate serves to keep it in a position that enhances catalytic activity. Mutations of the aspartate in trypsin (D102A, Craik et al., 1987) and subtilisin (D32A, Carter and Wells, 1988) cause a reduction of k_{cat} by about 10^4-fold. This has been reproduced quantitatively (Warshel et al., 1989) in calculations that consider both flexibility and electrostatic effect, demonstrating that it is probably due to electrostatic effects alone.

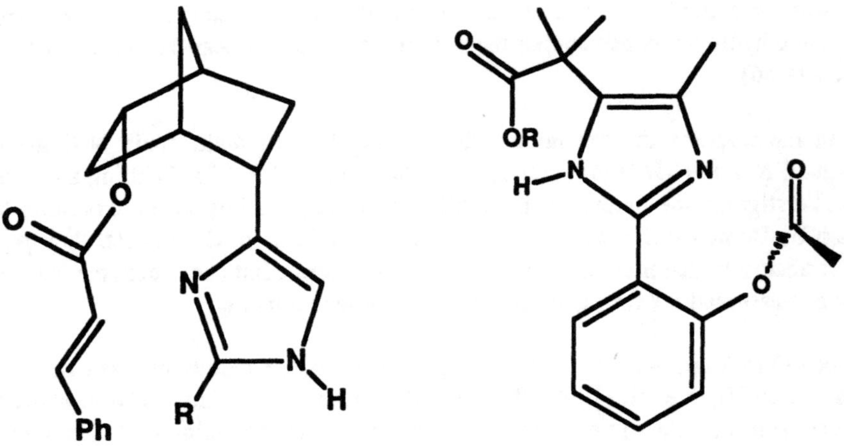

R = o-benzoic acid
R = H

R = H
R = CH_3

Figure 2. Two small-molecule models for the catalytic triad of serine proteinases

For the mechanism of SP, an advanced level of relationship between theory and experiments has been achieved by now. The results for some of the effects of residue replacements are close to experimental, and in other cases, they demonstrate the relative importance of the catalytic steps. Predictions by theory have encouraged experiments, with some success. Still, many more studies are required which in particular compare alternative pathways in a similar system and by the same methodology. Such studies, if employing QM methods, should desire to compare the basic components of their models to experiments in gas phase (such as proton affinities) and scale them properly in their final results for the SP model. Also, an increased representation of environment may be feasible, at least in parts, so that some idea of the contribution of more remote residues and of solvent molecules can be obtained. In MM and MD studies, that are most adapted to structural problems and structural design, some parameters have been found that may relate to reaction probabilities. This is also a direction that should be pursued, with constant refinement of the methods and longer simulations. QM/MM studies have a great potential of combining the benefits of the two approaches and hopefully, more of those should be pursued. Another important aspect is to calculate the pKa values of the catalytic triad residues as they change along the binding of substrates and along the reaction pathway (Finucane and Malthouse, 1992). Also, studies of the positions and roles of water molecules in the native structure of SP and in the acylated enzymes are necessary for improving our understanding of SP activity.

3. Cysteine Proteinases

Cysteine proteinases have been studied by theoretical methods less than the serine proteinases. Sulfur has more electrons than oxygen and this is restrictive for ab-initio computations. For semiempirical or classical computations, parametrization depends on the size of the data base, which is smaller for sulfur compounds than for the first row elements.

In cysteine proteinases (CP) , activity depends upon a cysteine thiol group. Papain, actinidin, and a few lysosomal cathepsins are the best known members of this family. They hydrolyze peptides and esters in a generally similar fashion to serine proteinases. Two residues are directly involved in the catalytic process, Cys and His, with apparent pKa values of 4.2 and 8.6, respectively. This is the reverse of their normal pKa order, with His being more acidic than Cys. A bell-shaped relation of activity vs. pH for papain is spread out, with maximal activity around pH=6.5 and about half of the activity is retained near 4.5 and 8.5, dropping fast below and above these values. Thus, the active species has a zwitterion state, with a cysteine anion (thiolate) and a histidine cation. CP were discussed in a few reviews (Baker and Drenth, 1987; Fersht, 1985; Polgar and Halasz, 1982)

The first step in peptide cleavage catalysis is believed to be a nucleophilic attack by thiolate on the peptide's carbonyl, forming an intermediate oxyanion, which is stabilized by hydrogen bonding from an oxyanion hole. This is comprised of NH from Cys-25 and of NH2 from the side chain of Gly-19. The tetrahedral intermediate that is formed in this step is analogous to the TI of SP. Proton transfer from NE of His is accompanied by C-N bond cleavage and formation of the intermediate acyl enzyme (thioester). This group is attacked by water, while His abstracts a proton and increases the basicity of the nucleophile. Despite this similarity to SP, papain has a low selectivity compared to SP. The oxyanion hole of papain is less stringent in its spatial demands. The rate of cleavage by SP is larger than that of papain by a factor of 10^2-10^3. The structure of papain shows that the catalytic Cys-25 is located at the N-terminal of an alpha-helix (residues 24-43, "L-I helix"). His-159 is hydrogen bonded to Asn-175 on the far side of Cys, instead of Asp in the SP. The ND nitrogen of histidine faces the sulfur, while it is NE in SP.

How similar are the two classes of proteinases? How do the structural differences affect the activity of CP ? Are the relative rates for acylation and deacylation similar in CP and SP ? How does the more polarizable and less electronegative sulfur compare to oxygen along the formation of the TI, and during hydrolysis of a thioester vs. an ester ?

Rullman et al. (1989) studied the initial proton transfer of Cys to His with a Hartee-Fock SCF direct reaction field (DRF) method, based on the refined X-ray structure of papain (Kamphuis et al., 1984). Parts of the active site residues were represented quantum mechanically and the environment was represented by partial charges and polarizabilities. The "QM motif" consisted of methanethiol (modeling Cys-25), imidazole (for His-159) and formamide (for Asn-175). All atoms at the vicinity of the active site were included, except for atoms that are too close to the active site atoms, which were deleted from the structure to prevent convergence problems. In addition, solvent effects were studied by means of a Monte-Carlo calculation of crystal water molecules that are located near the active site. In the initial structure of the active site residues, a planar arrangement was assumed (different than the X-rays), bringing the sulfur of Cys to a distance of 3.38Å from the nitrogen of His. PT energy was calculated as a function of the proton's position on a line joining N (His) and S (Cys). The proton was moved in small (0.1Å) steps.

In the gas phase, the zwitterionic S^----H-N^+ is higher in energy than the neutral S-H----N, but the protein environment stabilizes it by about 10 kcal/mol, and the two states become nearly equienergetic. Non-planarity, according to the X-ray structure, increases the difference between the two states by ~15 kcal/mol in favor of the neutral one. Asn-175 side chain stabilizes the zwitterion relative to the neutral by more than 4 kcal/mol. Similar values for the above differences between zwitterion and neutral were found, also, from classical interaction energies. This was not the case for the X-ray structure, which has the S atom out of the plane of His, and gave different results with the two methods. In addition to the large stabilizing effect of the L-I helix on the zwitterion relative to the neutral (~12.5 kcal/mol, most of it, 7.5 kcal/mol, is due to the main chain atoms of Cys-25) and the effect of Asn-175, also Asp-158 (~3.5 kcal/mol) and the solvent (~7 kcal/mol) contribute significantly. The largest energy factor contributing to the possible existence of an anionic Cys and protonated His is, however, the direct electrostatic interaction between them. Their binding energy amounts to ~100 kcal/mol compared to less than 2 kcal/mol for the neutral side chains. This compensates for the large difference in proton affinity, ~120 kcal/mol in favor of the neutral state. The electrostatic stabilization is very sensitive to the distance between the groups. At the planar arrangement, the difference between zwitterion and neutral states enables switching between the two, so that an equilibrium mixture is reached. This may change along substrate binding. It is also interesting to note that charged residues (Lys, Arg, Asp, Glu) in the protein did not affect the difference considerably: neutralizing these side chains reduced the stabilization of the zwitterion by only ~2.5 kcal/mol.

In a subsequent study, Dijkman and Van Duijnen (1991) reexamined the effect of Asp-158 on the stabilization of the zwitterion vs. neutral. As a negatively charged neighbor of His-159, it could affect proton transfer between Cys and His. Asp-158 is solvent accessible, so solvation could minimize the effect. A small part of the protein was included in this study, comprised of those residues that have an atom at a distance of 5Å from the sulfur of Cys-25. Solvent was introduced in the form of polarization charges located on the molecular Connolly surface. The integral form of the Poisson equation was solved by a two-dimensional integration technique, and the resulting polarization charges were employed in the DRF formalism (Thole and van Duijnen, 1980) to calculate the interaction between the active site and its environment. The conclusion is that a stabilizing effect of the solvent is operating, and it also masks the negative charge effect of Asp-158, which would otherwise be substantial. Thus, Asp-158 has no electrostatic effect on the proton stabilization or proton transfer between Cys-25 and His-159. One difficulty with the present calculation is the small size of the protein that is modeled, which is accessible to solvent much more than it would be if more parts of the protein were represented.

The results of an EVB/FEP for proton transfer from Cys to His were presented by Warshel (1991), based on a calibration of the protein reaction surface with solution results, which amount to 6 kcal/mol difference (in favor of SH relative to ImH+) due to the pKa values of the two conjugate acids. The protein is found to invert the stabilization of covalent and ionic state relative to their order in solution. This is a result of the stronger solvation in the enzyme, compared to water, and due to the orientation of protein dipoles.

To study the effect of Asp-158, Menard et al. (1991) mutated this residue to Asn, Glu, Ala and Gly, and found a small decrease in kcat/Km for the mutation to Asn, while other mutations caused larger effects. They attribute the effects to hydrogen bonding of Asp-158 (which remains with the mutation to Asn) and not to its charge.

Arad et al. (1990) simulated the reaction sequence of papain by constructing several enzyme-substrate models with molecular mechanics and following reaction paths with semiempirical quantum mechanics. AMBER force field (Weiner et al., 1986a) was employed for the construction. AM1 (Dewar et al., 1985) results for proton affinities of the modeled molecules were compared to 4-31G and to experiments. AM1 underestimates the proton affinities of methanethiol and of imidazole but overestimates the proton affinity of methanol. However, the proton transfer reactions from methanol to imidazole and from methanethiol to imidazole are overestimated by only 6 and 11 kcal/mol, respectively, and PT from imidazolium to formamide is underestimated by 6 kcal/mol.

Two minima were obtained from molecular mechanics. One is an "ion-molecule" (IM) complex, in which the peptide's N-H hydrogen bonds to the negatively charged sulfur (S---H of 2.2Å). The C=O is 4.0Å from S^- and forms an angle of 160° with it, so that the peptide is nearly in a parallel plane "above" the plane of the active site residues (figure 3). The other type, the conventional "oxyanion hole" bonding (where N-H from Gln-19 replaces the N-H of Gly-193 from pepsin, the other N-H donated by the backbone of Cys-25), is higher in energy than IM by ~30 kcal/mol, mostly due to better electrostatic interactions in the IM complex. The authors propose that the Michaelis complex may be better described by this IM structure and that it is unique to cysteine proteinases because of the charge separation in the native state, whereas in serine proteinases the native structure is neutral. The formation of two different low energy adducts is supported by nmr results (MacKenzie et al., 1986).

From this Michaelis complex, two routes are possible. One is bending the substrate's carbonyl towards the oxyanion hole so that the nitrogen gets closer to His ND-H. In the other, the carbonyl can bend with its oxygen closer to the histidine's ND-H. A short H-bond is formed in both cases between His-159 (donor) and either oxygen or nitrogen of the substrate. The two are judged by Arad et al., on the basis of their internal and interaction energies, to compete in the reaction pathway.

AM1 "gas phase" calculations on a smaller system, including Gln-19, Cys-25 and His-159 from the enzyme and Ace-Ser-Ile-Nme as the substrate, also find the IM structure more stable than the oxyanion one, but to a much lesser extent than in the MM results, only ~3 kcal/mol, following optimization. In the IM, the S---C=O distance is 2.9Å and the distance between the His N-H to the oxygen of the substrate is 1.9Å. The peptide remains nearly planar. In the oxyanion structure, S---C=O is 2.04Å and a TI is formed, with oxygen receiving H-bonds from both NH of Cys-25 and NH2 of Gln-19.

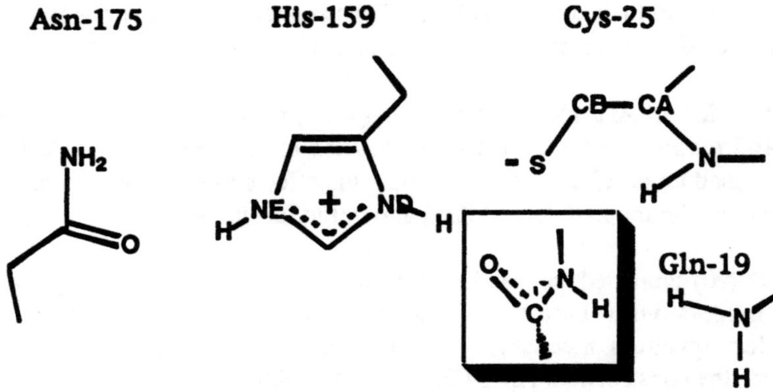

Figure 3. Active site residues of papain and the "parralel" position of the substrate

In the IM structure, the distance S---C=O was reduced in steps up to a distance of 2.0Å, where the energy was raised by ~29 kcal/mol, a similar value to previous results with ab-initio calculations (Howard and Kollman, 1988). PT from His to the carbonyl's oxygen was initiated at a S---C=O distance of 2.2Å, and a barrier for the PT of ~24 kcal/mol was found. It thus seems unlikely that PT from His to the oxygen occurs prior to sulfur attack on the carbonyl. However, tunneling effect could reduce this barrier.

In the oxyanion structure, PT from His to the substrate's nitrogen proceeds with no barrier if the imidazole (His model) is allowed to optimize. This structure is lower in energy compared to the oxygen-protonated species, by 5.5 kcal/mol. The accuracy of the model and of the calculations does not enable a decision which path is preferred, although it seems that the two could occur. An attack by S^- does not involve a stable anionic TI, as proton transfer from histidine to oxygen or nitrogen precedes or is concerted with the S---C=O bond formation. Due to the errors of AM1, the PT to the amide is favored. The authors suggest that the key role of His in CP is proton donation, while it is proton abstraction in the serine proteinases. Also, direct attack by sulfur without participation of histidine is judged as not likely to occur. To study the role of His as a proton donor in CP catalysis, the authors propose to mutate Asn-175 to Asp, thus probably increasing the basicity of His and reducing its ability to participate in the reaction.

Contribution of the side chain of Gln-19, in the purported "oxyanion hole" of papain, to stabilization of the transition state has been examined by mutating it to Ser and to Ala (Menard et al., 1991a). A reduction in k_{cat} was observed in both cases - ~20-fold for the Ala mutant and ~200-fold for Ser. K_m was affected to a much lesser degree, increasing about 3-fold in both. The small change in K_m suggests that hydrogen bonding to the substrate's carbonyl is absent or weak in the ground state Michaelis complex. Part of the reduced activity of the mutants could be due to conformational changes that are enabled by the shorter side-chains. It has been demonstrated earlier by these authors (Menard et al., 1991b) that mutating Ser-176 does not effect the catalysis, and thus the effect of mutating Gln-19 can not be due to its H-bond with Ser-176. Gln-19 is conserved in all the known cysteine proteinases, and thus this oxyanion hole interaction is important for the mechanism. In a further study of the oxyanion hole (Menard et al, 1995), Gln-19 was replaced by the isosteric glutamic acid and by histidine and asparagine. This can also probe electrostatic effects in the oxyanion hole, if the side chains of His and Glu are ionized at that position. K_m was, again, little affected, while kcat was reduced by ~25-fold for Q19E, ~850-fold for Q19H and ~35-fold for Q19N. The reason for the relatively small reduction by Glu has been demonstrated by molecular modeling, showing that the negatively charged residue can move away from the oxyanion hole and avoid some of the repulsion with the oxyanion TI. At low pH where the Glu side chain is neutral, the activity of the mutant is closest (~20-fold reduction) to wt.

Wang et al. (1994) analyzed by MD the roles of the "double catalytic triad" in papain catalysis, based on the structure of the enzyme, which is not completely known from crystallography (Kamphuis et al., 1984) due to the oxidation state of Cys-25 (present as cysteic acid in the crystal). Stochastic boundary MD (Brooks and Karplus, 1983) was carried out on the whole enzyme + ~350 water molecules. Three "layers" were treated according to their distance from the sulfur atom of Cys-25 - atoms within 12Å, atoms between 12-16Å and the more distant atoms were kept fixed. CHARMM forcefield was employed. The active site geometry was examined as a function of pH, for various mutual states of S-/SH and Im/ImH+. In addition, the mutations of Asp-158 (Menard et al., 1991) were studied.

Conclusions from this study were drawn on the basis of distances between atoms in the various side chains. This is a doubtful procedure and can not be employed to evaluate quantitative aspects of the activity. The average distances of sulfur to ring nitrogens in His-159 are longer in the D158N mutant compared to the D158E and D158G mutants. The authors suggest that sulfur would be more nucleophilic in the Asn mutant due to lesser hydrogen bonding. Indeed, in wt papain, the distances to His-159 nitrogens are even greater, and wt is 6-fold more reactive than the D158N mutant. The role of Asp-158 could be to keep the histidine ring, by H-bonding, at a greater distance from Cys-25 so that the thiolate is more flexible and nucleophilic, while histidine is more basic. In cathepsin-H, Asn replaces Asp-158, so that the H-bond to His can be maintained. In the simulation of the state (SH, ImH^+) His-159 may break its H-bond to Asp-158 in favor of hydrogen bonding to Asn-175. A role for Asn-175 is postulated, implying that a weaker H-bond of protonated His-159 to the carbonyl of Asn-175 (relative to the H-bond to ionized Asp-158) allows the protonated histidine to transfer its proton more readily to the amide nitrogen of the departing group and Asn may have an orientational effect on His with respect to this PT. The conclusion is that both Asn-175 and Asp-158 could assume alternative roles in catalysis, as their single separate effects are not very large, and much smaller than the effect of replacing the aspartate in serine proteinases (which leads to a reduction in activity of ~10,000-fold). A double catalytic triad, Cys-25-His-159-Asn-175 and Cys-25-His-159-Asp-158 is thus proposed by Wang et al.

Model substrate intermediates in the deacylation step of CP have been studied by AM1 (Dewar et al., 1985), on a series of dithioester intermediates (Duncan et al. , 1992). Model compounds were substituted N-benzoylglycine ethyl dithio esters. Their analogs in papain hydrolysis are the thiono esters, which showed high correlation with a positive coefficient between the deacylation rate constant and the Hammet σ constant of the substituents (reaction accelerated by electron withdrawing substituents). The substituents are quite remote from the reaction center, and so the transmission of their effect to the mechanism should be explained. The negatively charged thionoester intermediates (figure 4, after OH^- attack) are assumed by the authors to be close to the TS, and thus to reflect TS energies. They were divided into two "substructures" along the HN-CH_2 bond and molecular orbitals were calculated for the two separate parts, where the amide part varied due to different substituents. The energy of LUMO in the amide part is lowered by electron attractors, and thus the HOMO ("enzyme" part) - LUMO ("substrate" part) energy difference is reduced and correlates qualitatively with the rate of reaction. Orbital interactions along the attack of OH^- on a thiol ester (C(=O)-S) should be more pronounced compared to OH^- interaction with a dithio ester, since the σ* orbital of the thiol ester is lower than its π*. Thus, much of the donated density of the OH^- nucleophile goes into this antibonding σ* orbital, increasing the C-S distance. In the dithio ester (C(=S)=S) electron density is donated to the low lying π*. This explains, according to Duncan et al., the increased ~20-fold rate of deacylation of the thiol ester compared to the dithio ester (Carey et al., 1983).

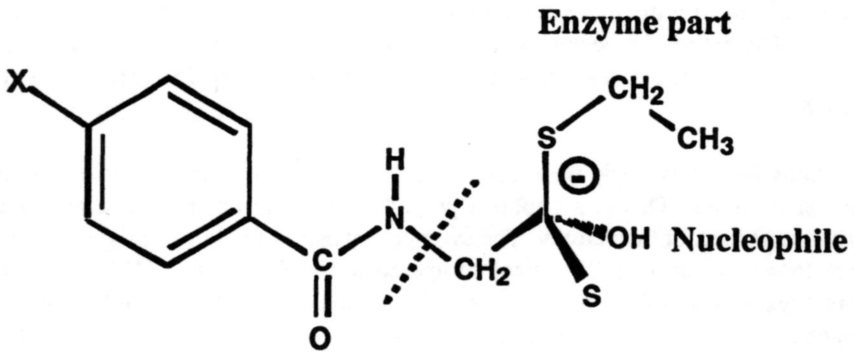

Figure 4. Model structure of thionoester intermediates

Thus, starting at the Michaelis complex and continuing through the various steps of the hydrolytic reaction by CP, there is still more to discover about the differences between cysteine and serine proteinases, with respect to the major and the minor effects on the mechanism. Concensus over the formation of the initial complex, the pathway(s) to formation of the intermediate thioester (including formation of an anionic TI) - is still missing. More experimental facts, kinetic and structural, are required, in particular of some mutants, such as the Asn-175 to Asp mutation, proposed by Arad et al. On the theoretical side, parameters for sulfur compounds of the type encountered in the catalysis by CP should be improved.

4. Aspartic Proteinases

Aspartic proteinases include some important cellular enzymes such as pepsin, renin, chymosin, cathepsin-D and others, as well as retroviral proteinases which are shorter and have a homodimeric structure. Most of the cellular enzymes of this family are largely homologous, especially around the active site where a sequence of three amino acids - Asp-Thr-Gly, is repeated twice along the polypeptide sequence in positions 32-34 and 215-217 (pepsin numbering) and these two triads face each other at the cleft of the active site. In the dimeric retroviral enzymes, the sequence appears once in each of the monomers and its final 3-dimensional arrangement is very similar to that of the cellular enzymes. The two hydroxylic residues are found to form a "fireman's grip" geometry, with each of them hydrogen bonding to the other at the "back" of the active site, more remote from the cleft where substrates and inhibitors bind. This cleft is filled with water in the native structures and has a "flap" (in cellular enzymes) or "flaps" (in the retroviral enzymes) structure facing the active aspartic acid pair.

Aspartic proteinases (AP) have been studied more thoroughly in recent years due to the discovery that HIV-1 proteinase (PR) belongs to this family and it became an important target for designing drugs against AIDS. Many crystal structures of HIV-1 PR were studied and those demonstrate the similarity of active site interactions, and possibly of its mechanism, to other AP. In the active site of aspartic proteinases, two aspartic acid moieties bear a nearly planar spatial arrangement with respect to each other. In the native enzymes, a water molecule is believed to reside at the center of the aspartic dyad, although other interpretations to a crystallographically discovered electronic density at this site were given. In complexes of AP with inhibitors such as pepstatin, the water molecule is absent and a hydroxyl group replaces it, with its oxygen nearly at the same position as the purported water molecule's oxygen. The profile of reaction rate vs. pH is bell-shaped so maximal activity is achieved when the site (two aspartates) is in the monoanionic state. In pepsins, the first pKa, near the low value of 1.5, is associated with Asp-32, while the higher pKa, ~4.5, is attributed to Asp-215. The assignment was based on reactions with irreversible active-site directed reagents. However, this has not been established by further experiments, and is of importance only in the cellular AP, while the retroviral AP are symmetrical. In the retroviral AP, "monoanionic" implies deviation from strict symmetry, and such deviation is not expected in these symmetric structures. It is more probable to expect that the single proton exchanges between the aspartic moieties, with the possible mediation of water, and so the pKa is of the "site" and not of a specific residue.

Kinetic experiments demonstrate that the extended active site cleft of pepsins (Tang et al., 1978) accommodates at least seven amino acids and that increasing chain length from two to six residues has a major effect on k_{cat} values and not on K_m (Allen et al., 1990). Energies of activation have been estimated to be at the range of 7-9 kcal/mol for good substrates and 13-14 for lesser ones, in a series of blocked peptides where the lysine-nitrophenylalanine peptide bond is cleaved.

Two mechanisms had been investigated for AP : a direct nucleophilic attack (Somayaji et al., 1988) and a general-acid general-base mechanism (GAGB, Fruton, 1976; Rich et al., 1982). The nucleophilic attack is initiated by an ionized aspartate on the peptide bond, and is promoted by polarization of the carbonyl by the protonated aspartate (Figure 5, A). The first product should be an anhydride, which is readily hydrolyzed in a second step. Attempts to trap the anhydride (or an amino-enzyme) by cryoenzymology failed (Hofmann and Fink, 1984; Hofmann et al., 1984) and there is no other type of evidence for the nucleophilic mechanism. There is, however, no clear-cut proof against it.

Based on crystallographic results, two main options for a GAGB mechanism were considered, depending on the position of the water molecule that participates. In the first (Figure 5, B), the water molecule remains at the center of the active site and is H-bonded primarily to the ionized aspartate (Pearl and Blundell, 1984; Suguna et al., 1987). The peptide bond is polarized by a proton residing on an outer oxygen of the neutral Asp. The polarization increases the positive charge on the carbonyl and the water molecule's ability to attack the carbonyl. This attack may be accompanied by one or two proton transfers: one from an outer oxygen of Asp to the oxygen of the peptide, the other from water to one of the oxygens of the aspartate. A new O-C bond is formed, two O-H bonds are broken and two other O-H's are formed. The carbonyl is transformed from trigonal to tetrahedral, and the tetrahedral intermediate should be a gem-diol. It is structurally close to a higher-energy structure which may be the TS. This TI is hydrogen bonded to one of the aspartates or to both, depending on the position of the single proton on the two aspartates. Another proton transfer to nitrogen is required to cleave the C-N bond, and that may be possible only after nitrogen inversion or rotation so that the nitrogen's lone-pair is in the direction of the proton. Following or concurrently with this PT to nitrogen, another proton has to be transferred from the gem-diol to an aspartate and the product, an acid, dissociates from the active site. In the second GAGB mechanism (figure 5, C), the water molecule in the center of the active site is replaced by the substrate's carbonyl, and the attacking water molecule dwells next to the ionized aspartate, at a position that is outside of the active site center (James and Sielecki, 1985). The GAGB mechanism of catalysis resides, thus, in the two aspartic acid moieties and the water molecule. However, one of the main features of the active sites of many AP is their high acidity, demonstrated by the low first pKa. Gly-35 and Gly-217 contribute to this acidity through hydrogen-bonding of their backbone N-H groups to the aspartates. This also serves to keep the rigidity of the active site. Details of AP studies and conclusions gained from them have been reviewed by Fruton (1976), Kostka (1985), Fersht (1985), Fruton (1987), and Polgar (1987). The crystallography of AP has been reviewed by Davies (1990).

Current knowledge about AP mechanism stills leaves much that is unknown: What is the state of the two aspartic acids in the native enzyme, what is the preferred direction of approach for the peptide and its positioning vis-a-vis the aspartates, is it a GAGB mechanism or a nucleophilic one, which step is rate determining, what are the steps of

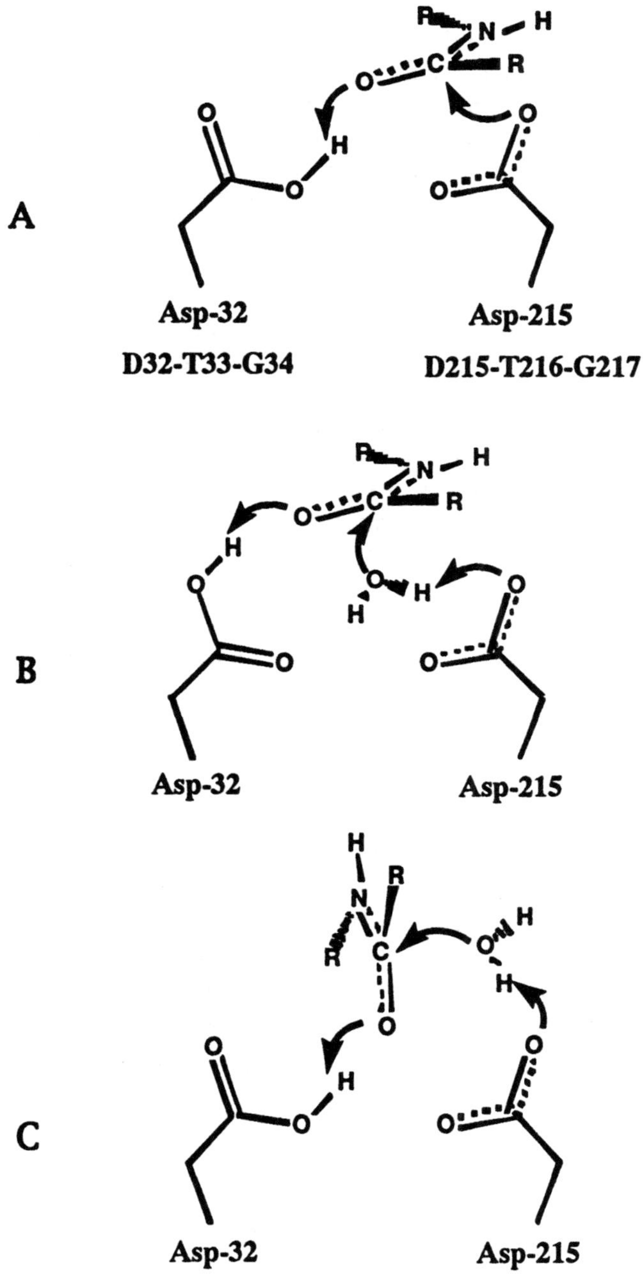

Figure 5. Nucleophilic attack (A) and General-Acid General-Base attacks (B,C)

bond making and proton transfers, what is the role of the flap(s) and of other remote residues. Only some of these questions have been addressed by theoretical studies to date.

It is assumed that the position of nucleophilic water is determined by the ionized aspartate more than by the neutral one and it influences the direction of approach of the peptide as well and thus much attention in mechanism studies has been devoted to theoretical determinations of pKa values and positions of the water molecule in the active site. Most studies assume a monoanionic site as a sole option, which requires to determine the preferred ionization between the two aspartic acids in the non-symmetrical AP. It was suggested that the density attributed to water at the center of the aspartates could be the result of averaging between two water positions due to charge delocalization between the aspartates (Pearl and Blundell, 1984).

The nature of the density at the center of the active site was studied by Hadzi et al. (1987) employing ab-initio computations on a small model of the active site, comprised of two formates representing the aspartic acids. The positions of the formates were derived from the X-ray results for endothiapepsin (Pearl and Blundell, 1984). In this study, it was assumed that a hydrogen bond connects the two closest oxygens of the aspartates, and neutral and protonated water, alternatively, were allowed to interact with the Asp models at the "center of symmetry" and with full minimization of their positions. With H_3O^+, a proton from the oxonium was transferred spontaneously to the apsartates in calculations with both STO-3G and with 4-31G basis sets. This is expected in this type of "gas phase" calculation. The hydrogen bond positioned between "inner" oxygens of the aspartates also imposes a set of hydrogen bonds of the active site residues with water. This is asymmetric if the oxygen of water is kept fixed at the crystallographic position, with one H-bond to an outer oxygen (of the neutral acid) and one to the inner oxygen (of the anionic acid). Optimization of the water position moved it a little out of the active site, with the two hydrogens pointing in the direction of the "outer" oxygens of the aspartates, and the oxygen is further outside, with an increased negative charge which makes it more ready as a nucleophile. In this geometry, the energy of interaction between water and the site is about 15 kcal/mol. The level of calculation can not clarify if H_3O^+ or H_2O is found at the center of the active site instead of water, but the authors suggest that with proper representation of the protein's electric and reaction fields, an energy minimum for H_3O^+ may be present and it could be the source of polarization of the substrate's carbonyl. It is unclear how the same molecule could be the source of the electrophile to C=O as well as the nucleophile that attacks it.

Additional active site residues were represented in the calculations by Hadzi et al.: serine-35 and threonine 218 (by water molecules), and two NH_3 were added to represent Gly-34 and Gly-217. These additions had minor effects on the nature of the interaction of water with the active sites of AP.

Goldblum (1988) studied the pKa of the active site of AP by constructing a stepwise model that included the two aspartates (modeled as formates), the peptide bonds of Thr-33-Gly-34 and Thr-216-Gly-217 (formamides) and the side chains of Ser-35 and Thr-218 (methanols), as well as a water molecule. Calculations by MNDO/H (Goldblum, 1987; Goldblum, 1988) demonstrate an enthalpy difference of ~2-3 kcal/mol between the acidities of Asp-32 and Asp-215 with the aspartate models alone, with coordinates extracted from three different pepsins (Subramanian et al., 1977; Bott et al., 1982; James and Sielecki, 1983). The study assumed that the single proton of the aspartates is bridging their "inner" oxygens. It was also assumed that entropy contributions to the difference between the acidities of the two aspartates are negligible. The additional neighboring residues (the peptide bond more than the hydroxylic side-chains) reduced the difference to about 0.4 kcal/mol in favor of Asp-32. A water molecule that was optimized for each ionization option did not change this difference. In its "best" position, with an interaction energy of ~14 kcal/mol, the water molecule is positioned with hydrogen bonds to the ionized aspartate and its oxygen directed towards the cleft. A general correlation of this "gas phase" pKa calculation to similar calculations for organic di-acids was found for the first (low) pKa and partially for the second pKa of the active site. In a subsequent study comparing the acidities of pepsin, renin and HIV-1 PR on the basis of small "gas phase" models (Goldblum, 1990), it was suggested that pepsin is more acidic than the others due to hydrogen bonding assistance by the additional fourth residue following the D-T-G triad (Ser-35 and Thr-218 in pepsins, Ser-35 and Ala-218 in h.renin and alanines in HIV-1 PR). The models included the 8 residues of the active site, but water molecules were missing. Experimental evidence does not agree with the above conclusions on the source of acidity variations in AP. Ido et al. (1991) mutated the fourth residue in the sequence of the active site of HIV-1 PR (D25-T26-G27-A28) To Ser. The mutant, A28S, has a first pKa of 3.4 compared to 3.3 for the native enzyme. The second pKa for the mutant is reduced to 5.6 from that of wt, which has 6.8. The opposite mutation, of Ser-35 and Thr-218 in pepsins to Ala (Lin et al., 1992), finds again a small effect, and the authors conclude that these residues contribute to structural rigidity and not to ionization properties.

The position and interactions of a water molecule in the center of the active site of AP has also been investigated by Turi and Naray-Szabo (1992) in order to complete the structural information obtained by X-rays. AM1 and PM3 semiempirical studies were employed for a relatively small model of the active site, including two formates and the water molecule. The methods were first compared to calculations with 4-31+G* (split valence and diffuse functions). On the basis of these calculations, PM3 was chosen to study the larger system. Heavy atom positions were extracted from Rhizopuspepsin (Suguna et al., 1987).

Two sets of calculations were performed, the first with optimization of added hydrogens only (X-ray positions frozen) and the second with additional optimization of oxygens, with only the two carbons fixed. Although energies were compared for the different structures, the assignment of the "best" structure was made on the basis of geometrical similarity to the active site atoms after the second type of optimization. Some of the optimized structures have lower energies than those that are closer to experimental geometry values. For the pair of acids in the monoanionic state (with no water) it was found that bridging the two acids is much lower in energy than having the single proton on an "external" oxygen (by more than 15 kcal/mol). With both acids neutralized, energy differences between the possible structures are much smaller. The position of a water molecule was studied with both the monoanionic and the neutral acid pair, while the possibility that H_3O^+ or NH_4^+ replace water were tested only with the anionic site.

The structure closest to experiment is a hydrated neutral pair of aspartates (figure 6), and it is higher in energy than one which has the water nearly inverted, with one of its protons H-bonding to the external oxygen of Asp-215 and the second proton pointing outwards, and two protons being carried by two adjacent oxygens of the two acids. The authors suggest that the lack of protein core stabilization may lead to the preference of the neutral form. No experimental evidence to support the suggestion of a neutral active site has been published. In the calculations with the protonated ions, O-H and N-H distances in the ions were frozen to disable spontaneous proton transfer. Based on the structures, the authors conclude that the active site in the crystal is hydrated and not ammoniated. The electrostatic potentials of more remote residues were found to strongly influence the propensity of Asp-32 or Asp-215 to ionize. Asp-307 in rhizopuspepsin stabilizes the ionization of Asp-32 if it is, itself, ionized. In the neutral state, Asp-307 stabilizes the ionization of Asp-215.

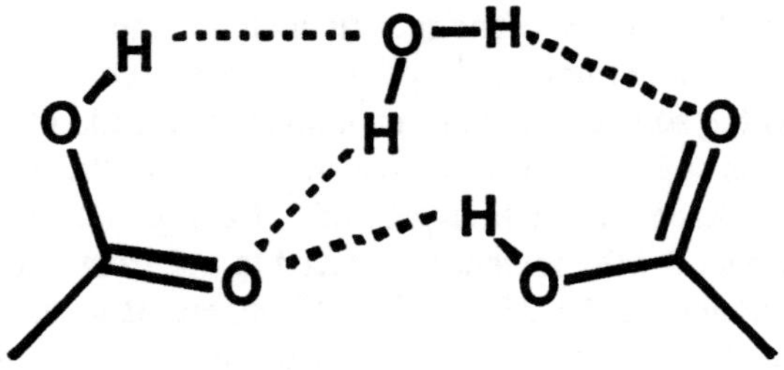

Figure 6. Proposed position of the water molecule in a neutral active site of AP

Beveridge and Heywood (1993) performed ab-initio calculations on a "gas phase" model with active site coordinates from pepsin (Sielecki et al., 1991) and endothiapepsin (Pearl and Blundell, 1984) to study the preferred position of the water molecule. Aspartates were modeled by formate/formic acid, Ser/Thr was modeled by methanol and the peptide bonds that donate hydrogen bonds (Gly34-Ser35, Gly217-Thr-218) were modeled by formamides. Geometry optimizations were performed by the 4-31G basis set. Second order configuration interaction (CI) wavefunctions were computed at the optimal positions with 4-31G and 4-31G* basis sets. Initially, four possible arrangements of water in the active site were tested, with several models starting from the bare two acids and water (model I) to additional methanols and formamides (model II) or point charges replacing methanols and formamides (model III). An interesting insight was provided by the large difference (about 6 kcal/mol) between alternative bifurcated arrangements (i.e., with D32 or D215 ionized). The authors suggest that the water molecule binds preferentially across the shorter diagonal of the aspartates. In the case of endothiapepsin, OD132--OD1215 distance is 4.43Å while the other diagonal is 4.76Å. They also suggest that the preference of ionization in all aspartic proteinases may be predicted from the diagonal distances: the aspartate which has the inner oxygen on the shortest O32--O215 diagonal is ionized at the low pKa. However, in an extended study of this system, Beveridge and Heywood (1994) examined additional positions of water and found that the lowest energy for the monoanionic state has a hydrogen bond between the two "inner" oxygens with D32 as donor, and this proton also forms an H-bond to the oxygen of water (figure 7, 3A). This was verified by MP2 calculations on the 4-31G optimized structure. A large binding energy for water (28 kcal/mol) was calculated in this position. This best position for the water molecule does not allow it to participate as a nucleophile and the authors suggest that it should undergo a prototropic rearrangement prior to catalysis, which could involve a hydronium ion. The productive position of water is found to be ~13 kcal/mol higher in energy than the stable structure, and the rearrangement must be assisted, possibly by the substrate. Thus, the outcome of this suggestion is that much of the binding energy of a substrate should be spent on activating a water molecule that is part of the native structure of the enzyme. It is probable that the differences between the water configurations in the active site are determined by more remote interactions that were not included in this study.

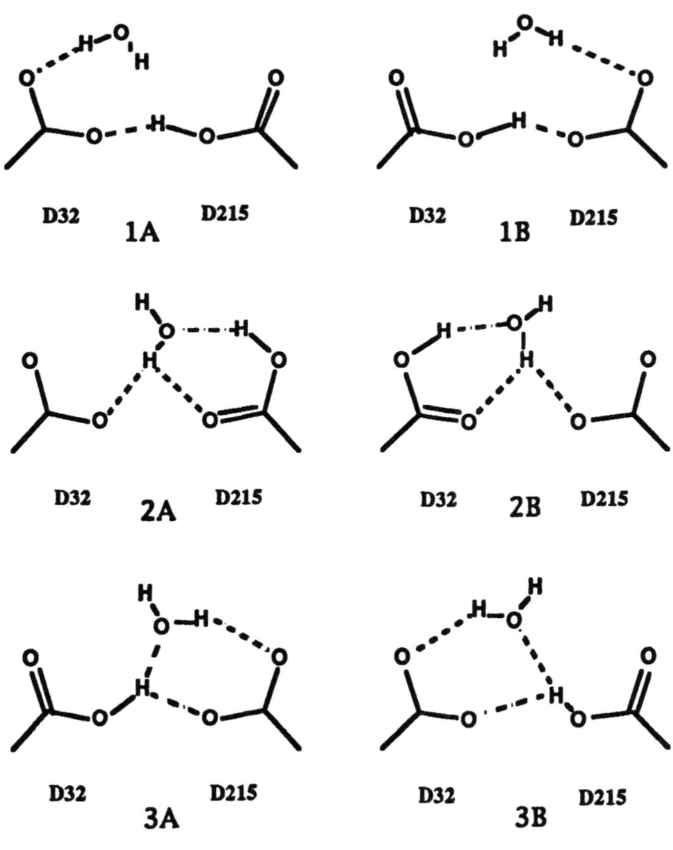

Figure 7. Possible water positions in a monoanionic site. Global minimum is 3A

The protonation state of the active site is assumed to be monoanionic in the native enzyme but, according to Harte and Beveridge (1993), it could change on interaction with inhibitors. In their study of complexes of HIV-1 PR with substrate-based inhibitors, all four possibilities of active site protonation (-1,-1; -1,0; 0,-1 and 0,0) were investigated. Monte Carlo and MD calculations were performed using the GROMOS86 force field (van Gunsteren and Berendsen, 1986) and SPC model (Berendsen et al., 1981) of water (7000 molecules with periodic boundary conditions in MD). With one inhibitor, only the diprotonated state (0,0) coincides well with the inhibitor's structure. With the other, the dianionic state (-1,-1) is best for reproducing the inhibitor's structure. The structural resemblance is not necessarily related to the energies, which were not evaluated. Both crystals were grown under similar conditions at pH 5.4. The (0,0) state is more questionable since the pH is above the assumed pKa of the aspartates. The authors suggest that the neutral (α-dihydroxy) structure of the inhibitor encourages a diprotonated active site, and also make the analogy with the purported mechanism where the transition state is diprotonated - one acid accepting a proton from the attacking water molecule and the other acid ready to donate a proton to the nitrogen of the amide. The second inhibitor has a reduced amide peptide isostere which, being a secondary amine, is protonated. Thus, it creates a different microenvironement than the other inhibitor. For substrates, the bell shaped pH-activity profiles for pepsins and for HIV-1 PR suggest that the monoanionic state is employed for catalysis, and so substrates may not be able to change the state of the aspartates and still be cleaved, while inhibitors may bind to the active site in different states of ionization.

Rao and Singh (1991) studied proton positions and movement between the two aspartates of the active site by ab-initio (6-31G). The "gas phase" calculations were performed on a planar structure of acetic acid models with coordinates from rhizopuspepsin. In this structure, PT between the inner oxygens of the aspartates requires an activation of ~15 kcal/mol, and this was not changed with water in the active site. With optimization of the planar arrangement of the acids, the inner oxygens get closer to each other and the PT activation energy is reduced to less than 1 kcal/mol. This is not necessarily relevant to the enzyme structure. Combined QM/MM was carried out for the enzyme, with active site residues calculated by ab-initio and with the appropriate modules of AMBER (Singh et al., 1986) for the interactions with the rest of the enzyme and with water molecules. In calculations of the active site it was assumed that the single proton in the active site resides on one of the "internal" oxygens. The energy of PT between the two acids was reduced in the presence of the environment. The authors find that their results are in agreement with those of Goldblum (1988) with respect to the low preference of acidity of one Asp vs. the other, the small effect of a water molecule and the influence of neighboring residues.

In the calculations of Rao and Singh, the planar arrangement of the aspartates was distorted and reached near perpendicularity in the TS and the final position along PT. This has no experimental verification. A similar distortion resulted with water in the active site. Rao and Singh conclude that the planar arrangement of the active site residues in the enzyme is not the most stable and that the distance between the internal oxygens should be shorter. They also comment on the difficulties of the particular force-field to deal with the special arrangement of the aspartic dyad and suggest to improve the force-field for this case of the aspartic dyad.

Another MD study was designed to determine the preferential position of the proton on the aspartates in the presence of a substrate (Harrison and Weber, 1994). This study demonstrates that the proton maintains coordination with all four oxygens of the catalytic aspartates and with the carbonyl oxygen of the scissile bond. A loosely bound hydrogen ion is therefore not exchanging rapidly with solvent and will rebond to either a catalytic aspartate or, possibly, to the substrate.

An important experimental contribution to the discussions of active site state has been recently presented. Iliadis et al. (1994) investigated the active site of pepsin by Fourier transform difference spectroscopy. They conclude that Asp-32 is present as an anion and Asp-215 is protonated, in the native enzyme. The two are not hydrogen bonded through their closest oxygens, even in the native structure. Also, water and not H_3O^+ is at the center of the active site.

Hadzi et al. (1988) extended their study of the initial steps in the mechanism by adding a substrate to their model. In this study they assume two alternative sources of polarization: oxonium ion and the proton bridging the two internal oxygens of the aspartates. A high stabilization energy is found (STO-3G basis set for initial optimization followed by 4-31G) for positioning the substrate model (N-methyl acetamide) with its carbonyl oxygen pointing away from the active site while N-H of the substrate forms a hydrogen bond to an "external" oxygen of one of the aspartates. In this model the water molecule was taken out, so only the spatial relation of the amide to the aspartates was tested.

For amide protonation by water , H_3O^+ was positioned in its optimal position near the aspartates, and an amide (acetamide) was allowed to approach with its carbonyl oxygen reaching hydrogen bonding distance from the water's proton. Transfer of this proton to the amide's oxygen is highly endothermic (~21 kcal/mol). Without the aspartate models, amide protonation was favored by nearly the same amount of energy. So, if H_3O^+ is the species at the active site, PT from it to the peptide should be unfavorable due to the effect of the negatively charged aspartic pair. A PT between a charged and neutral species that are at H-bonding distance should be less influenced by the environment represented in the calculation than PT between two neutral species that creates opposite charges, such as PT in SP (serine and histidine) and CP (cysteine and histidine). Another option for amide protonation is by the proton that bridges the two acids. This type of PT requires a different access ("head on") of the carbonyl into the aspartic dyad, as well as displacing the water molecule that resides in the center of the active site. This water molecule donates an H-bond to the anionic aspartate and is an acceptor of the N-H bond of the substrate. While binding is favorable in this position (-22 kcal/mol with STO-3G, -10 kcal/mol with 4-31G) - PT is again highly endothermic, about 30 kcal/mol with the larger basis set, and more than twice that amount with STO-3G. The conclusion of these studies is that there is no support from theory to proposed mechanisms in which protonation of the amide is a first step. However, some other options to such a protonation were not tested in this study.

The positioning of the peptide at the beginning of the catalytic cycle has been investigated by CNDO/2 as well. This method is not a good choice for hydrogen bonded structures, but was employed by Antonov and Alexandrov (1991) with a model comprised of acetate and acetic acid, water and N-methylacetamide. The two acids were hydrogen bonded between their closest ("inner") oxygens. The carbonyl of the substrate model was directed to the carboxyl groups, so that polarization was affected by the inner bridging proton. The oxygen of water was localized perpendicular to the carbonyl, above its carbon atom. By minimizing the potential energy with respect to structure, a short (1.31Å) H-bond was formed between an outer oxygen of the acetate (Asp-32) and a proton of water, and the charge on the water's oxygen was increased from -0.26 to -0.43. The distance of the carbonyl oxygen to the bridging proton (on Asp-215) is too large (~3Å) for it to be an effective electrophile. The amide nitrogen underwent pyramidalization and no inversion was required for the nitrogen's lone pair. Further approach of the water molecule to the carbonyl proceeded with no activation energy, but with a sharp decrease of the total energy (an enormous ~-190 kcal/mol). Synchronous PT from water to acetate (Asp-32) has been observed. At this stage, the two aspartates were protonated, and PT to nitrogen required another source or, alternatively, rotation of the two aspartates around their CB-CG bond, so that they "switched" roles - Asp-32 became the donor of the hydrogen bridge between the two close oxygens of the aspartic side chains, while Asp-215 rotated to donate a hydrogen bond to the substrate's nitrogen (with a short distance,1.3Å, between H and N). This state was calculated to be unfavorable. PT from Asp-215 to nitrogen was accompanied by synchronous proton transfer between the aspartate inner oxygens, so that Asp-32 anion was restored but the H-bond between the acids was broken, for unclear reasons. This state is also endothermic, by about 25 kcal/mol. At this stage, the C-N bond order was decreased to 0.77 from the initial 1.13 in the amide. In this study, many abnormally short H-bonds were encountered, probably due to artifacts of semiempirical QM methods that are based on neglect of differential overlap (Scheiner, 1984)

An important direction for studying the mechanism of AP emerged from crystallographic structures of the complexes of some pepsins with difluoro-ketone inhibitors. The groups of Blundell (Veerapandian et al., 1992) and of Davies (Parris et al., 1992) employed the same inhibitor (CP-82,218, figure 8, above) with endothiapepsin and rhizopuspepsin, respectively, while James et al. (1992) used a difluorostatone containing peptide in a complex with penicillopepsin. The difluoromethylene group next to a ketone increases the electrophilic character of the ketone and moves the balance of its reaction with water towards a gem-diol, closely related to the tetrahedral intermediate purported to participate in the hydrolytic reaction. The complexes of such compounds with the above pepsins demonstrate the spatial relationship of the alleged TI to the aspartic dyad, and distances of the two gem-diol oxygens to the aspartic oxygens suggests hydrogen bonding schemes between them. A few alternatives of these H-bonds are possible (figure 8, A-D). Based on coordinates for the complex of endothiapepsin and a tri-peptide analog that has the difluoro-ketone group near the C-terminal, a stronger support for one GAGB mechanism emerged. An ab initio STO-4G calculation of a few alternative schemes for hydrogen bonding were performed with a small model (formic acid residues for the aspartates, difluoroacetaldehyde hydrate for the inhibitor functions) and concluded that structure D (figure 8) is the preferred arrangement for the protons. It is about 16 kcal/mol more stable than an alternative configuration with Asp-32 protonated and Asp-215 ionized (A in figure 8).

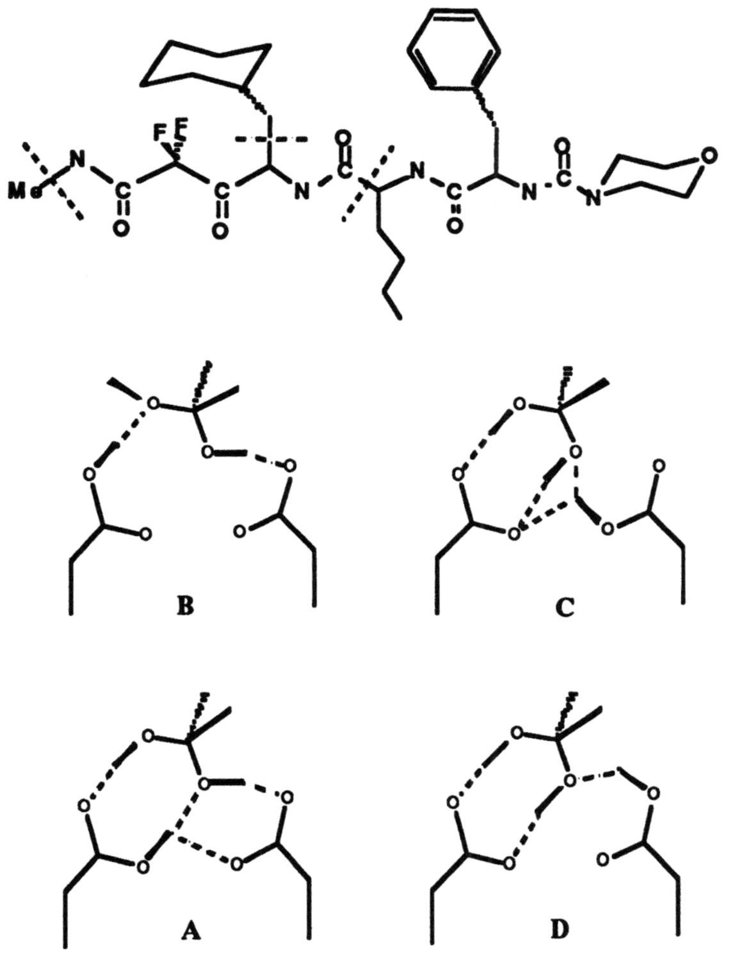

Figure 8. Tripeptide inhibitor (above) and possible H-bonding schemes of its hydrate with different monoanionic states of the active site in AP. The broken lines on the tripeptide indicate the model inhibitor

Calculation of the alternative schemes for the proton positions in the gem-diol complex with the active site were extended by Goldblum et al. (1993,1993a) to include the full eight residues of the active site (D32-T33-G34-S35 and D215-T216-G217-T218 of endothiapepsin) and a model of the inhibitor which contains a large part of its C-terminal (figure 9, A). Semiempirical MNDO/H calculations were employed to position the protons (two on the gem-diol oxygens and one proton on an oxygen of the aspartates, with all alternatives tested by conformational analysis). Throughout this study, all heavy atoms were maintained in their crystallographic positions. An equivalent study was done for HIV-1 PR, with the full inhibitor introduced to the active site by analogy to its spatial relations in the endothiapepsin complex. In this study, the full enzyme+inhibitor and a layer of water molecules (1150) were minimized by the cvff force field (Hagler et al., 1985) of DISCOVER for each of the four alternative arrangements of a single proton on the active site aspartates.

A

B

C

D

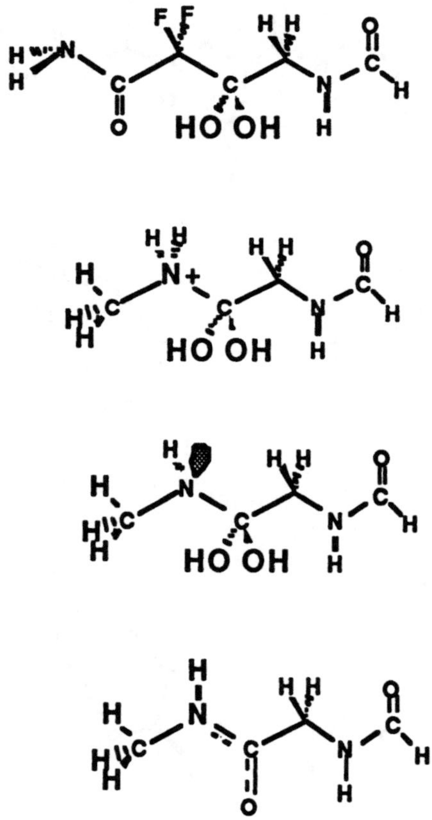

Figure 9. "Mutations" of a model inhibitor to a model substrate

In the most stable structure calculated, the two hydroxyls of the gem-diol form two parallel hydrogen bonds to the two oxygens of the ionized Asp-32, while Asp-215 is protonated on its external oxygen and donates a H-bond to the "inner" oxygen of the diol. The next stable structure, with a hydrogen bond donated by Asp-32 to the inner oxygen of Asp-215 and the inner oxygen of the gem-diol acting as an H-bond donor to the external oxygen of Asp-215, is ~16.5 kcal/mol higher in energy. The same structure was found to be more stable in MM calculation with the full, solvated, HIV-1 PR structure. A difference of ~20 kcal/mol was found between the two lowest energies, in favor of the one found by MNDO/H.

The gem-diol of the inhibitor was "computer mutated" in steps to form a model of the substrate's gem-diol (figure 9, D). Mutating the fluorines to hydrogens was used to verify that the fluorines are not responsible for the large preference of the low energy configuration. The -CH2- moiety was then mutated to -NH2- (figure 9, B). The same low energy configuration was found also for this molecule, but the difference to the next low-energy configuration was reduced to ~5 kcal/mol. In this "substrate" gem-diol (figure 8, D), one oxygen is derived from the carbonyl and the other from water. Thus, the previous step could be initiated by an "external" water, polarized by anionic Asp-32, or by an "internal" water, polarized by Asp-215. To achieve the preferred gem-diol structure directly with no further proton transfers, the "external" water must attack the carbonyl and transfer a proton in a strained 4-member transition state. A water molecule was not found to be stable in such a position in the calculations. The attack by water that resides at the center of the active site has a few options (figure 10). Three proton transfers were compared, and two of them (PT from Asp-32 to carbonyl or from water to the external oxygen of Asp-215) have low energy barriers at short (<1.75Å) distances of the water's oxygen to the carbonyl carbon. The possible sequence of reaction was studied by transferring the protons at different OW--C=O distances. It was found that the maximal gain in enthalpy is achieved close to a distance of 1.5Å. At this distance, all PT require energies of activation that are less than 3.0 kcal/mol, so that they should be concerted with the heavy atoms bond formation of the TI. At 1.5Å, PT from Asp to the carbonyl reduced the energy by ~9 kcal/mol, while the alternative PT from water to the ionized Asp stabilized the system by ~13 kcal/mol. TI (gem-diol) was more stable than the reactants by more than 20 kcal/mol. Despite the known restraining effect on charge separation in "gas phase" QM calculations, stable intermediates in this reaction sequence have lower energy than the reactants. It is thus possible that addition of the enzyme's environment would reduce the activation required for TI formation, which was found to be ~35 kcal/mol, much too high for such a reaction.

Studies by Ciarkowski et al. (1994) of the mechanism of AP by molecular mechanics with the MMX force field (Gajewski et al., 1990) and by semiempirical quantum mechanics methods AM1 (Dewar et al., 1985) and PM3 (Stewart, 1989) are hardly useful since energies are not reported for the first study while the model for a quantum mechanical study (constructed of two malonic acid molecules and water molecules that bridge the malonates) bears little resemblance to the X-ray structures of the active site of AP and includes effects that are not expected in the interaction of the peptide with the active site (Oldziej and Ciarkowski, 1994).

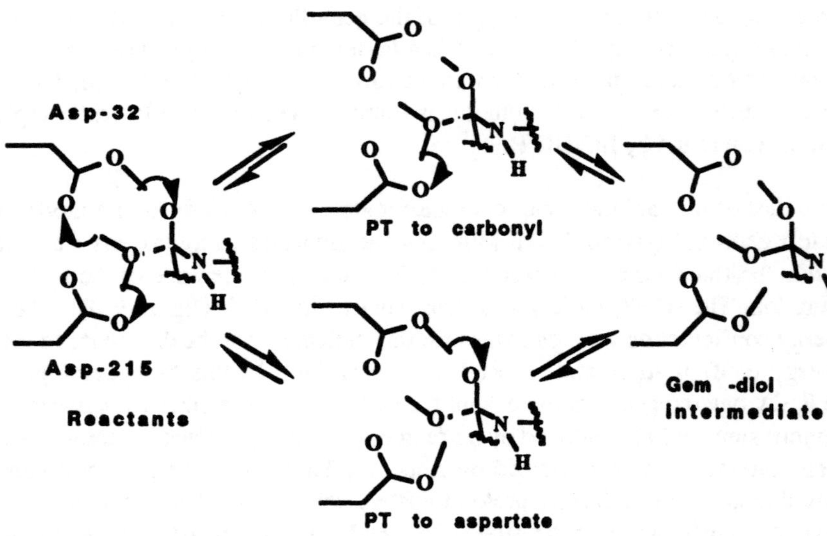

Figure 10. Alternative steps from reactants to gem-diol in a GAGB mechanism

A recent study by Chatfield and Brooks (1995) addresses the cleavage mechanism with MD simulations of HIV-1 proteinase. Three mechanisms were explored: GAGB, "zwitterionic" (protonated nitrogen of oxyanion) and direct nucleophilic attack. The first two included a water molecule in the simulations. The authors explored the conformations of the active site just prior to the first step of reaction and the MD runs were analyzed on the basis of "those that would be likely to initiate reaction", which they define as "productive conformations". A similar simulation was applied to the gem-diol intermediate in the first GAGB mechanism.

A seven residue substrate model was chosen on the basis of its similarity to an inhibitor that was crystallized in a complex with HIV-1 PR. It includes the Tyr-Pro cleavage site which is common to a few sites of the gag fusion protein, the substrate of HIV-1 PR. This model was constructed from the inhibitor by "mutating" only two of the inhibitor's residues. MD simulations were carried out with CHARMM (Brooks et al., 1982) and water molecules were represented by a modified (Steinbach and Brooks, 1993) TIP3P model (Jorgensen et al., 1983). Five hundred water molecules were employed (minimal solvation).

Productive conformations were defined for each mechanism. For direct nucleophilic attack, a close approach (less than 3.5Å) of an aspartate to the scissile carbon should be concurrent with formation of a hydrogen bond between the protonated Asp and the carbonyl oxygen or the amide carbon. For GAGB catalysis, a close approach of the lytic water to the scissile carbon, while two hydrogen bonds are forming - between the aspartate and water, and between the protonated Asp and the carbonyl oxygen or nitrogen. For the gem-diol intermediate, a productive conformation is defined by the hydrogen bond between the protonated Asp and the scissile nitrogen.

All the criteria were not energy based but employed only distances and did not even include angular/dihedral relations. A problem of charge distribution on the Asp residues had to be solved empirically, since with all known methods the simulations could not keep the two aspartates in a coplanar arrangement and did not result in productive conformations. Modest perturbation in the Asp charges resulted in major changes of the Asp conformations and H-bonding relations.

Reliability of simulations was examined by RMS deviations from the crystal structure. Those were found to be generally small. Substrate simulations were run for all possible monoprotonated states. None of the simulations supports the formation of a hydrogen bond between the scissile nitrogen and protonated Asp, not even transiently. With one of the acids protonated (25) - no productive structures were formed but, by switching the proton to the other acid (125), conformations consistent with both direct nucleophilic attack and GAGB were found. On the basis of the MD simulations, the authors suggest that both mechanisms are plausible. However, no assessment of energy barriers has been proposed.

Water-301, which was found in the structures of many complexes of HIV-1 PR with inhibitors, resides at the "back" of the substrate and hydrogen bonds to the flaps and to inhibitors. It is believed to bind to substrates as well. The simulations did not indicate any possible role for W301 as a nucleophile - it did not get closer than 4Å to the scissile carbon for more than 0.1 ps. A cooperativity of the scissile carbonyl movement with that of the carbonyl oxygen at the P1' position was detected, and the latter is crucial for the binding of W301. The authors propose that W301 may stabilize productive reactant conformations. Simulations of the gem-diol intermediate were performed for all four alternative protonation states of the aspartates (and the nitrogen with a configuration that has its lone-pair in the general direction of the aspartates). The distance between the protonated Asp oxygen and the scissile nitrogen was the smallest, 3.5Å, when Asp 25 was protonated. This is the same Asp that acts, in the first step, as a base to deprotonate the lytic water. When the chirality of nitrogen was reversed (so that its lone pair direction was away from the aspartates, as it should be following the attack by water on the carbonyl, according to the stereoelectronic principle), than the distance between nitrogen and protonated Asp increased to 5.5Å on the average. Also, water 301 has less hydrogen bonds to the carbonyl oxygen of proline when the Pro nitrogen is in the R-configuration, thus suggesting that it may provide the stabilization necessary for nitrogen inversion. The value of this study is in some of its suggestions and ideas for future studies by theoretical methods that can follow bond-making and breaking, and it may also suggest some future experiments.

The role of W301 has been analyzed on the basis of crystallographic studies of a complex between an octapeptide inhibitor and HIV-1 PR. Its distance from the scissile bond is believed to be long (>3.5Å) and thus it is unlikely that it serves a direct mechanistic role. However, it might be serving to force the C-N bond into a conformation that facilitates cleavage. In this study, Jaskolski et al. (1991) propose a different mechanism, with no intermediate gem-diol. They suggest that symmetry of this enzyme is maintained in the native structure, where a hydrogen bond is donated by one of the two external oxygens of the aspartates to the water molecule, and a movement of this proton towards the oxygen of water supports PT from water to the external oxygen of the other aspartate, thus keeping a dynamical symmetry. This symmetry may be perturbed by an approaching substrate, which could "lock" the proton at one position. The attack of the nucleophilic water on carbon and electrophilic attack by a proton on nitrogen could proceed simultaneously, so that the cleavage process is a one-step reaction. They also suggest that at the beginning of the reaction, the proton resides on an outer oxygen of the aspartate that is closer to the nitrogen of the substrate. The structural components, nevertheless, were not employed to propose any energy data.

Other experimental results were also employed to suggest steps in the mechanism. Most of them have not been, yet, explored by theoretical approaches in a comprehensive manner. They raise important problems that should be addressed by future theoretical explorations.

Hyland et al. (1991) characterized the mechanism of HIV-1 PR by using pH rate profiles with several substrates. On the basis of their studies and previous results for ^{18}O exchange into the substrate (Hyland et al., 1991a), they suggest that the two oxygens in the gem-diol intermediate are equivalent and that the rate constant for reversal of the gem-diol formation is larger than the rate of product formation. Thus, they assume that the next step, PT from Asp to nitrogen of the substrate, is rate limiting. They also propose that it is concerted with PT from the diol to the ionized aspartate.

The pH activity profile of most eukaryotic AP shows maximal activity in the pH range of 2-4, while HIV-1 PR catalyzes maximally between 4-6. This is attributed to the acidity of the active site, being lower in the eukaryotic enzymes (except for h.renin, where Ala replaces Ser as the fourth residue following one D-T-G triad (the other is D-T-G-T). The mutation of A28S in HIV-1 PR (Ido et al., 1991) hardly effects the first, low pKa of the native mutant, but does change the second one. Substrate binding changes the pKa values to 5.1 (mutant) vs. 4.3 (wt) for the low pKa, and 6.9 vs. 6.0 for the higher pKa (mutant and wt, respectively). k_{cat} for the mutant is lowered by a factor of 200 approximately, while K_m remains nearly unchanged. Based on the pKa results, Ido et al. suggest that in the native structure, the water bridges the external oxygens of the aspartates in a symmetrical fashion, and it also creates a bifurcated hydrogen bond to the internal oxygens. The negative charge of the site resides mostly on the water's oxygen. They also propound that the TS is analogous to that of serine proteinases, with a negative charge that develops on the oxygen of the carbonyl, assisted by a hydrogen bond from the protonated aspartate, which is not transferred to the peptide. Also, the proton of water is not completely transferred to the ionized aspartate but remains hydrogen-bonded to it. This proton is now transferred to the nitrogen as C-N cleavage progresses. The main difference between this mechanism and the GAGB mechanism with a gem-diol TI is thus in the positions of two protons - the one that polarizes the carboxyl (in the gem-diol it is fully transferred to the carbonyl) and the proton of water (in the gem-diol it is fully transferred to the ionized Asp). Substrate binding raises the low pKa by ~2 pH units for unclear reasons. There is only a small difference in the high pKa due to substrate binding, and this is explained as a demonstration of the similarity in the arrangement of the three protons in the active site, with and without a substrate. Also, it is not clear why is k_{cat} reduced ~60-fold in the A28S mutant.

A reverse study, of mutating serine and threonine residues to alanines in pepsin and rhizopuspepsin, also affects strongly the k_{cat} values, with minor changes in K_m (Lin et al., 1992). Substrate binding affects the pKa values significantly. This is due, according to Lin et al., to a primary effect of the water molecules in the active site cleft. Also, both the pKa variations and the differences between the width of the pH activity profile for pepsin and rhizopuspepsin may be a result of active site accessibility of water molecules. Again , in the S35A and T218A mutants of pepsin and the equivalent ones of rhizopuspepsin, the decrease of k_{cat} is significant (20-150 fold), while K_m is practically invariable. The authors suggest that reductions of k_{cat} should be attributed to loss of active-site rigidity. However, in their own study of HIV-1 PR mutant A28S, which increases rigidity, a reduction of k_{cat} was observed.

HIV-1 PR has a high mutation rate which is the cause of its drug-resistance. Lin et al. (1995) extended their research of HIV-1 PR mutants to positions that are more remote from the active site and are associated with the binding pockets of side-chains of substrates and inhibitors. Mutations of Val-82 (to E,A,D,N,Q,and S), Asp-30 (to F and W), Gly-48 (to H, D and Y) and Lys-45 (to E) were constructed, and kinetic studies with a set of synthetic nona- and decapeptide substrates demonstrated that k_{cat} was hardly affected, except for the D30W mutant which cleaved at a slower rate, by 100-fold. Gly-48 (in the flap) mutations show rate enhancement with a few substrates. A model for the G48Y mutant suggests that this may be due to increased rigidity of the flap. Tyrosine is also present in the single flap of pepsins, and an interesting speculation on its function was raised by Tang and Koelsch (1995). They suggest that Tyr-75 occupies a "guard booth" to the entrance of the cleft and it is likely to bind hydrophobic side chains in a non-specific interaction and "lead" the substrate to more specific interactions in the appropriate substrate pocket. The increase in the size of the hydrophobic cluster of Tyr + side chain (Trp, Phe, Tyr, etc.) favors the association with the interior of the cleft and induces flap closure and "squeezing" out the water. Following the hydrolysis, the production of charged species reduces the hydrophobic contribution and favors the opening of the flap and release of products. It has been shown by MD studies (Collins et al., 1995), albeit in HIV-1 PR, that flap movement may be substantial, being closed and opened even with no substrate or inhibitor present.

Involvement of the "flap" in the catalysis has been explored by Baca and Kent (1993), by chemically replacing the -CONH- linkage between residues Gly49-Ile50 in each flap of HIV-1 PR by an isosteric -COS- bond. The engineered enzyme retained substrate affinity and specificity but catalytic activity was reduced nearly 3000-fold. This corresponds to an increase of ~5 kcal/mol in the activation energy for the hydrolysis. It may be that this is the result of losing the two hydrogen bonds of water to the peptide's carbonyls at P1' and P2, as this water molecule is no more restricted in the mutant by the (missing) N-H bonds from the two peptide units of the flaps. A similar reduction, but of binding, was found for an inhibitor, JG-365 (2500-fold). It is consistent with the loss of two H-bonds, but is an effect on Ki, which should have been reflected in a substrate by a large increase in K_m. In view of the recent results by Hyland et al. (1991) that assign the rate determining step to C-N bond cleavage and not to water attack, Baca and Kent propose that the reduction in k_{cat} may be attributed to the loss of correct orientation of the nitrogen's lone-pair by the carbonyls at P1' and P2. Crystallography of this enzyme is required to verify if structural changes occur upon this special mutation. Is the water molecule (W301) expelled from its position in the native structure and a closer approach of the flap to the substrate becomes possible ?

In an attempt to identify further the nature of the rate-limiting step in the hydrolysis by HIV-1 PR, Rodriguez et al. (1993) employed ^{15}N isotope effect on the kinetics. A slightly inverse isotope effect in both water (0.995) and D2O (0.992) implies that in the transition state, stiffening of the bonds to nitrogen is involved, which is consistent with protonation of nitrogen. This effect is enhanced in D2O, as expected. Combining their kinetic results with the known structural findings, they suggest that, following the initial formation of the enzyme-substrate complex (with a "weak" hydrogen bond between an outer -OH of Asp to the carbonyl), a first intermediate is formed by strengthening this H-bond ("low barrier H-bond"). This increases the negative charge on the peptide's oxygen, while the electron deficient carbon is assisted by lone-pair donation from its attached nitrogen. The electrophilic carbon promotes the attack of water (that looses a proton to the ionized aspartate) towards the formation of a second intermediate, an amine hydrate, which maintains the "low energy" H-bond between the (former) carbonyl oxygen and the aspartate. Two proton transfers: one from the "neutral" aspartic acid to nitrogen, the other across the "low energy" H-bond back to the originally protonated aspartate, form the third intermediate which is zwitterionic (N-protonated, carbonyl oxygen charged). Rearrangement of the zwitterion cleaves the C-N bond towards the release of products.

An additional contribution to our understanding of the mechanistic steps in HIV-1 PR was presented (Polgar et al., 1994) with different substrates, where one has Glu in the P2' position (The most preferred residue in substrates of this enzyme) and in the other, Glu is replaced by Gln. This variation of side-chain charge modulates the pH dependency of K_m and k_{cat} for the two substrates. A change in the rate-determining step between the two substrates was proposed to explain their different response.

In conclusion, a plethora of suggestions for mechanism steps has been proposed on the basis of kinetic studies and of crystallography, but have not yet been followed by theoretical methods. There is still much to expect from theoretical computations for better understanding of AP mechanistic aspects. It should also be borne in mind that mechanistic variance may exist between the different enzymes of this family. Nevertheless, a few agreements may be found among the studies of AP. Those are the large stabilization of a water molecule in the active site (~15 kcal/mol), a small difference in acidity between the two aspartates and the preference for proton bridging between the two closest oxygens of the aspartates in the monoanionic site of native structures of AP.

References

Allen, B., Blum, M., Cunningham, A., Tu, G.-C., and Hofmann, T. (1990) J. Biol. Chem. 265, 5060-5065
Antonov, V.K., and Alexandrov, S.L. (1991) in Structure and Function of the Aspartic Proteinases (Dunn, B., ed.) Plenum Press, New York and London
Arad, D., Langridge, R., and Kollman, P.A. (1990) J. Am. Chem. Soc. 112, 491-502
Baca, M., and Kent, S.B.H. (1993) Proc. Nat. Acad. Sci. U.S.A. 90, 11638-11642
Baker, E.N., and Drenth, J. (1987) in Biological Macromolecules and Assemblies (Jurnak, F.A., and McPherson, A., eds.), Wiley, New York, Vol.3
Bemis, G.W., Carlson-Golab, G., and Katzenellebogen, J.A., (1992) J. Am. Chem. Soc. 114, 570-578
Berendsen, H.J.C., Postma, J.P.M., van Gunsteren, W.F., and Hermans, J. (1981) in Intermolecular Forces (Pullman, B., Ed.) D.Reidel, Dordrecht
Berger, A. and Schechter, I. (1970) Philos. Trans. R. Soc. London, B257, 249-264
Beveridge, A.J., and Heywood, G.C. (1993) Biochemistry 32, 3325-3333
Beveridge, A.J., and Heywood, G.C. (1994) J. Mol. Structure (Theochem) 306, 235-247
Bott, R., Subramanian, E., and Davies, D.R. (1982) Biochemistry 21, 6956-6962
Braxton, S., and Wells, J.A. (1991) J. Biol. Chem. 266, 11797-11800
Brooks, B.R., and Karplus, M. (1983) J. Chem. Phys. 79, 6312-6323
Brooks, B.R., Brucolleri, R.E., Olafson, B.D., States, D.J., Swaminathan, S., and Karplus, M. (1983) J. Comput. Chem. 4, 187-217
Bryan, P., Pantoliano, W., Quill, S.G., Hsioa, H.-Y., and Poulos, T. (1986) Proc. Nat. Acad. Sci. U.S.A. 83, 3742-3745
Carey, P.R., and Storer, A.C. (1983) Acc. Chem. Res. 16, 455-463
Carter, P., and Wells, J.A. (1988) Nature 332, 564-568
Chambers, J. and Stroud R. (1977) Acta Crytallogr. B33, 1824-1833.
Chatfield, D.C., and Brooks, B.R. (1994) J. Am. Chem. Soc. 117, 5561-5572
Ciarkowski, J., Oldziej, S., and Liwo, A. (1994) Polish J. Chem. 68, 939-947
Collins, J.R., Bert, S.K., and Erickson, J.W. (1995) Nature Struct. Biol. 2, 334-338
Craik, C.S., Roczniak. S., Largman, C., and Rutter, W.J. (1987), Science 237, 909-913
Daggett, V., Schroder, S. and Kollman, P. (1991) J. Am. Chem. Soc. 113, 8926-8935
Davies, D.R. (1990) Annu. Rev. Biophys. Biophys. Chem. 19, 189-215
Dewar, M.J.S., Zoelisch, E.G., Healy, E.F., and Stewart, J.J.P. (1985) J. Am. Chem. Soc. 107, 3902-3909
Dijkman, J.P., and van Duijnen, P.Th. (1991) Int. J. Quant. Chem. Quant. Biol. Symp. 18, 49-59
Ding, X., Rasmussen, B.F., Petsko, G.A., and Ringe, D. (1994) Biochemistry 33, 9285-9293
Dixon, M.M., and Matthews, B.W. (1991) Int. J. Biol. Macromol. 13, 89-96
Dutler, H., and Bizzozero, S.A. (1989) Acc. Chem. Res. 22, 322-327
Duncan, G.D., Huber, C.P. and Welsh, W.J. (1992) J. Am. Chem. Soc. 114, 5784-5794
Fersht, A. (1985) Enzyme Structure and Mechanisms, W.H. Freeman and Co., New York
Finucane, M.D., and Malthouse, J.P.G. (1992) Biochem. J. 286, 889-900
Fitzpatrick, P.A., Steinmetz, A.C., Ringe, D., and Klibanov, A.M. (1993) Proc. Nat. Acad. Sci. U.S.A 90, 8653-7
Flanigan, M.C., Komornicki, A., and McIver, J.W., Jr. (1977) in Semiempirical Methods of Electronic Structure Calculation, Part B: Applications (Segal, G.A., ed.) Plenum Press, New York and London
Fruton, J.P. (1976) Adv. Enzymol. Relat. Areas Mol. Biol. 44, 1-36

Fruton, J.S. (1987) in Hydrolytic Enzymes (Neuberger, A., and Brocklehurst, K., eds.) Elsevier, Amsterdam
Gajewski, J.J., Gilbert, K.E., and McKevey, J. (1990) in Advances in Molecular Modeling, A Research Annual, Vol. 2 (D. Liotta, ed.) Jai Press Inc. Greenwich and London
Gilson, M.K., and Honig, B.H. (1988) Proteins 4, 7-18
Goldblum, A. (1987) J. Comput. Chem. 8, 835-849
Goldblum, A. (1988) J. Mol. Structure (Theochem) 179, 153-163
Goldblum, A. (1988) Biochemistry 27, 1653-1658
Goldblum, A. (1990) FEBS Lett. 261, 241-244
Goldblum, A., Rayan, A., Fliess, A., and Glick, M. (1993) J. Chem. Inf. Comp. Sci. 33, 270-274
Goldblum, A., Glick, M., and Rayan, A. (1993a) Theor. Chim. Acta 85, 231-247
van Gunsteren, W.F., and Berendsen, H.J.C. (1986) Gromos, University of Groningen
Hadzi, D., Hodoscek, M., Harb. V., and Turk, D. (1987) J. Mol. Structure (Theochem) 150, 241-250
Hadzi, D., Hidiscek, M., and Turk, D. (1988) J. Mol. Structure (Theochem) 181, 71-80
Hahn, K.W., Klis, W.A., and Stewart, J.M. (1990) Science 248, 1544-1547
Hagler, A.T., Osguthorpe, D.J., Dauber-Osguthorpe, P., and Hemple, J.C. (1985) Science 227, 1309-1315
Harrison, R.W., and Weber, I.T. (1994) Protein Eng. 7, 1353-1363
Harte, W.E., Jr., and Beveridge, D.L. (1993) J. Am. Chem. Soc. 115, 3883-3886
Homamadi, F., Richards, N.G.J., Guida, W.C., Liskamp, R., Lipton, M., Caufield, C., Chang, G., Hendrickson, T., and Still, W.C. (1990) J. Comp. Chem. 11, 440-451
Hofmann, T., Dunn, B.M., and Fink, A.L. (1984) Biochemistry 23, 5231-5247
Hofmann, T., and Fink, A.L. (1984) Biochemistry 23, 5247-5256
Howard, A., and Kollman, P.A. (1988) J. Am. Chem. Soc. 110, 7195-7200
Huber, R., and Bode, W. (1978) Acc. Chem. Res. 11, 114-122
Hyland, L.J., Tomaszek, T.A., Jr., and Meek, T.D. (1991) Biochemistry 30, 8454-8463
Hyland, L.J., Tomaszek, T.A., Jr., Roberts, G.D., Carr, S.A., Magaard, V.W., Bryan, H.L., Fakhoury, S.A., Moore, M.L., Minnich, M.D., Culp, J.S., Desjarlais, R.L., and Meek, T.D. (1991a) Biochemistry 30, 8441-8453
Ido, E., Han, H.-P., Kezdy, F.J., and Tang, J. (1991) J. Biol. Chem. 266, 24359-24366
Iliadis, G., Zundel, G., and Brzezinski, B. (1994) FEBS Lett. 352, 315-317
Jackson, S.E., and Fersht, A.R. (1993) Biochemistry 32, 13909-13916
James, M.N.G., and Sielecki, A.R. (1983) J. Mol. Biol. 163, 299-361
James, M.N.G., and Sielecki, A.R. (1985) Biochemistry 24, 3701-3713
James, M.N.G., Sielecki, A.R., Hayakawa, K., and Gelb, M.H. (1992) Biochemistry 31, 3872-3886
Jaskolski, M., Tomasseli, A.G., Sawyer, T.K., Staples, D.G., Heinrickson, R.L., Schneider, J., Kent, S.B.H., and Wlodawer, A. (1991) Biochemistry 30, 1600-1609
Jorgensen, W.L., Chandrasekhar, J., Medura, J.D., Impey, R.W., and Klein, M.L. (1983) J. Chem. Phys. 79, 926-935
Kamphuis, I.G., Kalk, K.H., Swarte, M.B.A., and Drenth, J. (1984) J. Mol. Biol. 179, 233-256
Kossiakof, A.A., and Spencer, S.A. (1981) Biochemistry 20, 6462-6474
Kostka, V. (1985) in Apsartic Proteinases and Their Inhibitors (Kostka, V., Ed.) Walter de Gruyter, Berlin
Kraut, J. (1977) Annu. Rev. Biochem. 46, 331-358
Lamotte-Brasseur, J., Dive, G., Dehareng, D., and Ghuysen, J.-M., (1990) J. Theor. Biol. 145, 183-198

Lin, Y., Fusek, M. Lin, X., Hartsuck, J.A., Kezdy, F.J., and Tang, J. (1992) J. Biol. Chem. 267, 18413-18418
Lin, Y., Lin. X., Hong, L., Foundling, D., Heinrikson, R.L., Thaisirvongs, S., Leelamanit, W., Raterman, D., Shah, N., Dunn, B.M., and Tang, J. (1995) Biochemistry 34, 1143-1152
MacKenzie, N.E., Grant, S.K., Scot, A.I., and Malthouse, J.G.G. (1986) Biochemistry 25, 2293-2298
Mallick, I.M., D'Souza, V.Y., Yamaguchi, M., Lee, J., Chalabi, P., Gadwood, R.C., and Bender, M.L., (1984) J. Am. Chem. Soc. 106, 7252-7254
McGrath, M.E., Vasquez, J.R., Craik, C.S., Yang, A.S., Honig, B., and Fletterick, R.J. (1992), Biochemistry 31, 3059-3064
Menard, R., Khouri, H.E., Plouffe, C., Laflamme, P., Dupras,R., Vernet, T., Tessier, D.C., Thomas, D.Y., and Storer, A.C. (1991) Biochemistry 30, 5531-5538
Menard, R., Carriere, J., Laflamme, P., Plouffe, C., Khouri, H.E., Vernet, T., Tessier, D.C., Thomas, D.Y., and Storer, A.C. (1991a) Biochemistry 30, 8924-8928
Menard, R., Plouffe, C., Khouri, H.E., Dupras,R., Tessier, D.C., Vernet, T., Thomas, D.Y., and Storer, A.C. (1991b) Protein Eng. 4, 307-311
Menard, R., Plouffe, C., Laflamme, P., Vernet. T., Tessier, D.C., Thomas, D.Y., and Storer, A.C. (1995) Biocemistry 34, 464-471
Mildvan, A.S. (1974) Annu. Rev. Biochem. 43, 357-399
Nakagawa, S., Yu, H.-A., Karplus, M., and Umeyama, H. (1993) Proteins: Str. Funct. Gen. 16, 172-194
Oldziej, S., and Ciarkowski, J. (1994) Polish J. Chem. 68, 949-956
Parris, K.D., Hoover, D.J., Damon, D.B., and Davies, D.R. (1992) Biochemsitry 31, 8125-8141
Pearl, L.H., and Blundell, T.L. (1984) FEBS Lett. 174, 96-101
Polgar, L., and Halasz, P. (1982) Biochem J. 207, 1-10
Polgar, L. (1987) FEBS Lett. 219, 1-4
Polgar, L. (1990) Biol. Chem. Hoppe-Seyler 371, 327-331
Polgar, L. (1992) FEBS Lett. 311, 281-284
Polgar, L., Szeltner, Z., and Boros, I. (1994) Biochemistry 33, 9351-9357
Pople, J.A., and Beveridge, D.L. (1970) Approximate molecular Orbital Theory, McGraw-Hill, New York.
Rao, S.N., Singh, U.C, Bush, P.A., and Kollman, P.A. (1987) Naturre 328, 551-554
Rao, B.G., and Singh, U.C. (1991) J. Am. Chem. Soc. 113, 6735-6750
Rauk, A., Hamilton, G., and Moore, G.J. (1987), Biochem. Biophys. Res. Comm. 145, 1349-1355
Rich, D.H., Bernatowicz, M.S., Schmidt, P.G. (1982) J. Am. Chem. Soc. 104, 3535-3536
Rodriguez, E.J., Angeles, T.S., and Meek, T.D. (1993) Biochemistry 32, 12380-12385
Rogers, G.A., and Bruice, T.C., (1974) J. Am. Chem. Soc. 96, 2473-2481
Rullman, J.A.C., Bellido, M.N., and van Duijnen, P.Th. (1989) J. Mol. Biol. 296, 101-118
Russel, S., and Warshel, A. (1985) J. Mol. Biol. 185, 389-404
Sandler, M., and Smith, H.J., eds. (1989) Design of Enzyme Inhibitors as Drugs, Oxford University, Oxford.
Scheiner, S. (1984) Theoret. Chim. Acta 57, 71-77
Schowen, R.L. (1988) in Principles of Enzyme Activity, J.F. Liebman and A. Greenberg (eds.), Molecular Structure and Energetics, Vol. 9, VCH Publishers, Weinheim, FRG
Schroder, S., Daggett, V. and Kollman, P. (1991) J. Am. Chem. Soc. 113, 8922-8925
Sielecki, A.R., Federov, A.A., Boodhoo, A., Andreeva, N.J., and James, M.N.G. (1991) J. Mol. Biol. 214, 143-170

Singer, P.T., Smalas, A., Carty, R.P., Mangel, W.F., and Sweet, R.M. (1993) Science 259, 669-673
Singh, U.C., Weiner, P.K., Caldwell, J.W., and Kollman, P.A. (1986) AMBER 3.0, University of California, San Francisco.
Slilaty, S.N., and Vu, H.K. (1991) Protein Eng. 4, 919-922
Somayaji, V., Keillor, J., and Brown, R.S. (1988) J. Am. Chem. Soc. 110, 2625-2629
Stamato, F.M.L.G., and Tapia, O. (1988) Int. J. Quant. Chem. 33, 187-194
Steitz, T.A., and Shulman, R.G. (1982) Annu. Rev. Biophys. Bioeng. 11, 419-444
Stewart, J.J.P. (1989) J. Comput. Chem. 10, 209-220
Steinbach, P.J., and Brooks, B.R. (1993) Proc. Nat. Acad. Sci. U.S.A. 90, 9135-9139
Subramanian, E., Swan, I.D.A., Liu, M., Davies, D.R., Jenkins, J.A., Tickle, I.J., and Blundell, T.L. (1977) Proc. Nat. Acad. Sci. U.S.A. 74, 556-559
Suguna, K., Padlan, E.A., Smith, C.W., Carlson, W.D., and Davies, D.R. (1987) Proc. Nat. Acad. Sci. U.S.A. 84, 7009-7013
Sumi, H., and Ulstrop, J. (1988) Biochim. Biophys. acta 955, 26-42
Tang, J., James, M.N.G., Hsu, I.-N., Jenkins, J.A., and Blundell, T.L. (1978) Nature 271, 618-621
Tang, J., and Koelsch, G. (1995) Prot. and Pept. Lett. 2, 257-266
Thiel, W. (1981) J. Am. Chem. Soc. 103, 1413-1425
Thole, B.T., and van Duijnen, P. Th. (1980) Theoret. Chim. Acta (Berlin) 55, 307-318
Tong, W., and D'Souza, V.T. (1995) J. Comput. Chem. 16, 705-714
Tonge, P.J., and Carey, P.R. (1992) Biochemistry 31, 9122-9125
Turi, L., and Naray-Szabo, G. (1992) Int. J. Quant. Chem. 42, 1537-1551
Veerpandian, B., Cooper, J.B., Sali, A., and Blundell, T.L. (1992) Protein Sci. 1, 322-328
Wang, J., Xiang, Y.-F., and Li, C. (1994) Protein Eng. 7, 75-82
Warshel, A., Sussman, F., and Hwang, J.-K. (1988) J. Mol. Biol. 201, 139-159
Warshel, A., Naray-Szabo, G., Sussman, F., and Hwang, J.-K. (1989), Biochemistry 28, 3629-3637
Warshel, A. (1981) Biochemistry 20, 3167-3177
Warshel, A. (1987) Nature 330, 15-16
Warshel, A., and Russel, S., (1986) J. Am. Chem. Soc. 108, 6569-6579
Warshel, A. (1991) Computer Modeling of Chemical Reactions in Enzymes and Solutions, John Wiley and Sons, Inc.
Weiner, S.J., Seibel, G.L., and Kollman, P.A. (1986) Proc. Nat. Acad. Sci. (USA) 83, 649-
Weiner, S.J., Kollman, P.A., Nguyen, D.T., and Case, D.A. (1986) J. Comput. Chem. 7, 230-252
Wells, G.B., Mustafi, D., and Makinen, M.W. (1994) J. Biol. Chem. 269, 4577-4586
Wells J.A., Cunningham, B.C., Graycar, T.P., and Estell, D.A. (1986) Philos. Trans. R. Soc. London A317, 415-423
Whiting, A.K., and Peticolas, W.L. (1994) Biochemistry 33, 552-561

MODELLING OF PROTON TRANSFER REACTIONS IN ENZYMES

JOHAN ÅQVIST
Department of Molecular Biology
Uppsala University, Biomedical Centre, Box 590
S-75124 Uppsala, Sweden

1. Introduction

The transfer of protons between chemical groups is one of the fundamental types of processes that is catalyzed by enzymes. Proton transfer (PT) reactions can either be part (e.g., constitute individual steps) of more complex overall reactions types such as hydrolysis etc., in which bonds between heavy atoms are made or broken, or they can themselves provide a route for changing bonding and geometry as well as redistributing charge in a given substrate molecule. The latter is the case, e.g., in isomerization reactions such as that catalyzed by the enzyme triosephosphate isomerase (TIM), which will be the central enzyme of this chapter. As we will emphasize in the following sections, the energetics of PT reactions is mainly determined by the pK_a's of the donor and acceptor groups. The actual rate constants for PT reactions *in solution* are often found to obey rather simple free energy relationships (e.g., linear free energy relationships or LFERs) with respect to the donor–acceptor pK_a difference. This fact can turn out to be very useful for simulation of PT reactions in enzymes, since calibration of the relevant potential energy surfaces then can be carried out against the corresponding solution reactions requiring only knowledge of the donor–acceptor pK_a difference.

How do enzymes increase the rate of proton transfer? This a basic question in enzymology, the answer to which could either be fairly simple or quite complicated. The former being that enzymes simply change the relevant pK_a's and that all the rest follows from that. A more complicated picture could, e.g., involve changes in the rate of proton tunneling or other

G. Náray-Szabó and A. Warshel (eds.), Computational Approaches to Biochemical Reactivity, 341–362.

dynamic properties, subtle geometric or stereoelectronic effects etc. We will try to address these issues using a couple of different enzyme systems as pertinent examples, hopefully demonstrating to the reader that the computer simulation approach can be quite useful in obtaining new insight into enzyme catalysis. We find that, for the main purpose of estimating PT rate constants, the most important issue is to be able to model the donor–acceptor pK_a difference correctly since this appears to be where enzymes really can cause substantial deviations from normal values in solution. We also observe a pure transition state stabilization effect in the cases examined, which is due to the reduction of reorganization energy for the given reaction step compared to the uncatalyzed situation.

The computer simulations reported in this chapter are based on the empirical valence bond (EVB) model that will be only briefly described here as it has been extensively discussed elsewhere [1,2]. The enzymic reaction of TIM is used as our main example to illustrate both the origin of catalytic effects on PT and methodological aspects of the computational strategy. A more detailed account of the TIM calculations discussed in the next section has recently been reported by Åqvist & Fothergill [3].

2. Simulation of the Reaction Catalyzed by TIM

Triose phosphate iosmerase catalyzes the interconversion between dihydroxyacetone phosphate (DHAP) and glyceraldehyde–3–phosphate (GAP) [4-10]. The current view of the functional mechanism of the enzyme that has emerged from numerous studies can be summarized as follows (Fig. 1). (1) After binding of DHAP one of the C1 protons is abstracted by the catalytic base Glu165, yielding a corresponding enediolate species. (2) This enediolate is then protonated at O2 by the neutral imidazole ring of His95, yielding a doubly protonated enediol and an imidazolate anion. (3) His95 recaptures a proton from the O1 oxygen again yielding an enediolate species, now lacking the proton at O1. (4) Protonation of this enediolate at C2 by the carboxyl group of Glu165, thus restoring the general base to give the product D–GAP which then is released from the enzyme. TIM is a fast enzyme that is limited by product release at physiological substrate and product concentrations and with internal rate constants on the order of $10^4 s^{-1}$ [6,10]. As a comparison the rate limiting step for non-enzymatic conversion of DHAP is about $10^{-5} s^{-1}$ in which case the mechanism involves intramolecular proton abstraction by the phosphate group [10]. All of the proposed elementary reaction steps cannot be discerned experimentally and one instead observes only one kinetic intermediate state, but its exact nature is not known. The peculiar use of a neutral imidazole as a proton donor in the second reaction step (cf. Fig.1) has also attracted much

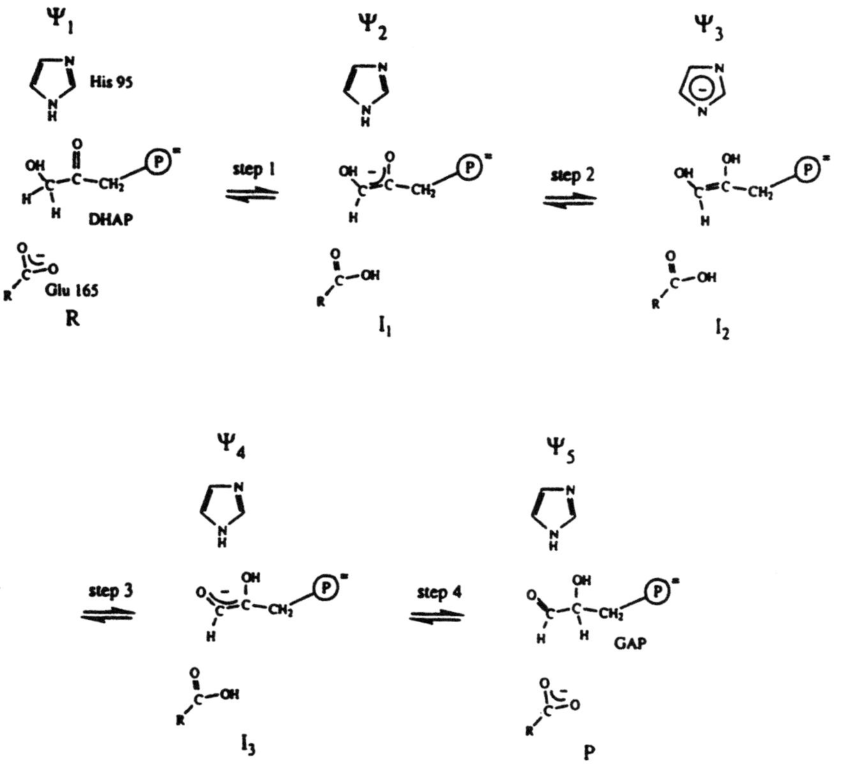

Figure 1. The catalytic mechanism of TIM is represented as a four-step process (excluding binding and release steps). Five VB configurations ($\Psi_1, \ldots, \Psi_5$) are used to describe the reaction.

attention but now seems to be established [11,12]. Since the pK_a of the neutral imidazole is better balanced with that of the substrate hydroxyl group this finding is not really so surprising.

A number of theoretical studies of TIM have been reported earlier dealing with, e.g., the effect of the Glu165→Asp mutation on binding and catalysis [13] as well as quantum and molecular mechanics calculations on the catalytic reaction [12,14]. The latter studies did, however, not reach a quantitative level and, in particular, no attempts to calculate *free energies* associated with the reaction were made. In fact, there have so far been only rather few reports of actual free energy calculations of enzymic reaction profiles [2]. Here, we will start our discussion with the calculation of a complete free energy profile [3] for the proposed reaction mechanism [7,9,12] of TIM using molecular dynamics (MD) free energy perturbation (FEP) simulations in combination with the empirical valence bond (EVB) method [1,2]. To our knowledge this is the first time that a free energy profile for *all chemical steps* of an enzymic reaction has been obtained by computer simulation. Although some controversy regarding the mechanism of TIM might still exist, our main goal here is to try to evaluate the detailed energetics implied by the currently favoured one in order to examine its feasibility. Furthermore, we would like to be able to rationalize differences in the energetics of the enzyme reaction compared to the uncatalyzed case in terms of the structure of the protein active site. The X-ray coordinates of the yeast enzyme in complex with the inhibitor phosphoglycolohydroxamate [8] were used as the starting point. This inhibitor closely resembles one of the proposed (high-energy) intermediates of the reaction and no major modelling is required to construct the different reaction species (cf. Fig.1) from it.

2.1. COMPUTATIONAL MODEL AND METHODS

We describe the TIM catalyzed reaction by the five bonding configurations, or VB resonance structures, depicted in Fig. 1. Each of these diabatic states $(\Psi_1, \ldots, \Psi_5)$ is represented by an analytical molecular mechanics force field [15,16]

$$\begin{aligned} \epsilon_i = H_{ii} &= \sum_j \Delta M_j^{(i)}(b_j^{(i)}) + \frac{1}{2}\sum_l \gamma_l^{(i)} k_l^{(i)} (\theta_l^{(i)} - \theta_{0,l}^{(i)})^2 \\ &+ \sum_m \gamma_m^{(i)} K_m^{(i)} [1 + cos(n_m^{(i)} \phi_m^{(i)} - \delta_m^{(i)})] \\ &+ V_{nb,rr}^{(i)} + \alpha^{(i)} + V_{nb,rs}^{(i)} + V_{ss} \end{aligned} \quad (1)$$

Here, all the energy terms except the last one (V_{ss}) represent interactions involving the reacting fragments, and these terms thus usually look different for different bonding configurations (the subscript r is used to denote the reacting groups, while s denotes the surroundings). The first three terms in eq. 1 are those for bonds (Morse potentials), bond-angles and torsional angles (including improper ones) involving the reacting fragments. Bond-angle and torsional energies can also depend on the degree of existence of certain bonds through the factor $\gamma_l^{(i)} = |M_j^{(i)}/D_j^{(i)}|$, where $D_j^{(i)}$ is the dissociation energy of the particular bond (j) to which the lth angle or torsion is coupled. $V_{nb,rr}$ and $V_{nb,rs}$ are the non-bonded electrostatic and van der Waals interactions among the reacting atoms themselves and with the surrounding atoms, respectively. These terms will, of course, differ between resonance structures as the charges and Lennard-Jones parameters can change between the states. In the cases where standard force field parameters [15], viz. partial charges, were not available they were derived from AM1/PM3 calculations using the AMSOL program [17] and merged with the GROMOS group parameters in order to achieve consistency with these. Fig. 2. shows the charge distributions that were used for the substrate in

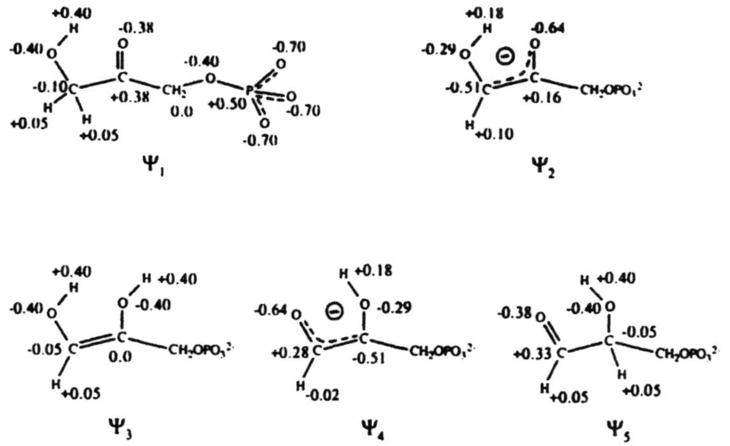

Figure 2. Partial atomic charges used in the calculations for the substrate in the various resonance forms. The standard GROMOS van der Waals parameters [15] were used except for the oxygens bearing a large negative charge ($|q| > 0.5$) for which the parameters derived by in [16] were used (these have been calibrated to reproduce hydration free energies for carboxylate ions).

each of the five VB states. The last term in eq. 1, i.e. the "invariant" part of the five force fields pertaining to interactions involving the surrounding atoms, is the normal GROMOS potential [15] that involves harmonic bond and angle terms, proper and improper torsions, electrostatic and Lennard-Jones terms.

While the force fields above can describe the interactions between the fragments of each state and with the surrounding medium, as well as the energies involved in distorting geometries from their equilibrium, they do not contain information about the relative energies of the fragments in vacuum. That is to say, that there is a real energy difference between any two resonance structures that is related to their heats of formation in vacuum, which is not included in the molecular mechanics force field. This is reflected by the constant $\alpha^{(i)}$ in eq. 1, which will then only enter as the difference ($\Delta\alpha_{ij} = \alpha^{(j)} - \alpha^{(i)}$) in energy between Ψ_j and Ψ_i with the reacting fragments at infinite separation in the gas-phase. Furthermore, classical force fields do not describe the adiabatic coupling between states reflected by off-diagonal terms (H_{ij}) of the hamiltonian. The above information can, however, be obtained either by quantum mechanical calculations or by gas-phase or solution experiments. The latter alternative, to use experimental data to calibrate the above mentioned gas-phase parameters of the system hamiltonian is the essence of the EVB method [1,2] (these parameters are listed in Table 1 below). In particular, it has proven useful to calibrate the potential energy surface with respect to solution experiments by simulating suitable reference reactions in water. The resulting parameters are then directly transferred to simulations of the corresponding reaction steps in the solvated enzyme. The advantage with this approach is that the energetics in an aqueous environment can be reproduced exactly and one then "only" has to be able to model the *substitution* of water by protein (note, however, that the choice of reference reactions corresponds to the assumed mechanism *in the enzyme* and not necessarily to that of the uncatalyzed reaction, which may proceed by a different mechanism). The alternative would be to rely on a (e.g., semi-empirical or ab initio molecular orbital) model to correctly reproduce the gas-phase energetics and then "substitute" vacuum for the actual environment in the enzyme, by coupling to a molecular mechanics force field [12]. This may involve more serious transferability problems and the idea with the EVB reference reaction calibration is thus to diminish these.

In the present case some of the data needed for calibration is not directly available from experiment due to the "high-energy" nature of the proposed intermediates, e.g. the solution pK_a's of hydrogens on C1 and C2 as well as the associated activation barriers. Starting from the pK_a of acetone in water of 19.2 [18] these pK_a's can, however, be estimated by the substituent effects of a hydroxyl group and the phosphate dianion. The latter interaction, i.e. between the enediolate and phosphate anions, has been determined to be about one pK unit in water [10]. In order to estimate the additional effect of a hydroxyl group on the pK_a of acetone we carried out semi-empirical SCF calculations in a homogeneous dielectricum ($\epsilon = 80$) using the AMSOL program [17]. Calculations with the AM1–SM2 and PM3–SM3 hamiltonians yield a pK_a shift of -3.8 and -5.4 pK-units, respectively. Taking into account the observed effect of the phosphate group [10] and after correcting for the number of equivalent protons we arrive at a pK_a value of 17.2 for abstracting a C1 proton from DHAP, using the AM1–SM2 result, and 15.6 using the PM3–SM3 result. The former of these values seems most reasonable in that it gives a free energy relationship that nicely fits experimental data on acetone and DHAP deprotonation from independent sources [10,19], as shown in Fig. 3 (our estimate of the pK_a at 17.2 is also rather close to the value inferred by Richard [10]). The free energy relationship in Fig. 3 can also be used to deduce the activation barriers for proton exchanges with the carboxylate moiety of glutamic acid in aqueous solution, which are used as reference reactions for the first and fourth step (cf. Fig. 1) of the enzymic conversion. The second and third steps of the enzyme reaction, according to the mechanism of Fig. 1, involve proton transfer between the substrate oxygens and the imidazole ring of His95. The pK_a difference between donor and acceptor in solution is about 2 pK-units, with the neutral imidazole ionising at the higher pK_a of 14 and the

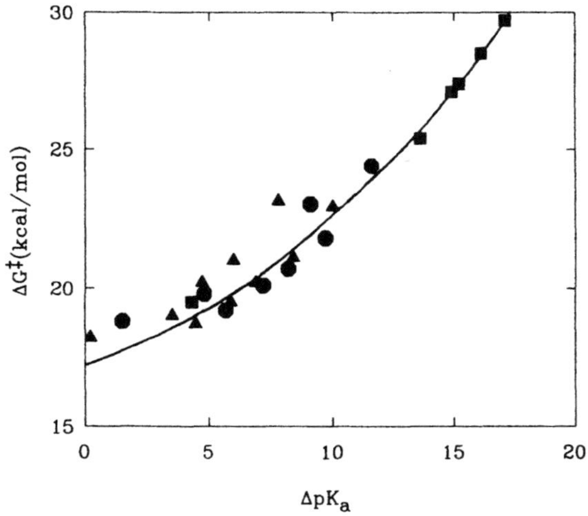

Figure 3. Free energy relationship (activation barrier versus ΔpK_a) comprising experimental data on acetone deprotonation by various bases ([19], squares) and on base catalyzed phosphate elimination reactions [10] of DHAP (circles) and LGAP (triangles), using the estimated pK_a values of DHAP and LGAP. The latter pK_a was related to the former by the observed equilibrium constant of 1/22 between LGAP and DHAP in solution [10]. All values have been corrected for the number of equivalent protons.

enediol at 12 [19]. The proton transfer barriers for these reference reactions can again be determined quite accurately from a free energy relationship based on experimental data of Eigen & Hammes [20].

Hence, the energetics of the four reaction steps as they would occur in aqueous solution, assuming a non-concerted mechanism, can be determined as described above and the resulting free energy diagram is shown as the upper curve in Fig. 4. Each of these steps is simulated in a separate MD calculation with the reacting fragments immersed in a sphere of water. The FEP technique [2,21-23] is used to drive the system, in a number of discrete steps, between the different VB configurations while mapping out the actual ground-state free energy profile, or potential of mean force (PMF) along the reaction coordinate [1,2,16]:

$$\Delta g(X_n) = \Delta G(\xi_m) - RT \ln < e^{-(E_g(X_n)-\epsilon_m(X_n))/RT} >_m \tag{2}$$

Here, ξ_m denotes a particular value of the FEP mapping parameter and $\Delta G(\xi_m)$ is the accumulated free energy change from the starting point of the perturbation. ϵ_m is the energy measured on the given mapping potential, determined by ξ_m [$\epsilon_m = (1 - \xi_m)\epsilon_i + \xi_m \epsilon_j$], and E_g is the ground-state energy of the system. The latter quantity can be calculated for any given configuration (set of coordinates) of the system by mixing the VB states and solving the corresponding secular equation

$$\boldsymbol{H}\,\boldsymbol{C} = E_g \boldsymbol{C} \tag{3}$$

The reaction coordinate X_n in eq. 2 is defined in terms of the energy gap $\Delta\epsilon = \epsilon_i - \epsilon_{i+1}$ between adjacent VB states (the subscript n only indicates that the variable X is discretized). These simulations of the uncatalyzed reference reactions are used to obtain values of the unknown gas-phase free energy differences, $\Delta\alpha_{ij}$, between the VB states and of the off-diagonal hamiltonian matrix elements. The latter are represented by functions of the form $H_{ij} = A_{ij}e^{-\mu_{ij}r_{XY}}$, where r_{XY} denotes the distance between the two heavy atoms that exchange the proton [1,2]. Table 1 lists the EVB parameters obtained from this calibration procedure together with the corresponding reaction free energies and barriers that were estimated for the solution reactions. After performing the initial calibration simulations in water the obtained parameters are kept fixed and the corresponding reaction steps are simulated inside the enzyme. Here, it should be pointed out that we neglect possible *differences* in barrier heights and reaction free energies, between the solution and enzyme reactions, that are due to quantum mechanical effects such as proton tunneling and zero-point energies. That is, the calibration in water yields an effective potential surface that includes such effects, but, to the extent that quantum corrections to the classical free energies differ between the enzyme and solution reactions they are neglected. Furthermore, we only examine the non-concerted mechanism here and assume that off-diagonal matrix elements H_{ij} for which $|i-j| \geq 2$ (indicative of simultaneous proton tranfers) are zero.

TABLE 1. EVB parameters and estimated energetics for the uncatalyzed reference reactions in water used for calibration.[a]

Step	$\Delta\alpha$	A_{ij}	μ_{ij}	ΔG°_{aq}	$\Delta G^{\ddagger}_{aq}$
1	+37.5	129.0	0.50	+17.8	+25.2
2	−0.5	110.5	0.50	+2.7	+7.2
3	+13.7	103.5	0.50	−2.7	+4.5
4	−45.2	160.0	0.50	−16.0	+8.1

[a] All quantities are in kcal/mol, except μ_{ij} which is given in Å^{-1}. Note that the gas-phase energy differences (see text), $\Delta\alpha$, cannot directly be equated to the true free energy difference between the reactant and product fragments of each step (as, e.g., in [16]), since they in this case impicitly include nonbonded interactions within each fragment that may be non-zero in the equilibrium conformation.

The simulation characteristics were the same as in [16] and [24]. For each step in water and in the protein a 70 ps trajectory was calculated using about 30 values of the FEP coupling parameter ξ, yielding a total simulation time of 560 ps excluding equilibration. At each ξ-point the first

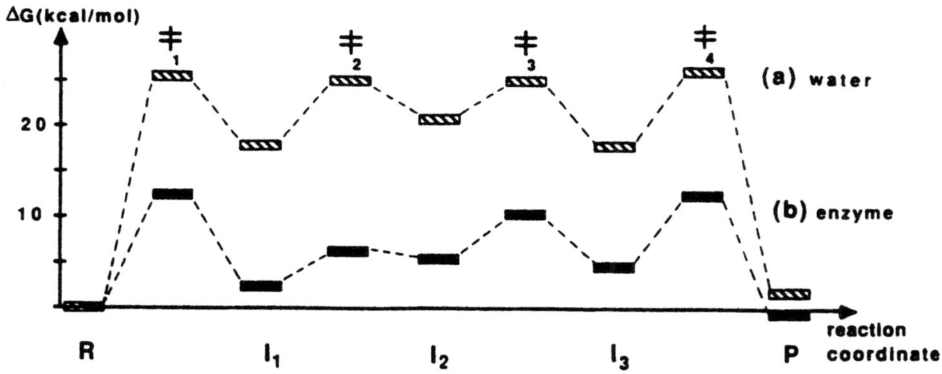

Figure 4. (a) Free energy diagram for the TIM reaction mechanism in aqueous solution based on experimental data [10,18,19,20] and semiempirical AMSOL [17] calculations. EVB/FEP/MD calibrations of each of these steps in water were carried out so that the resulting free energy profile reproduces the depicted energetics [13,14]. (b) Free energy diagram for the corresponding reaction steps in the enzyme.

40% of the data was discarded. The simulations were carried out at approximately 300K with an MD time step of 0.002 ps. Spherical systems of radius 15Å were used with the boundaries restrained by the SCAAS model [25] (water molecules) and by harmonic restraints to crystallographic positions (protein atoms). The spheres were centered on the C1 carbon atom both in the protein and water simulations. The substrate and charged protein groups interacted with everything within the simulation sphere, while cutoffs as specified in [24] were applied to other interactions. The convergence error bars obtained by forwards and backwards integration of the same trajectories were examined and found to range between 0.5 and 1.5 kcal/mol. In addition, some extra runs with different initial conditions were also carried out to try to assess the accuracy of the results and these gave very similar error ranges. Although the above values give an estimate of the MD convergence errors (per reaction step), it should be noted that an equally important source of error for the resulting enzyme energetics is the accuracy of the solution data used for calibration. These errors are unfortunately more difficult to estimate as they are determined by the accuracy of the pK_a's for the different species as well as the free energy relationships used to extract barrier heights (Fig.3 and [20]).

2.2. FREE ENERGY PROFILE AND SOURCES OF CATALYSIS

The results from the EVB/FEP/MD simulations of the TIM reaction are summarized in the diagram of Fig. 4 (lower curve). The large catalytic effect of enzyme is evident and it can be seen that all three intermediates along with their flanking transition states are stabilized relative to the reactant and product by some 15 kcal/mol, compared to the uncatalyzed profile. Of the four intervening free energy barriers one finds that those associated with PT to and from the substrate carbons limit the (internal) rate of conversion. The simulations place both I_1 and I_3 at lower energy than the doubly protonated species, with I_1 as the most stable intermediate along the path. This should thus be the main contributor to the experimentally observed kinetic intermediate. The calculated free energy profile appears to be rather compatible with available kinetic data [6,10]. Both the magnitude of the kinetically significant internal activation barriers (12.5 kcal/mol) and the corresponding equilibrium constants are fairly close to the experimental estimates. The theoretical calculations thus reproduce the catalytic effect of the enzyme remarkably well, within the accuracy limits that we can estimate (see discussion above).

The origin of the catalytic power of TIM has been the subject of much discussion [7,10,12,13]. We find here that there are two general effects at work: (1) *stabilization of charged intermediates* and (2) *reduction of reorganization energy.* The first of these favours the transfer of negative charge from Glu165 to the substrate as well as His95. The proximity of Lys12 (which also forms an ion-pair with Glu97) to the substrate together with hydrogen bonds from His95 and Asn10 are found to be important for the stability of the enediolate species, in agreement with earlier conjectures [9,12]. Knowles has also pointed out the presence of the "well-aimed" α-helix directed at His95 which presumably contributes to the stability of the imidazolate ion [7]. Furthermore, the present simulations indicate that an active site water molecule, located approximately between Glu165 and His95 (denoted 626 in the PDB entry 7tim), plays an important role during catalysis (Fig.5). This water has sufficient rotational and translational freedom so as to follow the "negative charge" as it is translocated within the active site, and it can provide stabilization by hydrogen bonding both to Glu165, the substrate and His95 when these are negatively charged. In fact, the H-bonding network in the active site appears to be particularly well suited for *adaptation* to the movement of negative charge during catalysis. Based on the above observation regarding the role of water-626 one would expect that a mutation which displaces this water molecule should hamper catalysis. Judging from the 3D structure, mutations of Ser96, Cys126 or Ile170 might cause such an effect. Interestingly, we note that Blacklow and Knowles [26] have examined the S96P mutant and found that it considerably reduces the catalytic efficiency. This mutant would indeed appear to displace water-626 and our results suggest that this can be the reason for its reduced activity. Furthermore, we observe that the hydroxyl group of

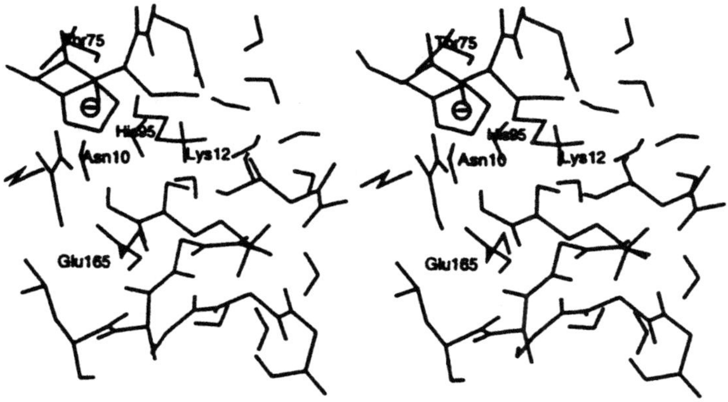

Figure 5. Stereo snapshot of the TIM active site in the intermediate state (I_2) with His95 negatively charged. The water molecule corresponding to number 626 in the PDB entry 7tim can be seen between His95 and Glu165.

Thr75 of the *second* subunit can interact with the deprotonated histidine in the I_2 intermediate state (Fig.5). As in the crystal structure, Thr75 donates a hydrogen bond to the carboxylate group of Glu97 when His95 is neutral, but in the second reaction step the hydrogen bond switches to the negative imidazolate ring of His95 thereby providing additional stabilization of this group. This behaviour could thus provide a new clue to the low enzymatic activity of monomeric TIM [27], since Thr75 belongs to the second subunit.

Our simulations show that the mechanism proposed by Bash et al. [12], involving the imidazolate species, is energetically and structurally very resonable. The distinction between this mechanism and that proposed by Alagona et al. [14], which involves *intramolecular* proton transfer between the substrate oxygens, is, however, rather subtle. The difference only pertains to whether a proton is relayed via His95 or whether the –NH dipole on this residue merely provides a hydrogen bond during intramolecular transfer. The barrier obtained by Alagona et al. in vacuum for this process is only about 14 kcal/mol, so one cannot exclude the possibility that the enzyme could reduce this activation energy enough to allow the intramolecular process. We find, however, that the barriers for the examined mechanism are low enough not to be rate determining in comparison with substrate binding and release, in accordance with experiment. This is in contrast with the calculations in [12] where the calculated internal barriers were too high to be compatible with experiment. It has also been shown that the mutant enzyme with His95 replaced by Gln, rather than to employ intramolecular proton transfer, uses Glu165 to effect all the proton tranfers [28].

The ability to reduce reorganization energies seems to be an important property of enzymes [29,2]. For a given chemical reaction step, part of the energy barrier can be viewed as an intrinsic contribution that does not depend on ΔG° (the reaction free energy) but on the curvature of the free energy functions of the reactant and product states, which reflects the reorganization, e.g. of dipolar groups, involved in the reaction [30]. In the case of large energy gaps between the reactant and product surfaces at their respective minima there will be a considerable reorganization of the system as the reaction proceeds, which gives rise to the intrinsic barrier. This is typically the case for charge separation and transfer reactions in aqueous solution, where water molecules must change their average polarization direction. In enzymes, polar groups are attached to a relatively rigid framework and are therefore less reorientable than in water. This fact, together with a clever design of the framework in which dipoles are pointing in a favourable direction, can lead to a considerable reduction of the reorganization energy compared to solution reactions. In the present calculations we observe this effect in *all* the steps of the catalyzed reaction. Fig. 6 compares the diabatic free energy curves (i.e. the free energy curves of the pure VB states without considering their mixing) in water and in the enzyme for the states Ψ_2 and Ψ_3, corresponding to the second reaction step. These free energy curves are obtained by the same potential of mean force formulation as that used to calculate the ground-state profile [31]. The reorganization energy upon going from Ψ_2 to Ψ_3 can be defined as $\lambda_{2\to3} = |\Delta\epsilon| - \Delta G^\circ_{2\to3}$, where $\Delta\epsilon = \epsilon_1 - \epsilon_2$ is the energy gap between the two curves at the minimum of Ψ_2. Vice versa, $\lambda_{3\to2} = |\Delta\epsilon| - \Delta G^\circ_{3\to2}$ for the opposite reaction, $\Delta\epsilon$ now being measured at the miniumum of Ψ_3. In the case of the second reaction step, the absolute value of ΔG° is small (about 3 kcal/mol both in solution and enzyme) and the reduction of reorganization energy can therefore be seen directly from the fact that the energy gaps at the minima of the diabatic "reactant" and "product" free energy surfaces for the reaction step are clearly smaller in the enzyme than in solution. In this case, as can be seen from Fig. 4, which gives the actual *adiabatic* energies (i.e. the ground state free energy surface after mixing the VB states), the enzyme leaves ΔG° close to its uncatalyzed value and mainly affects the barrier height.

Another convincing proof of the reorganization energy reduction effect can be obtained by artificially adjusting ΔG° (by varying the gas-phase energy difference $\Delta\alpha_{ij}$ discussed above) for a given reaction step in water so that it coincides with the calculated value in the enzyme, and then compare the resulting hypothetical barrier height with that in the enzyme. For the first reaction step, e.g., one then obtains a barrier of 17.3 kcal/mol in water with the same reaction free energy as in the protein. The fact that the barrier obtained in this way is considerably higher than in the protein shows that the enzyme does more than simply change relative pK_a's (the same conclusion is reached for the other steps as well). In a sense,

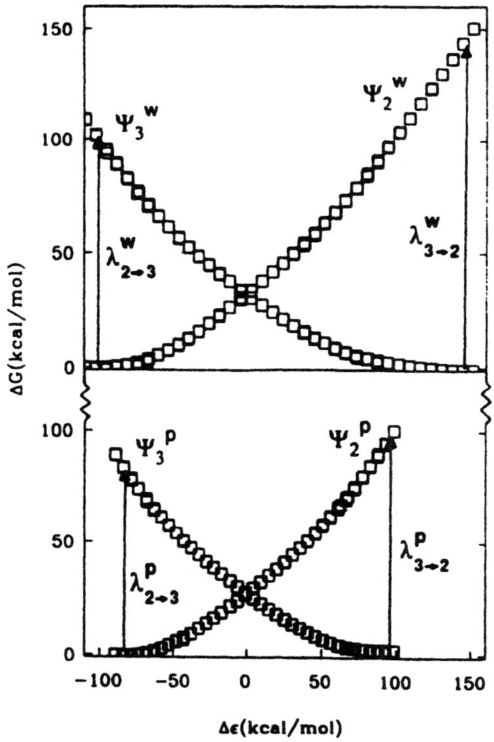

Figure 6. **Diabatic free energy functions for the states Ψ_2 and Ψ_3 in water (upper curve) and in the enzyme (lower curve) as a function of the energy gap between the two states. The reorganization energies (see text for definition) upon transfer from the "reactant" to the "product" state and vice versa for this (the second) reaction step are indicated. The reduction of reorganization energy in the protein is reflected by the energy gaps being significantly smaller than in the uncatalyzed reaction.**

the reorganization energy reduction effect corresponds to a "pure" form of the transition state stabilization paradigm since what it means is that, due to a smaller reorganization requirement than in water, the reactants and products "look more alike" and consequently the transition state energy will be lower. The reaction free energy is, however, not affected by reorganization energy reduction in itself, but by the concomitant stabilization of intermediates (which also, of course, contributes to the lowering of activation barriers). Hence, while the notion of "transition state stabilization" is very general and does not give any hint as to the actual origin of catalytic power, the two effects discussed here identify sources of such stabilization, namely the ability of the catalyst lower the energy of intermediates and to reduce the above-mentioned energy gaps. One can, in principle, imagine several ways of achieving the latter, but at least when the reaction involves polar states it is clear that a protein matrix with properly oriented dipoles (or electric fields) can provide an microenvironment that responds with a smaller reorganization than water, while still providing a substantial solvation of polar states.

3. Metal-Containing Enzymes

The observed electrostatic stabilization of charged high-energy intermediates (e.g., I_1,I_2,I_3 in TIM) also seems to be a common feature of enzyme reactions. The more popular view, perhaps, is that the effect of an enzyme is solely the stabilization of transition states. Our results suggest that a large portion of that is mainly a consequence of the stabilization of intermediates, although a more "orthodox" transition state stabilization also indeed does occur and originates from the reduction of reorganization energy. Earlier EVB/FEP calculations on the catalytic reactions of the metallo-enzymes staphylococcal nuclease (SNase) [32,33] and human carbonic anhydrase I (HCAI) [16,24], that will be briefly discussed below, show essentially the same result. These two enzymes contain Ca^{2+} and Zn^{2+}, respectively, as essential cofactors for catalysis. For stabilizing negatively charged intermediates, such as those arising from the abstraction of a proton from the reactants, an obvious strategy is to furnish the active site with a metal cation that can provide a "counter charge" to the high-energy intermediate. In the previous section, we noted that TIM effectively achieves this by various dipolar groups as well as by the positive charge of Lys12. However, it is clear that the strong electric field from a divalent metal ion should be (at least) equally efficient for this purpose.

SNase catalyzes the hydrolysis of both DNA and RNA at the 5′ position of the phosphodiester bond [34]

$$H_2O + 5'-OP(O_2)^{-}O-3' \rightleftharpoons 5'-OH + (OH)P(O_2)^{-}O-3' \qquad (4)$$

The proposed mechanism of the enzyme first involves a PT step from an active site water molecule to Glu43, that acts as a general base, followed by the nucleophilic attack of OH^- on the phosphorous atom in line with the 5′–O–P ester bond, leading to the formation of a trigonal bipyramidal (i.e. pentacoordinated) transition state or metastable intermediate [35,36]. The formation of this pentacoordinated transition state is the rate limiting chemical step, and products are then formed by subsequent breakage of the 5′–O–P bond and protonation of the leaving group [35,36].

Fig. 7 summarizes the EVB simulations of the SNase catalyzed reaction, in the enzyme and in water. It can immediately be seen that the enzyme excerts a substantial stabilization of the configurations with OH^- and the doubly charged phosphate group, as well as the intervening transition state, relative to the ground state where one unit of negative charge is residing on

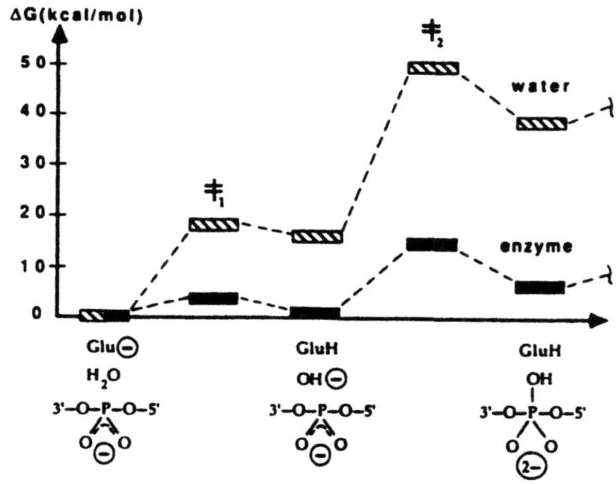

Figure 7. Calculated free energy profiles for the first two steps of the SNase reaction [32].

Glu43. To a first approximation this can be understood simply as an effect of concentrating all the negative charge (−2) closer to the metal, as Glu43 is well outside of the ion coordination sphere. Here, in addition to affecting the energetics of the PT step, the metal of course has an important second role, namely to facilitate the nucleophilic addition. This is apparently done in much the same way as the cost of PT is reduced, i.e. by a strong interaction with the accumulating negative charge on the phosphate group. Note, however, that as discussed in [32,33], there are also two positively charged arginines in the neighbourhood of the phosphate that contribute to the stabilization of the pentacoordinated state. In particular, one of these was found to exclusively interact with the doubly (but not the singly) negatively charged phosphate [32]. It has also been shown how the dual role of the metal ion in reactions of the hydrolysis type (PT + nucleophilic attack) can lead to an optimization of the catalytic rate with respect to the choice of metal [33].

In carbonic anhydrases, which catalyze the hydration/dehydration of carbon dioxide/bicarbonate ion:

$$H_2O + CO_2 \rightleftharpoons HCO_3^- + H^+ \qquad (5)$$

generation of the essential nucleophilic zinc-bound hydroxide ion proceeds through protolysis, or splitting, of a water molecule ligated to the metal. This hydroxide ion is then added to CO_2 in the actual interconversion step. The inital proton transfer step:

$$H_2O + H_2O \rightleftharpoons H_3O^+ + OH^- \qquad (6)$$

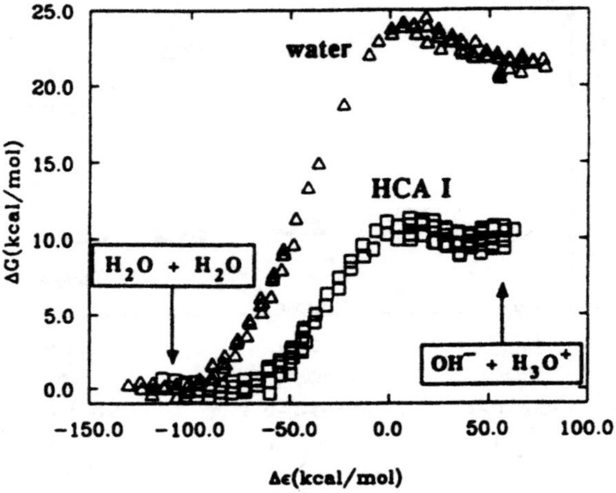

Figure 8. Calculated free energy profiles for the reference reaction (eq. 6) in water after calibration of the $\Delta\alpha$ and H_{12} parameters (upper curve, open triangles) and for the proton transfer step in HCA I (lower curve, open squares). The reaction coordinate $\Delta\epsilon$ denotes the energy gap between the two diabatic VB states [24].

between first and second coordination shell waters, or subsequent ones in which the excess proton is further translocated out into solution, have been found to limit the rate of catalysis in isozymes I and II [37]. It is thus not yet known where on the proton translocation path the transition state is situated. But, that this part of the reaction rather than the CO_2/HCO_3^- interconversion step limits the overall reaction rate. EVB/FEP calculations on the initial water splitting step as well as the CO_2/HCO_3^- interconversion have been carried out [24,16] and Fig. 8 shows the resulting catalyzed and uncatalyzed free energy profiles for the first of these steps. Again, it can be seen that the high-energy intermediate state, with OH^- ligated to Zn^{2+} in the enzyme active site, is stabilized by some 10 kcal/mol in comparison with the solution energetics. In fact, one also observes some stabilization of OH^- relative to HCO_3^- on the enzyme in the subsequent conversion step (not shown) [16]. This is again an example of the dual role of the metal, and one can note that the reaction would otherwise be too exothermic to allow efficient reversibility (which is indeed a physiological requirement). Also in this enzyme a significant reduction of reorganization energies is observed [2,16,31] which contributes to the lowering of transition state energies. In the first water splitting step, e.g., the value of the reorganization energy was found to be some 20 kcal/mol lower than the corresponding uncatalyzed water reaction. One should perhaps point out here that the Marcus rate equation for diabatic reactions

$$\Delta G^{\ddagger} = \frac{(\Delta G^0 + \lambda)^2}{4\lambda} \tag{7}$$

has sometimes been applied directly to experimental data for PT reactions, without taking into account the strongly adiabatic nature of PT [30,38]. The values obtained for the reorganization energy, λ, from eq. 7 will then be considerably smaller than those obtained from actual diabatic energy curves, such as those in Fig.6. This is because the lowering of the activation energy that occurs as a consequence of the adiabatic mixing is not taken into account by eq. 7. It can be shown that for two harmonic free energy functions of equal curvature one obtains in the range where $|\Delta G_0| < \lambda$ the expression

$$\Delta G^{\ddagger} \approx (\Delta G^0 + \lambda)^2/4\lambda - \bar{H}_{12}(X^{\ddagger}) + \bar{H}_{12}^2(X_0)/(\Delta G^0 + \lambda) \tag{8}$$

$$\bar{H}_{12}(X_0) < |\Delta G^0 + \lambda|/2$$

where X_0 and $X^{\ddagger}$ denote the reaction coordinate at reactant and transition state configurations, respectively, and $\bar{H}_{12}$ is the average value of H_{12} at the given X. The first term of eq. 8 is simply the Marcus expression [30] for the diabatic case where $H_{12}(X) = 0$ and the two last terms reflect the effect of the adiabatic coupling of the two surfaces on the transition and reactant states, respectively. Obviously, the validity of eq. 8 as usual requires that the free energy functions would be quadratic or, in other words, that the system will follow the linear response approximation. It has been shown earlier that eq. 8 indeed can describe the adiabatic case rather accurately [2,31].

4. Free Energy Relationships for PT in Solution

From the two metallo-enzyme examples above we thus find that the metal ion contributes both to increasing the acidity of the reacting water molecule, by stabilizing the negatively charged hydroxide ion, and to controlling the energetics of subsequent steps involving addition of the nucleophile to the substrate. Accompanying the stabilization of negatively charged "intermediate states" is also a reorganization energy reduction effect, just as found in TIM. Since a lot of data exists for base catalyzed water dissociation in solution, it is interesting to compare the calculated energetics of PT in SNase and HCAI to experimental free energy relationships [39,40] for this type of reaction. Fig.9 depicts such relationships between ΔG° and $\Delta G^{\ddagger}$ for the general reaction

$$H_2O + B \rightleftharpoons OH^- + BH^+ \tag{9}$$

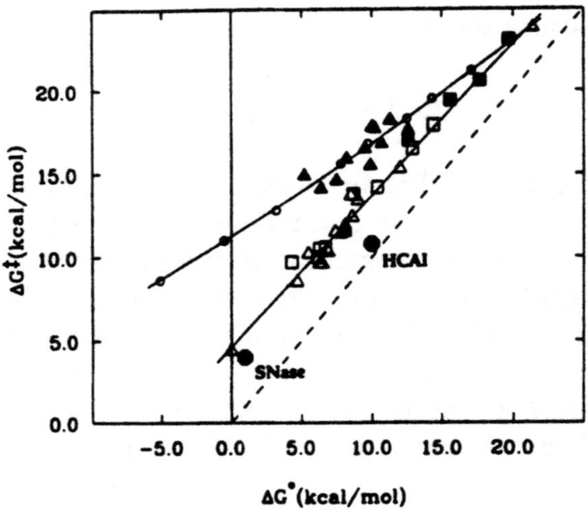

Figure 9. Activation vs. reaction free energy for reactions of the type in eq. 9. Experimental data for various bases (triangles) are taken from [39]. These fall into two main classes denoted by solid and open triangles (see text). Experimental data on the effect of metal ions on water dissociation when B is another water molecule (eq. 10, open squares) are from [40]. Solid squares denote theoretical calculations on some metals in eq. 10 [41]. Open circles show the calculated effect of changing ΔG° while keeping λ constant for the reaction of eq. 6 in water. The (upper) line through these points is a parabolic fit. The lower solid line is a linear fit ($\rho = 0.99$) to the experimental data (excluding the solid triangles). The dashed line is the limit $\Delta G^{\ddagger} = \Delta G^{\circ}$.

where the base B need not necessarily be uncharged. It can be seen from the Figure that the experimental data suggest that two different classes of proton abstracting bases can identified. Interestingly, these generally correspond to cases where (1) no intramolecular hydrogen bond can be formed as a result of the base accepting a proton and (2) an intramolecular H-bond can be formed within the base after PT. The former case corresponds to the lower curve in Fig. 9 (open triangles), while the H-bond forming bases are denoted by solid triangles (upper curve). Also shown in the Figure (open circles) is the result of parametrically varying ΔG° in the EVB model for PT between two water molecules in solution (i.e. when the base B is another water molecule). This can be done as described in [31] by simply changing the gas-phase energy difference ($\Delta\alpha$) between the two VB states used to describe the reaction. It is important to point out here, that by this procedure the reorganization energy of the reaction does *not* change since the two diabatic free energy functions (for reactants and products) do not change. It is therefore only the (hypothetical) ΔpK_a or reaction free energy that varies along this calculated curve. This can also be seen from the fact that the Brönsted factor β is 0.5 for the curve at zero reaction free energy, in agreement with the theoretical expectation for the "constant reorganization energy" case (eqs. 7,8). What is interesting with this exercise is that one finds that the "constant reorganization energy" curve appears to coincide with the experimental case where the bases form internal H-bonds. Conversely, this would suggest that along the lower LFER, with no H-bonds in the protonated base, the reorganization energy of the reaction actually changes as a function of the pK_a of the abstracting base (we use the term LFER here, even if the relationship is not exactly linear).

How can this behaviour be understood? Well, it seems logical that the reorganization energy in general should be larger when an internal H-bond forms within the base as it is protonated, compared to the case where the donor and acceptor are more likely to form such an H-bond after PT (i.e. in the "product state"). That is to say that there would be an extra reorganization contribution associated with making/breaking of the internal H-bond in BH^+, or conversely, that the reorganization is smaller if the interaction between OH^- and BH^+ is stronger and more specific (e.g. H-bonding). Furthermore, the phenomenon that the reorganization energy changes, i.e. decreases as $\Delta G^\circ \to 0$, along the lower LFER should simply reflect an increasing interaction strength or specificity between the water donor and the acceptor (i.e. inter- rather than intramolecular H-bonding), as the pK_a difference between them decreases. An interesting case here is the comparison between trimethylamine and trimethylphosphine, where the former compound falls on the lower LFER with the latter conforms to the upper one. Here, semiempirical AM1 and PM3 as well as ab initio 6–31G** HF calculations (Hansson & Åqvist, unpublished) indeed suggest that the charge distribution on the trimethylphosphonium ion is such that strong H-bonding between its hydrogen atom (which appears to have an excess of negative charge rather than positive) and OH^- would not occur, in contrast to trimethylamine. Although $(CH_3)_3PH^+$ does not, of course, have any internal H-bond, it apparently behaves in the same way as bases that do, since like them it does not interact by specific H-bonding with OH^-. Accordingly, this base should belong to the upper "LFER" which is indeed also the case.

Also shown in Fig. 9 are the results of calculations [41] on the effect of metal cations on the energetics of eq. 6, corresponding to the reaction

$$M(OH_2)_x^{n+} + H_2O \rightleftharpoons M(OH_2)_{x-1}(OH)^{(n-1)+} + H_3O^+ \qquad (10)$$

where M denotes the metal ion. Here, we find that the theoretical calculations yield a nice agreement with the lower experimental LFER and that the reorganization energy decreases somewhat as the aquo-cation becomes more acidic. Experimental data [40], in fact, show that divalent, trivalent and tetravalent metal aquo-cations indeed obey the lower LFER of Fig. 9 (open squares). There does, however, not seem to exist any experimental data on rates for the corresponding monovalent cases. As an example, our predicted values of ΔG° and $\Delta G^\ddagger$ for $Zn^{2+}(H_2O)$ (eq. 10) are 12.0 and 17.0 kcal/mol, respectively [41]. The corresponding experimental values are 12.9 and 16.5 kcal/mol, respectively [40].

In view of the discussion in the previous sections where we have concluded that enzymes can considerably reduce PT reorganization energies, it comes as no surprise that both the SNase and HCAI catalyzed PT steps lie well below both solution LFERs of Fig. 9. This demonstrates that they are more efficient catalysts than a given base in solution with a corresponding pK_a value would be. It is also interesting to note that in the initial PT step, HCAI is considerably more efficient than a Zn^{2+} in water and that it, in fact, operates close to the limit $\Delta G^\ddagger = \Delta G^\circ$.

5. Concluding Remarks

In this chapter, we have discussed the results of simulations of proton transfer reactions in a couple of different enzymes. The picture emerging from such calculations is that two main effects contribute to the efficiency of enzyme catalyzed PT. These are (1) the ability of the enzyme to modulate pK_a differences by electrostatic stabilization of charged high-energy intermediates and (2) the ability to significantly reduce reaction reorganization energies. The latter effect, which was already conjectured by Albery in 1980 [38], is mainly due to a strategic preorientation of dipolar groups [42] that does not incur high reorganization costs during the reaction. Both of the above effects depend, of course, in a rather complicated way on the dielectric properties of the protein, which must be such that (certain) electrostatic interactions can be considerably enhanced. This can be understood by comparing the energetics of enzyme reactions to the effects that the individual components of their catalytic machinery (e.g., general acids/bases, metals etc.) are known to have on solution reactions. One then finds that enzymes do more than simply assemble these components with their given energetic contributions. Rather, they can add further to the efficiency of these components by embedding them in specific active site microenvironments. We believe that the two effects discussed above provide a rather simple answer to how enzymes increase the rates of PT. More exotic explanations involving, e.g., increased tunneling rates [43] or single well (low barrier) hydrogen bonds [44] can also be given. The issue is, however, not really to what extent such quantum phenomena exist in enzymes, but rather whether they contribute significantly to catalysis or not. We have not tried to address those questions explicitly here, but we do find that the computational models employed above seem quite sufficient for explaining the origin of catalysis of proton transfer in enzymes.

Acknowledgement. Support from the Swedish Natural Science Research Council (NFR) is gratefully acknowledged.

References

1. Warshel, A. (1991) *Computer Modeling of Chemical Reactions in Enzymes and in Solutions*, Wiley, New York.
2. Åqvist, J. and Warshel, A. (1993) *Chem. Rev. 93*, 2523-2544.
3. Åqvist, J. and Fothergill, M. (1995). Submitted.
4. Rieder, S.V. and Rose, I.A. (1959) *J. Biol. Chem. 234*, 1007-1010.
5. Knowles, J.R. and Albery, W.J. (1977) *Acc. Chem. Res. 10*, 105-111.
6. Nickbarg, E.B. and Knowles, J.R. (1988) *Biochemistry 27*, 5939-5947.
7. Knowles, J.R. (1991) *Nature 350*, 121-124.
8. Lolis, E., Alber, T., Davenport, R.C., Rose, D., Hartman, F.C. and Petsko, G.A. (1990) *Biochemistry, 29*, 6609-6618.
9. Davenport, R.C., Bash, P.A., Seaton, B.A., Karplus, M., Petsko, G.A. and Ringe,D. (1991) *Biochemistry, 30*, 5821-5826.
10. Richard, J.P. (1984) *J. Am. Chem. Soc. 106*, 4926-4936.
11. Lodi, P.J. and Knowles, J.R. (1991) *Biochemistry 30*, 6948-6956.
12. Bash, P.A., Field, M.J., Davenport, R.C., Petsko, G.A., Ringe, D. and Karplus, M. (1991) *Biochemistry, 30*, 5826-5832.
13. Daggett, V., Brown, F. and Kollman, P. (1989) *J. Am. Chem. Soc. 111*, 8247-8256.
14. Alagona, G., Desmeules, P., Ghio, C. and Kollman, P.A. (1984) *J. Am. Chem. Soc. 106*, 3623-3632.
15. van Gunsteren, W.F. and Berendsen, H.J.C. (1987) *Groningen Molecular Simulation (GROMOS) Library Manual*, Biomos B.V., Nijenborgh 16, Groningen, The Netherlands.
16. Åqvist, J., Fothergill, M. and Warshel, A. (1993) *J. Am. Chem. Soc. 115*, 631–635.
17. Cramer, C.J. and Truhlar, D.G. (1992) *Science 256*, 213-217.
18. Chiang, Y., Kresge, A.J., Tang, Y.S. and Wirz, J. (1984) *J. Am. Chem. Soc. 106*, 460-462.
19. Guthrie, J.P. (1979) *Can. J. Chem. 57*, 1177-1185.
20. Eigen, M. and Hammes, G.G. (1963). *Adv. Enzymol. 25*, 1-38.
21. Jorgensen, W.L. (1989) *Acc. Chem. Res. 22*, 184-189.
22. Beveridge, D.L. and DiCapua, F.M. (1989) *Ann. Rev. Biophys. Biophys. Chem. 18*, 431-492.
23. Straatsma, T.P. and McCammon, J.A. (1992) *Ann. Rev. Phys. Chem. 43*, 407-435.
24. Åqvist, J. and Warshel, A. (1992) *J. Mol. Biol. 224*, 7-14.
25. King, G. and Warshel, A. (1989) *J. Chem. Phys.91*, 3647-3661.
26. Blacklow, S.C. and Knowles, J.R. (1990) *Biochemistry 29*, 4099-4108.
27. Borchert, T.V., Abagyan, R., Jaenicke, R. and Wierenga, R.K. (1994) *Proc. Natl. Acad. Sci. U.S.A. 91*, 1515-1518.
28. Nickbarg, E.B., Davenport, R.C., Petsko, G.A. and Knowles, J.R. (1988) *Biochemistry 27*, 5948-5960.
29. Yadav, A., Jackson, R.M., Holbrook, J.J. and Warshel, A. (1991) *J. Am. Chem. Soc. 113*, 4800-4805.
30. Cohen, A.O. and Marcus, R.A. (1968) *J. Phys. Chem. 72*, 4249-4256.
31. Warshel, A., Hwang, J.-K. and Åqvist, J. (1992) *Faraday Discuss. 93*, 225-238.
32. Åqvist, J. and Warshel, A. (1989) *Biochemistry 28*, 4680-4689.
33. Åqvist, J. and Warshel, A. (1990) *J. Am. Chem. Soc. 112*, 2860-2868.
34. Tucker, P. W., Hazen, E. E., Jr. and Cotton, F. A. (1979) *Mol. Cell. Biochem. 23*, 67-86.
35. Cotton, F. A., Hazen, E. E., Jr. and Legg, M. J. (1979) *Proc. Natl. Acad. Sci. U.S.A.*

76, 2551-2555.
36. Sepersu, E. H., Shortle, D. and Mildvan, A. S. (1987) *Biochemistry 26*, 1289-1300.
37. Silverman, D.N. and Lindskog, S. (1988) *Acc. Chem. Res. 21*, 30-36.
38. Albery, J.W. (1980) *Ann. Rev. Phys. Chem. 31*, 227-263.
39. Eigen, M. (1964) *Ang. Chem. Int. Ed. Eng. 3*, 1-72.
40. Burgess, M. A. (1978) *Metal Ions in Solution*, Ellis Horwood Ltd., Chichester, England.
41. Åqvist, J. (1991) *J. Phys. Chem. 95*, 4587-4590.
42. Warshel, A. (1978) *Proc. Natl. Acad. Sci. U.S.A. 75*, 5250-5254.
43. Klinman, J.P. (1989) *TIBS 14*, 368-373.
44. Cleland, W.W. and Kreevoy, M.M. (1994) *Science 264*, 1887-1890.

PROTEIN-LIGAND INTERACTIONS

TERRY P. LYBRAND
University of Washington
Center for Bioengineering
Molecular Bioengineering Program
Box 351750
Seattle, WA 98195-1750

1. Introduction

Protein-ligand interactions are at the heart of much of biochemistry. Most physiologic and pharmacological responses are mediated by specific receptor-ligand interactions, and most biological receptors are proteins or multi-protein complexes. Enzyme catalysis, signal transduction, transcription, and translation are all intimately dependent on specific, selective protein-ligand interactions. Nonspecific protein-ligand interactions are also common, but nonspecific interactions are rarely involved in precise, well-regulated biological processes.

Specific protein-ligand interactions are controlled by detailed molecular recognition processes. Ligands may range from small organic molecules to large biomolecules such as proteins, nucleic acids, or carbohydrates. In all cases, however, the fundamental nature and basis of molecular recognition and selectivity is the same. Specific protein-ligand interactions are characterized by size, shape, and physicochemical property complementarity between the protein receptor site and the ligand. For example, a hydrophobic receptor site will bind preferentially hydrophobic ligands that fit wellištericly, being neither too large (excessive steric repulsion) nor too small (suboptimal van der Waals interactions). In many cases, specific protein-ligand complexes are also characterized by formation of distinct hydrogen bonds and complementary ionic and polar interactions. Ligands that form fewer specific hydrogen bonds and ionic or polar interactions with the target receptor site typically bind with lower affinity and less selectivity than those ligands whose interactions with the receptor are optimized. Because so many biochemical processes are triggered and/or regulated by specific protein-ligand interactions, it is of great fundamental and practical interest to understand in atomic detail the nature of these interactions, and the molecular attributes that lead to high receptor binding affinity and selectivity for a given ligand. In therapeutic applications, high affinity usually correlates with high potency and receptor site specificity. High receptor specificity, in turn, usually correlates with diminished side effects that are often produced by cross-reaction with other, similar receptor molecules. Therefore, a detailed understanding of the molecular attributes necessary for specific, high affinity protein-ligand interactions should be useful in structure-based drug design applications.

G. Náray-Szabó and A. Warshel (eds.), Computational Approaches to Biochemical Reactivity, 363–374.

A wide range of biophysical experimental tools have been used to characterize protein-ligand interactions. X-ray diffraction experiments can provide detailed, atomic resolution images of a protein-ligand complex. However, diffraction experiments are possible only when suitable crystals of a protein-ligand complex are available. Diffraction experiments also yield a time-averaged picture of the complex, and it may be difficult to deduce important dynamical details essential to complex formation and selectivity from these static images. A variety of spectroscopic techniques can often be used to provide detailed structural, and even dynamical, information about protein-ligand complexes. However, these techniques usually do not provide as detailed and complete structural information as x-ray diffraction. Other experimental techniques, such as site-directed mutagenesis and ligand binding assays can often provide indirect structural information, but these approaches are most powerful when used in combination with high-resolution structural information from diffraction or spectroscopic studies.

Molecular modeling techniques are used extensively to supplement biophysical experimental methods for characterization of protein-ligand complexes. Modeling methods are often an essential tool for refinement of detailed structures from raw experimental data, and modeling techniques can provide additional information not readily available from experimental approaches. An overview of some modeling techniques suitable for investigation of protein-ligand interactions will be presented here.

2. Computational Methods

2.1 INTERACTIVE MOLECULAR GRAPHICS

Interactive molecular graphics modeling tools are among the oldest, and most widely used, molecular modeling techniques for the study of protein-ligand interactions. Molecular graphics software that incorporates full color display with depth cueing, hidden line and surface removal, stereo viewing capabilities, and the ability to manipulate structures interactively, provides a powerful tool for investigation of receptor-ligand complexes.(Olson & Morris, 1993; Weber at al., 1994) Interactive molecular graphics modeling tools are generally relatively simple and easy to use, and they capitalize on a user's knowledge and intuition about protein-ligand complexes. On the other hand, interactive graphics model building can be quite tedious, and is subject to personal biases. As a result, graphics modeling of protein-ligand complexes can be somewhat limited in scope, and many viable and important interactions or binding modes may be missed. Furthermore, graphics modeling tools depend on the availability of a three-dimensional structure or model for the complex. Usually, molecular graphics modeling tools are used in conjunction with other experimental or computational methods, as they are one of the best ways to analyze and interpret complex structural data.

2.2 LIGAND-DERIVED INSIGHTS

As mentioned above, detailed knowledge of protein-ligand interactions is of great interest in drug design applications. However, there is no 3D structural information for most therapeutically interesting drug receptor targets. Therefore, chemists have devised methods to extract some general information about receptor site structure and characteristics from the ligands that bind to those receptor sites. The underlying rationale is simple: the receptor site must represent a "negative image" of ligands that bind selectively and with high affinity.(Lybrand, 1984) This information can be deduced by examining a collection of ligands known to bind to the receptor site of interest for patterns in physicochemical properties, such as size, shape, and 3D spatial distribution of specific functional groups. A subset of physicochemical attributes common to all ligands that bind a particular receptor site with good affinity is termed a pharmacophore. The pharmacophore represents the essential set of molecular characteristics necessary for high affinity binding (and often, for biological activity). Success in a pharmacophore mapping study requires that all ligands included for analysis indeed bind the target receptor site (except for ligands included in the study as negative controls), and that all ligands bind with comparable, or at least predictable, binding modes. If the ligands in question are flexible, appropriate conformers must also be identified and considered. Usually, only relatively low energy conformers are considered(Marshall, 1987), as highly strained conformations are less likely to bind to a target receptor site with high affinity, i.e., the strain energy must be compensated by the receptor-ligand interactions. Once a pharmacophore has been defined, the importance of key ligand attributes for receptor binding can be quantitated in many cases, yielding a 3D quantitative structure-activity relationship (3D-QSAR). One commonly used 3D-QSAR technique is called Comparative Molecular Field Analysis , or CoMFA.(Marshall & Cramer, 1988) This method attempts to correlate 3D-spatial properties, such as molecular electrostatic potentials and molecular volumes, with binding affinity or other properties for a series of ligands. 3D-QSAR methods like CoMFA utilize a technique known as partial least-squares(Stahle & Wold, 1986) to derive correlations between the limited number of observations (e.g., binding affinity or activity of each ligand) and the huge data sets, e.g., fields are defined at multiple points on a grid surrounding the ligands, with each grid point as a data point or variable in the fitting procedure, so the data set is highly overdetermined. In 3D-QSAR, the validity and predictive power can be tested by cross-validation. Each ligand used to derive the QSAR is removed from the data set in sequence and treated as an unknown. A 3D-QSAR is then rederived for the remaining ligands. If the new 3D-QSAR can predict correctly the correlated properties of the "unknown" ligands, the QSAR has some predictive capability. Cross-validation for 3D-QSAR is especially important, due to the large number of data points used to generate the correlation equations. 3D-QSAR suffers from some similar limitations as 3D pharmacophore searches. In particular, use of unsuitable conformers for flexible molecules or inappropriate alignment of all the ligands may alter the QSAR. These limitations also make it much more challenging to handle highly dissimilar molecules in a pharmacophore mapping study or CoMFA analysis.

2.3 HOMOLOGY MODELING

While 3D structural data is often unavailable for a receptor site of interest, there may be structural information for a related protein or group of proteins. In these cases, a model

3D structure for the receptor of interest can be generated using the related protein(s) as a template. This procedure is called homology model building, and is quite powerful when structures are available for suitable reference proteins.(Johnson, et al., 1994) The higher the sequence homology between the protein receptor of interest and the reference protein, the greater the odds that the 3D homology model will be a good representation of the correct structure.(Chothia & Lesk, 1986) Once a 3D model is derived for the receptor site, a wide variety of computational tools are available to investigate receptor-ligand interactions.

2.4 DOCKING METHODS

Weaknesses cited above for manual exploration of ligand-receptor interactions using interactive molecular graphics techniques include the laborious nature of the task and the likelihood that important binding modes or interactions will be missed. To address these limitations with manual ligand docking approaches, automated procedures have been developed. These techniques can be used to explore numerous possible binding orientations and interactions at a well-defined target site, or can be used to probe an entire protein structure for plausible binding pockets. Some methods probe potential binding sites via docking of molecular fragments, such as typical functional groups.(Goodford, 1985; Gillet, at al., 1993; Lauri & Bartlett, 1994; Eisen, et al., 1994) A complete molecular docking is achieved by connection of the molecular fragments with suitable linking groups to form an intact molecule. Other programs perform formal molecular docking with specific molecules of interest.(Meng, at al., 1992) Many of these programs can also be used to scan databases of small molecules, to select compounds with apparent appropriate characteristics to bind a target receptor site. This strategy is often utilized in structure-based drug design applications, in an attempt to identify molecules that may be suitable starting points for optimization of binding interactions, i.e., "lead" compounds in drug development.(Rutenber, at al., 1993; Ring, at al., 1993)

Automated docking procedures make it possible to perform exhaustive searches for all plausible binding orientations for a receptor-ligand complex, or to examine huge data sets of small molecules to identify potentially suitable ligands. However, the computational power and efficiency of automated docking searches offers little improvement over manual exercises if each possible docking orientation has to be evaluated manually. Therefore, a variety of scoring schemes have been implemented in docking programs to rank the quality of each complex generated. These scoring schemes may simply measure the quality of a steric fit, or they may incorporate a simplified empirical potential energy function that evaluates the quality of other contacts, such as hydrogen bonds and polar interactions. In principle, a user would select only the top scoring complex from a search for further consideration. In practice, it has been shown that the scoring schemes in use at present are not completely reliable. In controlled tests, it is often the case that the docking scores for a series of molecules at a defined target site do not correlate well with the relative binding affinities of the molecules, for example. A molecule that rates best in the docking score will often bind notably less well than another molecule that scores much lower in the automated search. At present, the lack of quantitative predictive power of the scoring schemes remains one of the major weaknesses of automated docking procedures, and dictates that a good deal of manual evaluation using molecular graphics techniques is necessary.

Most automated docking programs have performed rigid docking searches, i.e., both the protein receptor and test ligands are kept in a single, fixed conformation as potential binding orientations and sites are explored. While this enhances the computational efficiency of a docking search significantly, it can seriously bias results. Rigid docking makes no allowance for any type of induced fit between ligand and receptor, but protein crystal structures with and without ligands show clearly that some conformational rearrangement of both receptor site and ligand often accompanies complex formation. Since rigid docking algorithms cannot account for induced fit, they may miss important docking possibilities. Lack of allowance for induced fit may also account for some of the shortcomings in scoring schemes, as even minute conformational adjustments may be sufficient to alter the predicted binding preferences in a series of ligands.

Several strategies are possible to introduce conformational flexibility in automated docking searches. The simplest approach, but also the most expensive computationally, is to perform a systematic search for all rotatable bonds in the ligand and receptor site. In a systematic search application, the torsion angle associated with each rotatable bond is varied in small increments (e.g., 10-30°) through a full 360° range. Hence, the problem size grows combinatorially with the number of rotatable bonds. This method is usually computationally prohibitive for the full complex, but is used often to introduce ligand flexibility in the docking searches. Several recent studies have used genetic algorithm techniques to select low-energy ligand conformations during docking exercises with a rigid receptor site(Oshiro, et al., 1995; Judson, et al., 1995). This approach appears to be more efficient for ligand conformational searches during an automated docking exercise than either systematic search or energy minimization with traditional algorithms. An alternate method to allow ligand flexibility involves the generation of a conformational database.(Miller, et al., 1994) A conformational search is performed for all ligands to be examined in the docking exercise. All low-energy conformers of each ligand are included as separate "molecules" in the ligand database, and then a traditional rigid docking search is performed for each ligand. This approach retains the computational efficiency of rigid ligand docking, on a per-molecule basis. However, the ligand database can be quite large if each molecule has several viable conformations, and there is extensive preprocessing necessary to generate this database. Systematic conformational search for the ligands can be coupled with a restricted conformational search for binding pocket side chains to introduce limited receptor site flexibility.(Leach, 1994) In this strategy, the conformational search for amino acid side chains in the receptor site is restricted to allowable conformers, as tabulated in protein side chain rotamer libraries (Ponder & Richards, 1987; Dunbrack & Karplus, 1993). Another powerful, but extremely computationally expensive, approach to include ligand and receptor site flexibility in docking exercises is the performance of limited energy minimization after the initial rigid docking stage (Mizutani, at al., 1994) It is generally impossible to perform energy minimization for every docking orientation generated in an automatic search, so scoring schemes are normally used to select a limited number of the "best" complexes for subsequent energy minimization. The primary weakness of this approach, besides the enormous computational expense, is the possibility that important docking orientations will be eliminated after the rigid docking step, due to inadequate scoring functions.

2.5 ENERGY FUNCTION METHODS

2.5.1 *Potential Energy Function Models*

Numerous computational techniques are available to study receptor-ligand interactions that are based on an explicit evaluation of the interaction energies between receptor and ligand molecules. These techniques generally utilize an empirical potential energy function to describe the energy hypersurface for receptor-ligand interactions, such as that shown in Equation 1:

$$V(\mathbf{r}_1,\mathbf{r}_2,\ldots,\mathbf{r}_N) = \sum_{\text{bonds}} \frac{1}{2}K_b\left(R - R_0\right)^2 + \sum_{\text{angles}} \frac{1}{2}K_a\left(\theta - \theta_0\right)^2 + \sum_{\text{dihedrals}} K_d\left[1 + \cos\left(n\phi - \gamma\right)\right] + \sum_{i,j} 4\varepsilon_{ij}\left[\left(\frac{\sigma_{ij}}{r_{ij}}\right)^{12} - \left(\frac{\sigma_{ij}}{r_{ij}}\right)^{6}\right] + \frac{q_i q_j}{\varepsilon r_{ij}} \qquad (1)$$

where K_b, K_a, K_d, are force constants and R_0, θ_0 and γ are equilibrium structural parameters for bonds, bond angles, and torsion angles, respectively. The nonbonded parameters σ_i, ε_i, q_i are the Lennard-Jones radius, well depth and partial charge for each atom in the system. These relatively simple potential energy functions can generally be parameterized to represent well the properties and behavior of large biological molecules. In some cases, more complicated potential functions may be needed to model important properties of molecules (e.g., vibrational properties). Morse potentials may be used in place of harmonic potentials, cross terms to permit bond stretch-bond angle bend coupling or bond stretch-torsion angle rotation coupling may be introduced, and more complicated functional forms may be employed to represent nonbonded interactions. Most empirical potential energy functions represent nonbonded interactions as a sum of all pairwise, or two-body, interactions. However, nonadditive contributions (three-body and higher order interactions) are often important in biomolecular systems. Many potential energy functions used for calculations on biomolecular and aqueous systems are calibrated to reproduce condensed phase properties, i.e., the nonbonded terms are adjusted to reproduce the nonadditive effects present in condensed phase systems in an average way. {Weiner, at al., 1986; Nilsson and Karplus, 1986; Jorgensen and Tirado-Rives, 1988} These potential energy functions are classified as effective pair potentials. In recent years, many potential energy functions have been developed that include nonadditive effects explicitly.(e.g., Cieplak, et al., 1990; Rullman and van Duijnen, 1988; van Belle, at al., 1987; Barnes, et al., 1979; Warshel & Levitt, 1976) These potential functions often include inducible point dipoles and involve self-consistent field calculations. Improved representations for long-range electrostatics interactions have also been adopted in recent years for biomolecular calculations. The Lennard-Jones and Coulombic terms in Equation 1 are normally truncated at some intermediate particle separation distance in most calculations, and Ewald summation is now used frequently to make corrections for long-range electrostatic interactions in large biomolecular complexes.(Essmann, et al., 1995) The simple Coulomb term in Equation 1 may also be replaced with the Poisson-Boltzmann equation (Equation 2) for a more accurate treatment of electrostatic interactions.(Gilson, 1995)

$$\nabla \cdot \varepsilon \nabla \phi - \varepsilon \kappa^2 \sinh(\phi) + 4\pi\rho = 0 \qquad (2)$$

$$\kappa^2 = \frac{8\pi N e^2 I}{1000 \varepsilon k_B T}$$

where ϕ is the electrostatic potential, ε is the dielectric constant, ρ is the charge density, κ is the inverse Debye-Hückel screening length, I is ionic strength, e is the unit electric charge, k_B is Boltzmann's constant, and T is the absolute temperature. Potential functions with explicit nonadditive terms, Poisson-Boltzmann electrostatics calculations, and Ewald or other long-range interaction energy corrections are computationally far more expensive than simple effective pair potential functions. However, they often yield more accurate and physically reasonable interaction energies.

Force constants for bond, bond angle, and torsion angle terms may be determined from vibrational and other spectroscopic methods or from quantum mechanical calculations for small reference molecules. Parameters for Lennard-Jones terms may be derived from scattering, crystal packing, or liquid structure data as well as quantum mechanical calculations. Atomic partial charges are generally obtained from *ab initio* quantum mechanical calculations. The molecular electrostatic potential for a molecule is calculated over a grid of points, and an atom-centered point charge model is derived that reproduces the electrostatic potential at each grid point.{Cieplak, et al., 1995; Reynolds, et al., 1992)

2.5.2 *Energy Minimization and Molecular Dynamics*

Given a suitable potential energy function, a variety of techniques are available to calculate ligand-receptor interaction energies.(Brooks, at al., 1988; McCammon & Harvey, 1987) Energy minimization, where the potential energy for the system is minimized with respect to the atomic coordinates, is the oldest and most straightforward method used to study ligand-receptor interactions. This procedure can yield useful insight and may reproduce qualitative ligand binding trends in some cases, but provides only interaction energies, or enthalpies, and is generally rather limited in scope. It is desirable to obtain instead free energies of binding for ligand-receptor complexes, and this requires methods which can sample representative configurational states for the molecular systems. In principle, molecular dynamics (MD) simulations could be run with a ligand and receptor molecule, immersed in a solvent system, until ligand-receptor encounter and complexation occurred. However, this is completely impractical, as the events involving in the complexation reaction, including diffusional encounter of ligand and receptor, possible conformational rearrangements of ligand and receptor, and desolvation of ligand and receptor site, all occur on simulation time scales that are much too long to be practical, i.e., many events in a ligand binding reaction are extremely rare events in contemporary MD simulations. A variety of special techniques based on molecular dynamics simulations have been developed to compute binding free energies for ligand-receptor complexes. A potential of mean force calculation can be used in some cases to obtain this information. In this method, the ligand and receptor molecule are placed in a solvent box, and a reaction coordinate is defined that maps a pathway for the formation of the ligand-receptor complex, starting from the solvent-separated pair. A series of molecular dynamics simulations are run, with the system restrained to sample small, overlapping regions along the reaction coordinate. Results from each individual MD sampling run may then be pieced together to generate the full reaction coordinate profile. From this information, it is possible to calculate the free energy of binding for the ligand. Relative free energies of binding for several ligands to a common receptor, or for a single ligand to different receptors can be estimated as the difference between computed binding free energies for each individual reaction. This approach works well for relatively simple molecular systems, but is often impractical for complicated systems like ligand-protein complexes. It can be difficult to define a reasonable reaction coordinate to describe a ligand binding reaction with a protein molecule, and extremely long MD simulations are needed to sample overlapping regions of the reaction coordinate. There is also frequently a large statistical uncertainty in the computed free energy results, due to the normal thermal fluctuations in these simulations. In these cases, in can be impossible to estimate relative binding free energies for several different complexes, as the statistical uncertainty of each calculation may be larger in magnitude than the relative binding free energy difference.

2.5.3 *Free Energy Perturbation Methods*

An alternate set of techniques have been developed in recent years to compute relative binding free energies. These techniques are based on principles from perturbation methods(Beveridge & DiCapua, 1989), and enable direct calculations of free energy differences for ligand binding reactions.(Straatsma & McCammon, 1991) For example, the relative free energy of binding for two ligands, A and B, at a common receptor site, R, can be computed in a straightforward fashion with perturbation methods. Rather than simulate the physical binding processes by the horizontal reactions shown in Equation 3, the non-physical transformations shown in the vertical reactions in Equation 3 are simulated. In these processes, ligand A is transformed into ligand B while free in solution (ΔG_3), or when bound in the receptor site (ΔG_4). Since these four reactions comprise a closed (pseudo) thermodynamic cycle, the relative binding free energy can be computed as the difference in the free energy changes for the two non-physical transformation processes.

$$\begin{array}{ccccc} & & \Delta G_1 & & \\ A & + & R \rightleftarrows & A{:}R & \\ \Delta G_3 \uparrow\downarrow & & & \downarrow\uparrow \Delta G_4 & \\ B & + & R \rightleftarrows & B{:}R & \\ & & \Delta G_2 & & \end{array} \tag{3}$$

$$\Delta\Delta G = \Delta G_4 - \Delta G_3 = \Delta G_2 - \Delta G_1$$

Several conditions must be satisfied if a free energy perturbation approach is to be applied in a molecular simulation. The changes, or transformations, to be modeled in the simulations must be relatively small, i.e., ligands A and B must be rather similar, so that the perturbation to ligand B is a modest change relative to the reference state ligand A. Both ligands must also bind to the receptor R in the same location and general orientation, i.e., the binding mode must be the same for both ligands. If these conditions are not satisfied, free energy perturbation methods generally do not work well.

When these conditions are satisfied, the free energy perturbation approach offers several advantages over potential of mean force calculations for physical ligand binding reactions. The free energy change for the ligand transformation reactions is generally small (since the two molecules are similar), and the statistical uncertainty in the computed results is usually similarly small, if sufficient configurational sampling is performed. Thus, it is possible to calculate small relative binding free energy differences reasonably accurately. It is not necessary to define a complicated reaction coordinate to perform a free energy perturbation calculation, and these calculations do not normally require a series of individual simulations to sample the full ligand transformation, so free energy perturbation calculations are normally much less computationally expensive that potential of mean force calculations. The calculations as outlined in Equation 3 also provide useful insight into the relative importance of differential desolvation free energies and intrinsic ligand-receptor interactions. In some cases, ligand A may exhibit a favorable binding free energy relative to ligand B because it is more easily desolvated, rather than due to better interactions with the receptor.

Free energy perturbation methods have been used with good success to calculate relative binding free energies for ligands ranging from small ions and solutes(Lybrand, et al., 1986; Thomas & Kollman, 1994) to drug-protein complexes(Eriksson & Nilsson, 1995; Orozco, et al., 1993; Hirono & Kollman, 1991) to antigen-antibody complexes(Lee, et al., 1992). While configurational sampling can pose a problem and lead to unreliable results in some complex systems(Ewing & Lybrand, 1994), these methods are among the most powerful and promising techniques for quantitative prediction and detailed molecular insight of ligand binding free energies.

3. Conclusions

A few computational methods useful for study of ligand-receptor complexes have been described briefly here. Some of these computational tools provide accurate quantitative results for ligand-receptor complexes in favorable situations, while other methods provide at best only qualitative information and trends. However, it is important to note that computational tools provide much more than numbers. Computational methods can reveal structural and thermodynamic details of ligand-receptor complexes that are often not attainable with experimental techniques alone. As such, computational tools yield detailed, atomic-resolution insight and understanding of ligand-receptor complexes. Coupled with dramatic advances in computer hardware technology and capabilities, these computational and modeling tools play an increasingly important role in the study of ligand-receptor interactions, and in applications areas such as structure-based drug design.

4. References

Barnes, P., Finney, J.L., Nicholas, N.D., and Quinn, J.E. (1979) Cooperative effects in simulated water, *Nature* **282**, 459-464.

Beveridge, D.L. and DiCapua, F.M. (1989) Free energy via molecular simulation: applications to chemical and biomolecular systems, *Ann. Rev. Biophys. Biophys. Chem.* **18**, 431-492.

Brooks, C., Karplus, M., and Pettitt, B.M. (1988) Proteins: A theoretical perspective of dynamics, structure, and thermodynamics, *Adv. Chem. Phys.* **71**.

Chothia, C. and Lesk, A.M. (1986) The relation between the divergence of sequence and structure in proteins, *EMBO J.* **5**, 823-826.

Cieplak, P., Kollman, P., and Lybrand, T. (1990) A new water potential including polarization: application to gas-phase, liquid, and crystal properties of water, *J. Chem. Phys.* **92**, 6755-6760.

Cieplak, P., Cornell, W.D., Bayly, C., and Kollman, P.A. (1995) Application of the multimolecule and multiconformational RESP methodology to biopolymers: charge derivation for DNA, RNA, and proteins, *J. Comput. Chem.* **16**, 1357-77.

Dunbrack, R.L., Jr. and Karplus, M. (1993) Backbone-dependent rotamer library for proteins. Application to side-chain prediction, *J. Mol. Biol.* **230**, 543-574.

Eisen, M.B., Wiley, D.C., Karplus, M., and Hubbard, R.E. (1994) HOOK: A program for finding novel molecular architectures that satisfy the chemical and molecular requirements of a macromolecule binding site, *Proteins* **19**, 199-221.

Eriksson, M.A. and Nilsson, L. (1995) Structure, thermodynamics and cooperativity of the glucocorticoid receptor DNA-binding domain in complex with different response elements. Molecular dynamics simulations and free energy perturbation studies, *J. Mol. Biol.* **253**, 453-472.

Essmann, U., Perera, L., Berkowitz, M.L., Darden, T., Hsing, L., and Pedersen, L.G. (1995) A smooth particle mesh Ewald method, *J. Chem. Phys.* **103**, 8577-8593.

Ewing, T.J.A. and Lybrand, T.P. (1994) A comparison of perturbation methods and Poisson-Boltzmann electrostatics calculations for estimation of relative solvation free energies, *J. Phys. Chem.* **98**, 1748-1752.

Gillet, V., Johnson, A.P., Mata, P., Sike, S., and Williams, P. (1993) SPROUT: A program for structure generation, *J.Comput-Aided Mol. Des.* **7**, 127-153.

Gilson, M.K. (1995) Theory of electrostatic interactions in macromolecules, *Curr. Opin. Struct. Biol* **5**, 216-223.

Goodford, P.J. (1985) A computational procedure for determining energetically favorable binding sites on biologically important macromolecules, *J. Med. Chem.* **28**, 849-857.

Hirono, S. and Kollman, P.A. (1991) Relative binding free energy calculations of inhibitors to two mutants (Glu46 -- Ala/Gln) of ribonuclease T1 using molecular dynamics/free energy perturbation approaches, *Prot. Eng.* **4**, 233-243.

Johnson, M.S., Srinivasan, N., Sowdhamini, R., and Blundell, T.L. (1994) Knowledge-based protein modeling, *Crit. Rev. Biochem. Mol. Biol.* **29**, 1-68.

Jorgensen, W.L. and Tirdao-Rives, J. (1988) The OPLS potential functions for proteins. Energy minimizations for crystals of cyclic peptides and crambin, *J. Am. Chem. Soc.* **110**, 1657-1666.

Judson, R.S., Tan, Y.T., Mori, E., Melius, C., Jaeger, E.P., Treasurywala, A.M., and Mathiowetz, A. (1995) Docking flexible molecules: A case study of three proteins, *J. Comput. Chem.* **16**, 1405-1419.

Lauri, G. and Bartlett, P.A. (1994) CAVEAT: A program to facilitate the design of organic molecules, *J. Comput.-Aided Mol. Des.* **8**, 51-66.

Lee, F.S., Chu, Z-T., Bolger, M.B., and Warshel, A. (1992) Calculations of antibody-antigen interactions: microscopic and semi-microscopic evaluation of the free energies of binding of phosphorylcholine analogs to McPC603, *Prot. Eng.* **5**, 215-228.

Leach, A.R. (1994) Ligand docking to proteins with discrete side-chain flexibility, *J. Mol. Biol.* **235**, 345-356.

Lybrand, T.P., McCammon, J.A., and Wipff, G. (1986) Theoretical calculation of relative binding affinity in host-guest systems, *Proc. Natl. Acad. Sci. USA* **83**, 833-835.

Lybrand, T.P. (1984) Ph.D. dissertation, *Univ. of California, San Francisco*

Marshall, G.R. (1987) Computer-aided drug design, *Ann. Rev. Pharmacol. Toxicol.* **27**, 193-213.

Marshall, G.R. and Cramer, R.D., 3rd. (1988) Three-dimensional structure-activity relationships, *Trends Pharmacol. Sci.* **9**, 285-289.

McCammon, J.A. and Harvey, S.C. (1987) *Dynamics of Proteins and Nucleic Acids*, Cambridge University Press

Meng, E.C., Shoichet, B.K., and Kuntz, I.D. (1992) Automated docking with grid-based energy evaluation, *J. Comput. Chem.* **13**, 505-524.

Miller, M.D., Kearsley, S.K., Underwood, D.J., and Sheridan, R.P. (1994) FLOG: A system to select quasi-flexible ligands complementary to a receptor of known three-dimensional structure, *J. Comput-Aided Mol. Des.* **8**, 153-174.

Mizutani, M.Y., Tomioka, N., and Itai, A. (1994) Rational automatic search method for stable docking models of protein and ligand, *J. Mol. Biol.* **243**, 310-326.

Nilsson, L. and Karplus, M. (1986) Empirical energy functions for energy minmization and dynamics of nucleic acids, *J. Comput. Chem.* **7**, 591-616.

Olson, A.J. and Morris, G.M. (1993) Seeing our way to drug design, *Perspect. Drug Disc. Design* **1**, 329-344.

Orozco, M., Tirado-Rives, J., and Jorgensen, W.L. (1993) Mechanism for the rotamase activity of FK506 binding protein from molecular dynamics simulations, *Biochemistry* **32**, 12864-12874.

Oshiro, C.M., Kuntz, I.D., and Dixon, J.S. (1995) Flexible ligand docking using a genetic algorithm, *J. Comput-Aided Mol. Des.* **9**, 113-130.

Ponder, J.W. and Richards, F.M. (1987) Tertiary templates for proteins. Use of packing criteria in the enumeration of allowed sequences for different structural classes, *J. Mol. Biol.* **193**, 775-791.

Reynolds, C.A., Essex, J.W., and Richards, W.G. (1992) Atomic charges for variable molecular conformations, *J. Am. Chem. Soc.* **114**, 9075-9079.

Ring, C., Sun, E., McKerrow, J.H., Lee, G.K., Rosenthal, P.J., Kuntz, I.D., and Cohen, F.E. (1993) Structure-based inhibitor design by using protein models for the development of antiparasitic agents, *P.N.A.S. USA* **90**, 3583-3587.

Rullman, J.A.C. and van Duijnen, P.T. (1988) A polarizable water model for calculation of hydration energies, *Mol. Phys.* **63**, 451-475.

Rutenber, E., Fauman, E.B., Keenan, R.J., Fong, S., Furth, P.S., Ortiz de Montellano, P.R., Meng, E., Kuntz, I.D., DeCamp, D.L., Salto, R. (1993) Structure of a non-peptide inhibitor complexed with HIV-1 protease. Developing a cycle of structure-based drug design, *J. Biol. Chem.* **268**, 15343-15346.

Stahle, L. and Wold, S. (1986) On the use of some multivariatestatistical methods in pharmacological research, *J. Pharmacol. Methods* **16**, 91-110.

Straatsma, T.P. and McCammon, J.A. (1991) Theoretical calculations of relative affinities of binding, *Adv. Enzymol.* **202**, 497-511.

Thomas, B.E., IV and Kollman, P.A. (1994) Free energy perturbation calculations of the relative binding affinities of an 8-subunit cavitand for alkali ions in methanol, *J. Am. Chem. Soc.* **116**, 3449-3452.

van Belle, D., Couplet, I., Prevost, M., and Wodak, S. (1987) Calculation of electrostatic properties in proteins. Analysis of contributions from induced protein dipoles, *J. Mol. Biol.* **198**, 721-735.

Warshel, A. and Levitt, M. (1976) Theoretical studies of enzymic reactions: dielectric, electrostatic, and steric stabilization of the carbonium ion in the reaction of lysozyme, *J. Mol. Biol.* **103**, 227-249.

Weber, J., Deloff, A., and Flukiger, P. (1994) Visualization in computational chemistry, *SPEEDUP* **8**, 63-70.

Weiner, S.J., Kollman, P., Nguyen, D.T., and Case, D. (1986) An all atom force field for simulations of proteins and nucleic acids, *J. Comput. Chem.* **7**, 230-252.

SUBJECT INDEX

Understanding Chemical Reactivity

1. Z. Slanina: *Contemporary Theory of Chemical Isomerism.* 1986 ISBN 90-277-1707-9
2. G. Náray-Szabó, P.R. Surján, J.G. Angyán: *Applied Quantum Chemistry.* 1987 ISBN 90-277-1901-2
3. V.I. Minkin, L.P. Olekhnovich and Yu. A. Zhdanov: *Molecular Design of Tautomeric Compounds.* 1988 ISBN 90-277-2478-4
4. E.S. Kryachko and E.V. Ludeña: *Energy Density Functional Theory of Many-Electron Systems.* 1990 ISBN 0-7923-0641-4
5. P.G. Mezey (ed.): *New Developments in Molecular Chirality.* 1991 ISBN 0-7923-1021-7
6. F. Ruette (ed.): *Quantum Chemistry Approaches to Chemisorption and Heterogeneous Catalysis.* 1992 ISBN 0-7923-1543-X
7. J.D. Simon (ed.): *Ultrafast Dynamics of Chemical Systems.* 1994 ISBN 0-7923-2489-7
8. R. Tycko (ed.): *Nuclear Magnetic Resonance Probes of Molecular Dynamics.* 1994 ISBN 0-7923-2795-0
9. D. Bonchev and O. Mekenyan (eds.): *Graph Theoretical Approaches to Chemical Reactivity.* 1994 ISBN 0-7923-2837-X
10. R. Kapral and K. Showalter (eds.): *Chemical Waves and Patterns.* 1995 ISBN 0-7923-2899-X
11. P. Talkner and P. Hänggi (eds.): *New Trends in Kramers' Reaction Rate Theory.* 1995 ISBN 0-7923-2940-6
12. D. Ellis (ed.): *Density Functional Theory of Molecules, Clusters, and Solids.* 1995 ISBN 0-7923-3083-8
13. S.R. Langhoff (ed.): *Quantum Mechanical Electronic Structure Calculations with Chemical Accuracy.* 1995 ISBN 0-7923-3264-4
14. R. Carbó (ed.): *Molecular Similarity and Reactivity: From Quantum Chemical to Phenomenological Approaches.* 1995 ISBN 0-7923-3309-8
15. B.S. Freiser (ed.): *Organometallic Ion Chemistry.* 1996 ISBN 0-7923-3478-7
16. D. Heidrich (ed.): *The Reaction Path in Chemistry: Current Approaches and Perspectives.* 1995 ISBN 0-7923-3589-9
17. O. Tapia and J. Bertrán (eds.): *Solvent Effects and Chemical Reactivity.* 1996 ISBN 0-7923-3995-9
18. J.S. Shiner (ed.): *Entropy and Entropy Generation.* Fundamentals and Applications. 1996 ISBN 0-7923-4128-7
19. G. Náray-Szabó and A. Warshel (eds.): *Computational Approaches to Biochemical Reactivity.* 1997 ISBN 0-7923-4512-6

Kluwer Academic Publishers – Dordrecht / Boston / London

GPSR Compliance

The European Union's (EU) General Product Safety Regulation (GPSR) is a set of rules that requires consumer products to be safe and our obligations to ensure this.

If you have any concerns about our products, you can contact us on ProductSafety@springernature.com

In case Publisher is established outside the EU, the EU authorized representative is:

Springer Nature Customer Service Center GmbH
Europaplatz 3
69115 Heidelberg, Germany

Batch number: 10620143

Printed by Printforce, the Netherlands

GPSR Compliance

The European Union's (EU) General Product Safety Regulation (GPSR) is a set of rules that requires consumer products to be safe and our obligations to ensure this.

If you have any concerns about our products, you can contact us on ProductSafety@springernature.com

In case Publisher is established outside the EU, the EU authorized representative is:

Springer Nature Customer Service Center GmbH
Europaplatz 3
69115 Heidelberg, Germany

Batch number: 10620143

Printed by Printforce, the Netherlands